ENCYCLOPEDIA *of* DISCOVERY

Science
and History

CONSULTING EDITORS

Science
and History

FOG CITY PRESS

Published by Fog City Press
814 Montgomery Street
San Francisco, CA 94133 USA

Copyright © 2002 Weldon Owen Pty Ltd
Reprinted 2002, 2003, 2004, 2005, 2006

Chief Executive Officer John Owen
President Terry Newell
Publisher Lynn Humphries
Managing Editor Janine Flew
Design Manager Helen Perks
Editorial Coordinator Jennifer Losco
Production Manager Louise Mitchell
Production Coordinator Monique Layt
Sales Manager Emily Jahn
Vice President International Sales Stuart Laurence

ISBN 1 876778 92 X

Color reproduction by Colourscan Co Pte Ltd
Printed by SNP Leefung Printers Limited
Printed in China

A Weldon Owen Production

Project Managing Editor Rosemary McDonald
Project Editors Helen Bateman, Ann B. Bingaman,
Jean Coppendale, Kathy Gerrard, Selena Quintrell Hand
Text Claire Craig, Ian Graham, Terry Gwynn-Jones,
Neil Jameson, Anne Lynch, Judith Simpson, Richard Wood
Educational Consultants Richard L. Needham,
Deborah A. Powell
Text Editors Jane Bowring, Claire Craig,
Gillian Gillett, Tracy Tucker
Project Art Director Sue Burk
Designers Catherine Au-Yeung, Juliet Cohen,
Lyndel Donaldson, Gary Fletcher, Kathy Gammon,
Avril Makula, Kylie Mulquin, Mark Nichols, Jill Ryan
Assistant Designers Robyn Latimer, Janet Marando,
Angela Pelizzari, Regina Safro, Melissa Wilton
Visual Research Coordinators Jenny Mills, Esther Beaton
Photo and Visual Research Peter Barker, Karen Burgess,
Annette Crueger, Carel Fillmer, Kathy Gerrard,
Fran Meagher, Amanda Parsonage, Lorann Pendleton,
Kristina Sturm, Fay Torres-Yap, Amanda Weir

Contents

Science

History

Introduction

Encyclopedia of Discovery: Science and History is a dynamic, fact-filled book that explains how things work in our everyday lives, and how people lived their lives many centuries ago. Here, in one compact volume, is detailed information about such diverse topics as flight, computers, sound recording, the world's most remarkable buildings, medical marvels, and our favorite sports and games. There are descriptions and images of the first automobiles and today's sophisticated counterparts; of computers, clocks, and cameras; and of a multitude of other ingenious inventions from every period in history.

Also presented are insights into the lives of ancient peoples from around the globe. Take a trip into the past and discover how mummies were made in Ancient Egypt, what happened to the gladiators in Rome's Colosseum, and why a Chinese emperor was buried with a vast terracotta army.

Science and History comprises 11 chapters, each structured as a series of self-contained double-page spreads dealing with one aspect of the subject. Simple, direct language and detailed, atmospheric illustrations and photographs will engage young readers and encourage them to discover for themselves the world around them.

Science

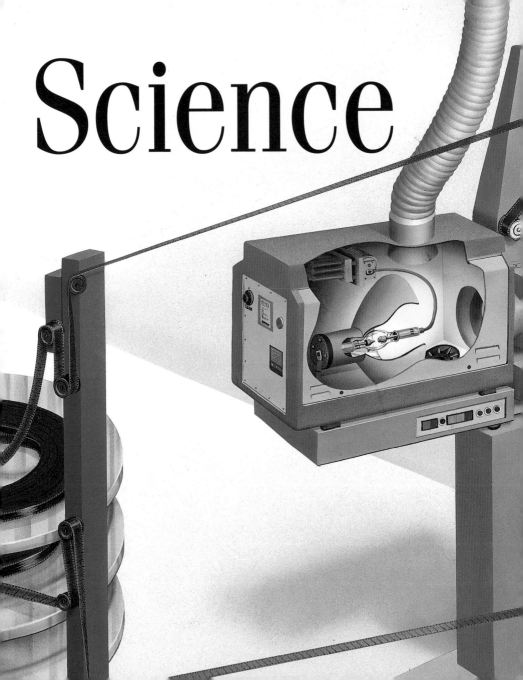

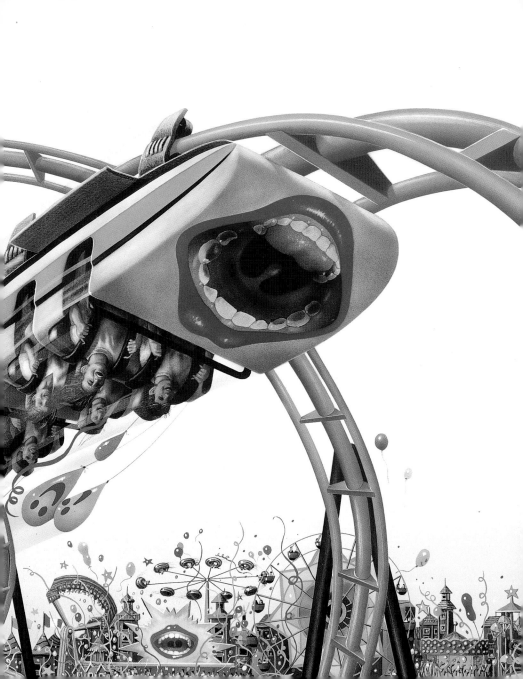

How Things Work

Why do space rockets need oxygen in their fuel?

How does a satellite stay in orbit?

How does a clock keep time?

Contents

• LEISURE AND ENTERTAINMENT •

• TRANSPORTATION •

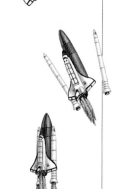

Wind Power

People have used the power of the wind for more than 5,000 years. It propelled their sailing boats over rivers, lakes and oceans; it turned the heavy blades of windmills to grind grain and pump water. Wind has energy because it is always moving in one direction or another. This energy can be caught, or harnessed, by large sails or blades. When electricity was developed in the nineteenth century, wind power did not seem as efficient as this marvelous new source of power, and most windmills disappeared. But wind power is making a comeback. Today, modern versions of windmills called wind turbines are used to generate electricity. Groups of wind turbines with long, thin metal or plastic blades, which look like airplane propellers on top of tall thin towers, are often erected together in wind farms that stretch across the landscape. By the middle of the twenty-first century, one-tenth of the world's electricity could be powered by wind turbines.

WIND FARMS
These are built in very windy areas and are controlled by computers that turn their blades into the wind. When the wind turns the blades, the spinning motion is converted into electricity.

Blades
The blades of the turbine are set at an angle that can be changed to suit the wind's speed or direction.

WIND-ASSISTED TANKER
This ship has stiff fiberglass sails as well as engines. It can save fuel by using sails whenever there is enough wind. Computers calculate the wind speed and indicate when it is time to unfold the sails.

Cables
Underground cables collect the electricity produced by the turbines at a wind farm.

Gearbox
The gearbox, driven by the turbine shaft, controls the speed of the generator.

Generator
The generator converts the spinning motion into electricity.

Turbine shaft
Wind turns the blades, which turn the central turbine shaft. The speed of the shaft varies according to the strength of the wind.

Nacelle
The nacelle (the part that contains the machinery) pivots to keep the blades pointing into the wind. The angle of the blades is set automatically to suit the wind speed.

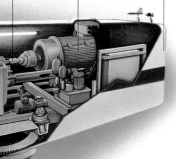

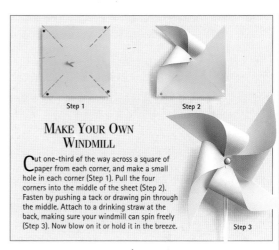

Step 1

Step 2

MAKE YOUR OWN WINDMILL

Cut one-third of the way across a square of paper from each corner, and make a small hole in each corner (Step 1). Pull the four corners into the middle of the sheet (Step 2). Fasten by pushing a tack or drawing pin through the middle. Attach to a drinking straw at the back, making sure your windmill can spin freely (Step 3). Now blow on it or hold it in the breeze.

Step 3

TIMES PAST
This kind of windmill was used many years ago to grind grain.

Cap
The cap carrying the sails could turn so that the sails faced into the wind.

Fantail
Wind blowing against the fantail made it spin and turned the mill cap until the sails faced the wind.

Tower
The tower holds the blades at a safe height above the ground and contains the cables that carry the electricity underground.

Grain hopper
Grain fell from a container, called a hopper, down to the two grindstones below.

Driveshaft
This used the turning motion of the sails to move the grindstones.

Canvas-covered sails
Canvas sheeting stretched over the wooden frame of the sails caught the wind and moved the sails around.

Grindstones
Two heavy stones rotated and crushed the grain beneath them.

Discover more in Riding on Air

The Ways of Water

Transmission lines
Strengthened electric cables called transmission lines carry electricity away from the power plant.

Spillway
The spillway gates are opened to release water when the level of water behind the dam is too high.

Water covers more than two-thirds of the Earth's surface and is constantly on the move. It rushes along rivers and streams; it flows into oceans. This endless movement of water creates energy that can be harnessed. For centuries, people have channeled flowing water into waterwheels that turn to grind grain. Hydroelectric power stations use water in a similar way, but to generate electricity. These enormous concrete constructions are usually found in mountainous regions where there is a high rainfall. Engineers build huge dams across steep-sided valleys. Turbines (modern versions of ancient wooden waterwheels) are placed in the path of the water that gushes with force through the dam. This torrent of water strikes the angled blades of the turbines, which begin to spin and extract an incredible amount of energy from the water. The process of producing hydroelectric power is set in motion.

Control room
The operation of the entire power plant is directed from the control room.

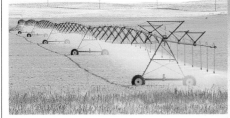

WATERING THE LAND
The water for this insectlike irrigation system is coming from the dam of a hydroelectric power station.

Drift tube
Water leaves the turbines through the drift tube.

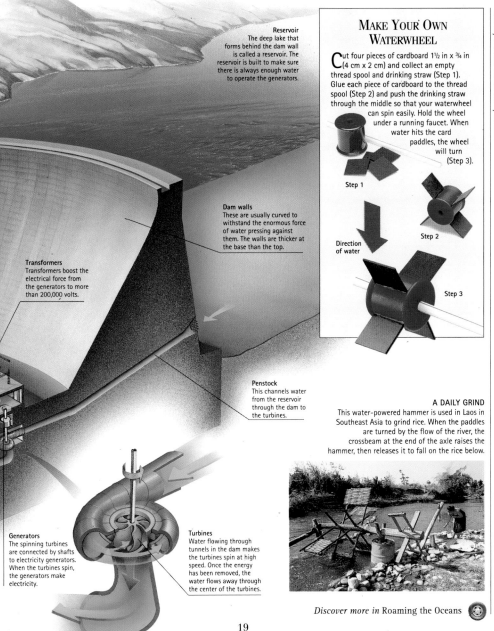

Reservoir
The deep lake that forms behind the dam wall is called a reservoir. The reservoir is built to make sure there is always enough water to operate the generators.

MAKE YOUR OWN WATERWHEEL

Cut four pieces of cardboard 1½ in x ¾ in (4 cm x 2 cm) and collect an empty thread spool and drinking straw (Step 1). Glue each piece of cardboard to the thread spool (Step 2) and push the drinking straw through the middle so that your waterwheel can spin easily. Hold the wheel under a running faucet. When water hits the card paddles, the wheel will turn (Step 3).

Step 1

Step 2

Direction of water

Step 3

Dam walls
These are usually curved to withstand the enormous force of water pressing against them. The walls are thicker at the base than the top.

Transformers
Transformers boost the electrical force from the generators to more than 200,000 volts.

Penstock
This channels water from the reservoir through the dam to the turbines.

A DAILY GRIND
This water-powered hammer is used in Laos in Southeast Asia to grind rice. When the paddles are turned by the flow of the river, the crossbeam at the end of the axle raises the hammer, then releases it to fall on the rice below.

Generators
The spinning turbines are connected by shafts to electricity generators. When the turbines spin, the generators make electricity.

Turbines
Water flowing through tunnels in the dam makes the turbines spin at high speed. Once the energy has been removed, the water flows away through the center of the turbines.

Discover more in Roaming the Oceans

Passing on the Power

TURBOGENERATOR
Electricity is made by a turbogenerator—a generator driven by a turbine. When a wire moves near a magnet, electricity flows along the wire. Inside the generator, strong magnets make electricity flow through coils of wire.

E lectricity has to be sent from the power station where it is made to the homes and businesses where it is used. Whether the power station is nuclear powered, hydroelectric or burns coal, the electricity it makes is distributed in the same way. Transformers at the power station boost the electricity to a very high voltage—hundreds of thousands of volts. The electricity is then carried by metal cables suspended from tall transmission towers, or pylons. It usually ends its journey by passing along underground cables. By the time it reaches your home, transformers have reduced its voltage to a level that depends on which country you live in. Electricity generated in one place can be sent to another part of the country if more power is needed.

Rotor
The rotor consists of coils of wire that rotate at high speed. Electric current flowing through the coils creates powerful magnetic fields around them.

Anode
A carbon rod acts as the positive electrode.

Electrolyte
This is a chemical paste.

Cathode
The zinc battery case forms the negative electrode.

ELECTRICITY DISTRIBUTION
Electricity generated at a power station is distributed through a network of cables above and below the ground.

Transformers
Transformers increase the voltage before electricity is transmitted.

BATTERIES
When a battery is connected to an electric circuit, a chemical reaction between the negative terminal (cathode) and a liquid or paste (electrolyte) creates a current. This current travels round the circuit and returns to the battery at the positive electrode (anode).

A BRIGHT IDEA

Most light bulbs contain a thin coiled wire filament that heats up and glows when an electric current flows through it. They are called incandescent bulbs (left). An energy efficient bulb (right) is a fluorescent tube that needs less electricity to produce the same amount of light as a normal bulb. When an electric current passes through mercury vapor inside the tube, the vapor releases invisible ultraviolet rays, and the coating on the inside of the bulb converts them into visible light.

DID YOU KNOW?

Power stations have to be ready to boost electricity production whenever demand suddenly increases. In many countries, television schedules help to predict power demands! At the end of films or major sporting events, the demand for electricity soars as millions of television viewers switch on their electric kettles to make tea or coffee.

Stator
The stator, which does not move, is made from coils of wire surrounding the rotor. As the rotor turns, its magnetic fields cut through the stator coils and make an electric current flow through them.

Power take-off cables
Thick cables lead electric current away from the generator.

Transmission towers
The transmission lines are held high above the ground by tall transmission towers. Glass or ceramic insulators between the metal towers and the cables stop the current from running down the towers into the ground.

Transmission lines
Cables strengthened by steel carry the current.

Street transformer
Before electricity reaches your home, its voltage is reduced by transformers. The voltage level depends on the country you live in.

Home
Electricity enters your home through a meter that measures how much electricity is used.

SAVING ENERGY
An energy-efficient house is designed to minimize
energy waste. It generates its own electricity, but it
is still connected to the national power grid. If it
generates more electricity than it needs, the excess
is supplied to the grid. If it needs more electricity,
this is supplied by the grid.

Keeping warm
Most of the heat lost by a house
escapes through the roof. The roof
of an energy-efficient house is
lined with insulation material to
stop heat from escaping.

Solar panels
When the sun shines on a solar
panel, solar energy is converted
into electricity to power electrical
appliances in the house, such as
water heaters or cooling fans.

• USING THE ELEMENTS •

Harnessing the Sun

The sun is an extraordinarily powerful form of energy. In fact, the Earth receives 20,000 times more energy from the sun than we currently use. If we used much more of this source of heat and light, it could supply all the power needed throughout the world. We can harness energy from the sun, called "solar" energy, in many ways. Satellites in space have large panels covered with solar cells that change sunlight directly into electrical power. Some buildings have solar collectors that use solar energy to heat water. These panels are covered with glass and are painted black inside to absorb as much heat as possible. Some experimental electric cars are even powered by solar panels. Solar energy is a clean fuel, but fossil fuels, such as oil or coal, release harmful substances into the air when they are burned. Fossil fuels will run out eventually, but solar energy will continue to reach the Earth long after the last coal has been mined and the last oil well has run dry.

The sunny side
The house is built with one
long side facing the sun so
that it can absorb as much
solar energy as possible
during the day.

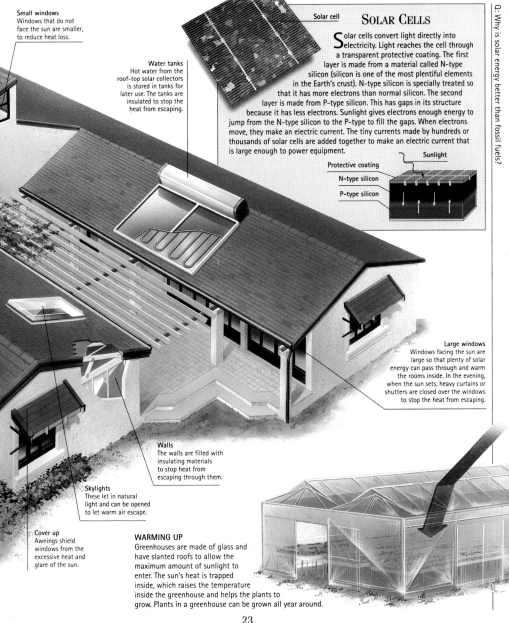

Small windows
Windows that do not face the sun are smaller, to reduce heat loss.

Water tanks
Hot water from the roof-top solar collectors is stored in tanks for later use. The tanks are insulated to stop the heat from escaping.

Solar cell

SOLAR CELLS

Solar cells convert light directly into electricity. Light reaches the cell through a transparent protective coating. The first layer is made from a material called N-type silicon (silicon is one of the most plentiful elements in the Earth's crust). N-type silicon is specially treated so that it has more electrons than normal silicon. The second layer is made from P-type silicon. This has gaps in its structure because it has less electrons. Sunlight gives electrons enough energy to jump from the N-type silicon to the P-type to fill the gaps. When electrons move, they make an electric current. The tiny currents made by hundreds or thousands of solar cells are added together to make an electric current that is large enough to power equipment.

Sunlight

Protective coating
N-type silicon
P-type silicon

Large windows
Windows facing the sun are large so that plenty of solar energy can pass through and warm the rooms inside. In the evening, when the sun sets, heavy curtains or shutters are closed over the windows to stop the heat from escaping.

Walls
The walls are filled with insulating materials to stop heat from escaping through them.

Skylights
These let in natural light and can be opened to let warm air escape.

Cover up
Awnings shield windows from the excessive heat and glare of the sun.

WARMING UP

Greenhouses are made of glass and have slanted roofs to allow the maximum amount of sunlight to enter. The sun's heat is trapped inside, which raises the temperature inside the greenhouse and helps the plants to grow. Plants in a greenhouse can be grown all year around.

Escapement
This regulates the speed of the clock. It consists of an anchor that rocks from side to side, and an escape wheel that is repeatedly caught and released by the anchor.

Hour hand
The hour hand makes one revolution every 12 hours.

Minute hand
The minute hand moves 12 times faster than the hour hand and makes one revolution every hour.

Pendulum
The swinging pendulum regulates the rocking motion of the anchor.

• MACHINES •

About Time

People have been keeping the time for thousands of years. The first time-keeping devices were very inaccurate. They measured time by the sun, or by the falling levels of water or sand. Mechanical clocks are much more accurate. They have three main parts: an energy supply, a mechanism for regulating the energy and a way of showing the passing of time. The energy is supplied by a coiled spring or a weight. The spring unwinds, or the weight falls, and turns a series of interlocking, toothed wheels. Hands linked to the wheels rotate around a dial. For the clock to be accurate, the hands must turn at a constant speed. In large clocks, a pendulum swings at a constant rate and regulates the movement of the escapement. Digital or electronic watches have a piece of quartz crystal that vibrates at 32,768 times a second. An electronic circuit uses these movements to turn the hands or change numbers on the watch face.

ON YOUR MARK, GET SET, GO!
Athletes often cross the finish line at exactly the same moment and it is difficult to decide who has won the race. Officials accurately record the athletes' race times so that very close finishes can be separated by degrees of a second.

Gears
These make sure that the minute hand goes around 12 times faster than the hour hand.

Weight
This hangs on a cord wound around a shaft so that the weight turns the shaft to move the gears.

KEEPING TIME
Athletes train hard for their events. Stopwatches can help them monitor their progress by measuring times to within 100th of a second. Some stopwatches can also store up to 100 laps in their memories and even print times using built-in printers.

THINGS IN COMMON

Pendulum clocks and digital watches are very different in size, but they are made from the same basic building blocks. Both have an oscillator that moves or swings at a regular rate (left), a device that turns these movements into time-keeping pulses (center) and a display for showing the time (right).

Pendulum

Escapement

Display

Crystal

Circuit

Display

Discover more in Computer Friendly

25

A vacuum cleaner works in a similar way to a straw. When you drink through a straw you suck out the air and this draws up the fluid. A vacuum cleaner creates a powerful flow of air that sucks up dust and dirt through a hose and traps it inside the machine.

FAST FOOD

A microwave oven uses powerful radio waves of a very short wavelength (microwaves) to cook food very quickly. These waves heat the inside as well as the outside of food immediately. In more traditional ovens, the heat takes longer to cook the inside of the food.

• MACHINES •

Saving Time and Effort

We use machines around the house every day. They make our lives easier and give us time to do other things. Hundreds of years ago, for example, household chores took most of the day. Water was carted from a well, food was cooked over an open fire, and houses were swept with branches. Today, most homes have labor-saving devices, which are designed to make jobs around the home less of an effort. Washing machines automatically wash and rinse clothes, then spin them to force out most of the water. Some even dry the clothes completely. Refrigerators and freezers keep food fresh longer so that we do not need to shop every day. Dishwashers, remote controls for televisions and videos, microwave ovens and vacuum cleaners are some of the appliances found in many homes throughout the world.

Dust bag
Air carries dust and dirt into the dust bag. The air then escapes through tiny holes in the bag and leaves the dust trapped inside. Some vacuum cleaners have a "micro-filter," with even smaller holes in it, to trap the tiniest dust particles.

Fan
A spinning fan sucks air and dust through the flexible hose into the vacuum cleaner.

Motor
The fan is driven by an electric motor. Some vacuum cleaners can vary the speed of the motor so that the suction power can be adjusted to clean different surfaces.

Insulation
The oven is double-walled and insulated. This stops heat from leaking out of the oven.

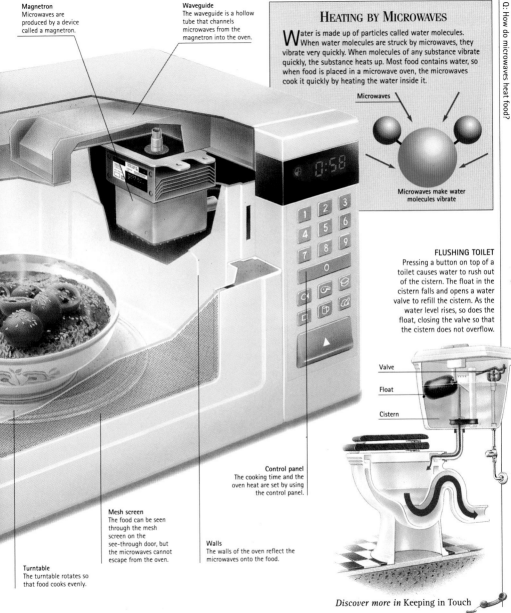

Magnetron
Microwaves are produced by a device called a magnetron.

Waveguide
The waveguide is a hollow tube that channels microwaves from the magnetron into the oven.

Heating by Microwaves

Water is made up of particles called water molecules. When water molecules are struck by microwaves, they vibrate very quickly. When molecules of any substance vibrate quickly, the substance heats up. Most food contains water, so when food is placed in a microwave oven, the microwaves cook it quickly by heating the water inside it.

Microwaves

Microwaves make water molecules vibrate

FLUSHING TOILET
Pressing a button on top of a toilet causes water to rush out of the cistern. The float in the cistern falls and opens a water valve to refill the cistern. As the water level rises, so does the float, closing the valve so that the cistern does not overflow.

Valve

Float

Cistern

Control panel
The cooking time and the oven heat are set by using the control panel.

Mesh screen
The food can be seen through the mesh screen on the see-through door, but the microwaves cannot escape from the oven.

Walls
The walls of the oven reflect the microwaves onto the food.

Turntable
The turntable rotates so that food cooks evenly.

Discover more in Keeping in Touch

27

Office Essentials

Modern businesses depend on being able to send and receive information quickly. Telephones enable people to talk to each other over long distances, but the worldwide telephone network carries much more than people's voices. Computers and fax machines, for example, use this network to send information to each other. "Fax" is short for "facsimile transmission" (facsimile means copy). A fax machine can transmit a copy of an image on paper—a printed document, handwritten message or drawing—to anywhere in the world within seconds. It does this by changing the information on the paper into electrical signals, then converting these into sounds that are sent along normal telephone lines. Another fax machine receives the sounds and changes them back into a printed copy of the original image. Computers exchange information by telephone in the same way. Some of the information exchanged by computers is called electronic mail or E-mail, because it is an electronic version of the ordinary postal system.

Drum
The rotating drum attracts black toner powder onto itself, then transfers the powder onto the paper. This creates an image on the page.

Print head
A row of lights flashes on and off as the charged drum rotates next to it. The electric charge on the drum is weakened wherever light strikes it. Black toner powder sticks only to the uncharged parts of the drum and forms an image of the transmitted document on it.

Numerical keypad
Telephone numbers are dialed by using this pad.

One-touch keys
Frequently used telephone numbers are keyed into the fax machine and stored in an electronic memory. When a number is selected from the memory, by pressing one of these buttons, the machine dials it automatically.

Image sensor
The electrical signal produced by the photosensor is changed into sounds that are transmitted down a telephone line.

FAXING A MESSAGE

The fax machine divides the image on the paper into a grid of tiny squares and detects whether each square is light or dark.

Each square is registered as either completely black or completely white.

The pattern of black and white squares is changed into an electrical signal. A pulse of electricity represents a dark square.

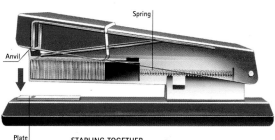

Spring

Anvil

Plate

STAPLING TOGETHER

A stapler fastens sheets of paper together with short lengths of wire called staples. The staples are glued in a row, but when the stapler's jaws are pressed down over the paper, one staple is separated from the rest and forced through the paper. A metal plate with specially shaped grooves in the base bends the ends of the staple inward so that it cannot fall out again.

STICK-ON NOTES

Stick-on notes can be stuck to almost any surface, peeled off again easily and stuck somewhere else. The secret lies in the gum on the back. It is not as sticky as the gum on adhesive tape, and the notes can be peeled off paper without damaging it.

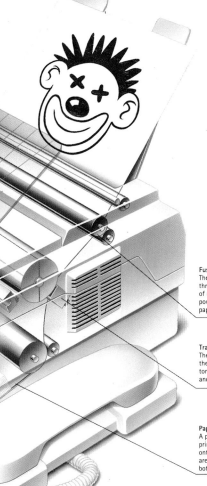

Fuser unit
The paper leaves the machine through the fuser unit, a pair of rollers that melts the toner powder and fuses it to the paper by heat and pressure.

Transfer unit
The transfer unit charges the paper so that it attracts toner powder off the drum and onto the paper.

Paper
A plain paper fax machine prints copies of documents onto sheets of paper that are stored in a tray at the bottom of the machine.

This simple electrical signal is changed into a complex code. As documents are usually printed on white paper, white is given a shorter code than black. The code is changed into sounds that can be sent down a telephone line.

The receiving fax machine changes the sounds into an electrical signal, which controls a printer.

By printing line after line of black spots, the receiving fax machine builds up a copy of the original document.

Jib
The cross arm, or jib, suspends the hook that lifts the load. The jib is suspended by cables or steel rods from the top of the tower.

Counterweight
Concrete slabs on one side of the tower balance the weight of the loads it lifts on the other side.

Operator's cab
The crane operator sits in a cab at the top of the tower and moves the load by operating controls. The front of the cab is made from glass to give the operator a clear view of the hook, from the ground up.

• MACHINES •

Building Upward

Machines are used in the construction industry to lift, move, cut, drill and connect the various materials used. A tower crane, for example, is used to lift heavy materials up to the workers. The horizontal arm of the tower crane is called the jib. This can swing around horizontally, but it cannot be raised or lowered. The crane's hook is raised and lowered by winding the cable it hangs from around a large motorized drum. Thousands of tons of materials have to be delivered to the construction site. One of these materials, concrete, is delivered by concrete mixers. Their drums rotate constantly to stop the concrete inside from setting hard before it is poured out wherever it is needed. Pulleys (wheels with grooves around their rims) are often used on building sites to help move very heavy loads. When a rope or chain is threaded around several pulleys, a pull on the rope or chain is enhanced, or magnified, by the pulleys to enable a small effort to move a heavy load.

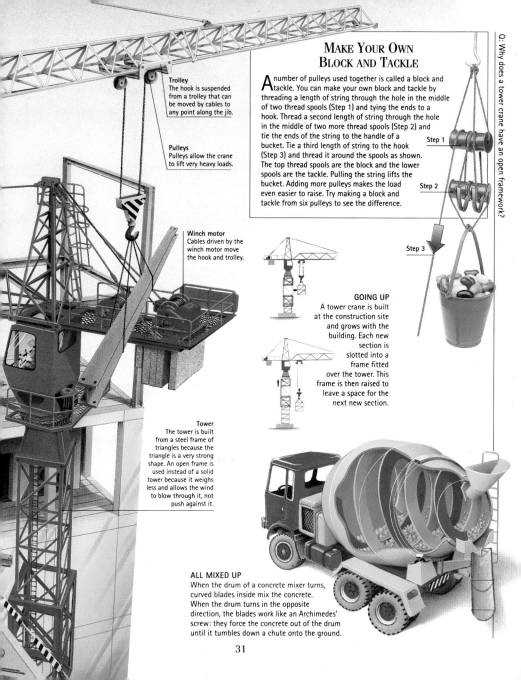

Trolley
The hook is suspended from a trolley that can be moved by cables to any point along the jib.

Pulleys
Pulleys allow the crane to lift very heavy loads.

MAKE YOUR OWN BLOCK AND TACKLE

A number of pulleys used together is called a block and tackle. You can make your own block and tackle by threading a length of string through the hole in the middle of two thread spools (Step 1) and tying the ends to a hook. Thread a second length of string through the hole in the middle of two more thread spools (Step 2) and tie the ends of the string to the handle of a bucket. Tie a third length of string to the hook (Step 3) and thread it around the spools as shown. The top thread spools are the block and the lower spools are the tackle. Pulling the string lifts the bucket. Adding more pulleys makes the load even easier to raise. Try making a block and tackle from six pulleys to see the difference.

Step 1

Step 2

Step 3

Winch motor
Cables driven by the winch motor move the hook and trolley.

GOING UP
A tower crane is built at the construction site and grows with the building. Each new section is slotted into a frame fitted over the tower. This frame is then raised to leave a space for the next new section.

Tower
The tower is built from a steel frame of triangles because the triangle is a very strong shape. An open frame is used instead of a solid tower because it weighs less and allows the wind to blow through it, not push against it.

ALL MIXED UP
When the drum of a concrete mixer turns, curved blades inside mix the concrete. When the drum turns in the opposite direction, the blades work like an Archimedes' screw: they force the concrete out of the drum until it tumbles down a chute onto the ground.

31

Making Shopping Easy

Technology has made shopping quick and convenient. Automated teller machines (ATMs) give us immediate access to our money and reduce the need to stand in long lines inside the bank. Personal identification numbers (PINs) and cash cards replace passbooks and withdrawal slips. All cash cards have a magnetic strip on their backs. When the code and information stored here match the information in the bank's computer, the machine gives you the requested amount of money. Computers and lasers speed up the service at checkout counters in stores. A laser scans the barcode of every product and tells the computerized cash register how much the product costs. It also records how many of a certain product have sold, so that more stock can be ordered when necessary. Some products have security tags attached to them. If anyone tries to take a tagged product out of the store, sensors at the door detect the tag and sound an alarm.

PRICE SCAN
As the laser beam scans the barcode, a light-sensitive sensor in the handset picks up its reflections. An electric current flows through the sensor and produces electrical pulses in a pattern that matches the barcode's pattern of black lines. A pair of thin lines in the middle divides the barcode in two. The first half of the code contains the manufacturer's name and the second half is the code for the product. The pairs of thin lines at each end of the pattern tell the computer where the code starts and finishes. A barcode computer can always "translate" the code lines because the same standard—the Universal Product Code—is used for barcodes all over the world.

> ### DID YOU KNOW?
> The printing on paper currency can be copied by forgers. In 1988, the Commonwealth Reserve Bank of Australia issued a plastic folding banknote that is very difficult to copy.

Currency cassettes
Banknotes are stored in boxes called cassettes.

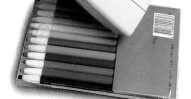

Barcode scanner
A laser in the handset scans the barcode. The sensor identifies the product and its price, then sends these details to the cash register.

BANKING MADE EASY
An automated teller machine (ATM) enables people to withdraw money from an account. Most ATMs can also show us how much money is in our account and can transfer money from one account to another.

Card reader
The card is drawn inside the machine by motorized rollers. Information recorded invisibly on a magnetic strip on its back is read in a way that is similar to a tape being played in a tape recorder.

Screen
The screen gives step-by-step instructions for using the machine.

Printer
The printer prints out a record of the cash withdrawal, and the ATM pushes out a receipt for the customer.

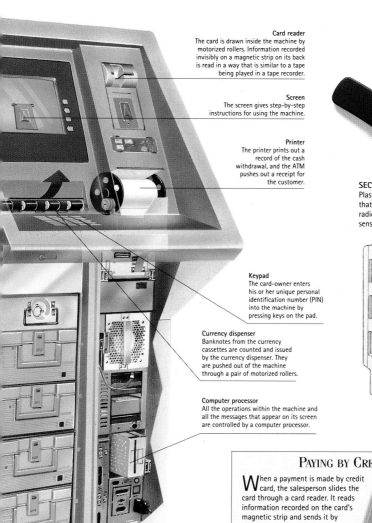

Keypad
The card-owner enters his or her unique personal identification number (PIN) into the machine by pressing keys on the pad.

Currency dispenser
Banknotes from the currency cassettes are counted and issued by the currency dispenser. They are pushed out of the machine through a pair of motorized rollers.

Computer processor
All the operations within the machine and all the messages that appear on its screen are controlled by a computer processor.

SECURITY TAGS
Plastic security tags contain coils of wire that can be detected magnetically or by radio waves when they pass through the sensors at the store entrance.

PAYING BY CREDIT CARD

When a payment is made by credit card, the salesperson slides the card through a card reader. It reads information recorded on the card's magnetic strip and sends it by telephone to a central computer. This checks the details and approves the payment. Later, the card's owner receives a bill for this payment. Payments are also made in this way with debit cards, but the amounts are transferred from the card-owner's bank to the store.

Discover more in Recording Sound

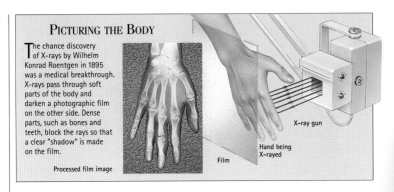

• MACHINES •

At the Hospital

In hospitals, medical staff use machines and instruments to help people. Ultrasound scanners, X-ray machines and other medical equipment can be used to diagnose an illness, treat an injury or monitor changes in the patient's condition. More complex scanners use the high-speed processing power of computers to create intricate pictures. These scanners can show cross-sections of a body, three-dimensional views of internal organs, and pictures of the brain showing which parts of it are active while the patient is thinking, seeing, hearing or moving. These pictures can be seen on the scanner's own special screen, and can also be printed out on paper or on film. Such a detailed view of a disease or an injury allows doctors to see all the angles of a medical problem and then figure out the best way to treat it.

Hand–held probe

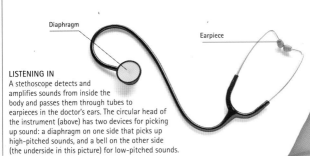

Diaphragm

Earpiece

LISTENING IN
A stethoscope detects and amplifies sounds from inside the body and passes them through tubes to earpieces in the doctor's ears. The circular head of the instrument (above) has two devices for picking up sound: a diaphragm on one side that picks up high-pitched sounds, and a bell on the other side (the underside in this picture) for low-pitched sounds.

34

AN INSIDE VIEW

This woman is having an ultrasound examination of her baby. The probe that is held on her stomach sends bursts of ultrasound down into her body. It also receives the reflections bouncing back again. Reflections from deeper inside the patient take longer to bounce back. The machine records the different "flight times" of the sound waves and produces a picture of a part of the body. Unborn babies are often examined in this way.

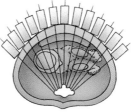

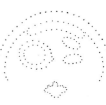

Scanning
The ultrasound probe is moved from side to side, sending ultrasonic vibrations down into the patient's body. When the ultrasound vibrations strike anything inside the body, some are reflected. Others pass through to be reflected by deeper layers.

Generating an image
The ultrasound reflections are received by the probe and combined by a computer to make a picture of the patient's internal organs. If the patient is a pregnant woman, an ultrasound scan shows a picture of her unborn baby and its internal organs.

Inside information
An ultrasound operator can tell from the picture on the machine's screen whether an unborn baby is a boy or a girl. He or she can also examine the baby's internal organs, especially the heart, to make sure that the fetus is developing normally. Ultrasound can also confirm the number of babies the mother is carrying.

DID YOU KNOW?

In the 1950s, doctors realized that an unborn baby in its mother's fluid-filled womb was like a submarine in the sea. Submarines use a system called sonar (from SOund Navigation And Ranging) to detect objects near them. Sonar sends out bursts of ultrasound and detects reflections that bounce back from solid objects. This system was adapted and used to examine people in hospitals.

Keeping in Touch

R adio waves are vibrating, invisible waves of energy. They are similar to light waves, and are very useful for carrying information across great distances. Radio and television programs, for example, travel from transmitters all over the world to our homes. As radio waves can travel through outer space, astronauts' voices and information collected by satellites can also be transmitted by radio. Many natural objects in the universe send out radio signals that radio telescopes on Earth can receive. Whatever a radio receiver is used for, it always has the same parts. An aerial, or receiving antenna, picks up the radio signals and feeds them down a cable to the receiver. A tuner selects particular signals and discards all the rest, and an amplifier strengthens the selected signals. A radio telescope receives data, which can be displayed as perhaps pictures or charts; while a radio at home changes the radio signals it receives into sound.

RADIO TELESCOPE
A radio telescope forms images of the sky from very faint radio signals. They are so weak that they cannot be used until a series of amplifiers makes them 1,000 million million times larger. The telescope scans an object in the sky from side to side and builds up a picture of it from a series of horizontal lines.

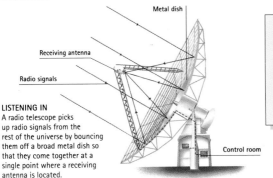

Metal dish

Receiving antenna

Radio signals

LISTENING IN
A radio telescope picks up radio signals from the rest of the universe by bouncing them off a broad metal dish so that they come together at a single point where a receiving antenna is located.

Control room

DID YOU KNOW?
The world's largest radio telescope is in Arecibo, Puerto Rico. It was built by lining a natural hollow in the ground with wire mesh to form a reflecting dish 1,000 ft (305 m) across. The Arecibo telescope cannot be moved, but most radio telescopes can be tilted and turned to point at any part of the sky.

When you switch on your radio, you might listen to a tape or a disk, or to someone speaking into a microphone at a studio.

An engineer sitting at a mixing desk adjusts the signals from the studio so that no signal is too large or too small.

Mixing desk

Studio signal

Carrier wave

The studio signal is mixed with a high-frequency (rapidly vibrating) signal that is called a carrier wave (it carries the studio signal), and then beamed across the country from a transmitter.

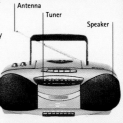

THE RADIO SKY
Pictures made by radio telescopes are not like normal pictures of the sky. Radio waves have no color, so the colors in a radio telescope picture are added by computer.

Frequency modulation (FM)

Amplitude modulation (AM)

The studio signal changes either the frequency (speed of vibration) of the carrier wave, which is called frequency modulation; or the carrier wave's amplitude (size), which is called amplitude modulation.

Various signals flow around a radio antenna and create tiny electrical currents. The radio is tuned into just one of these signals, which is separated from its carrier wave. The signal is then amplified (made more powerful) by the radio, and the speaker changes it back into sound.

Antenna

Tuner

Speaker

• COMMUNICATIONS •

Messages from Space

In the sky
A low-flying satellite orbits at a height of about 155–186 miles (250-300 km) just outside most of the Earth's atmosphere. It can dip down to as low as 74 miles (120 km) to take close-up photographs of interesting places. Its advanced camera systems can see details as small as 2 in (5 cm) across.

S atellites circling the Earth send us pictures of the weather and relay telephone calls and television programs around the world. They also study vast areas of the Earth and its oceans, taking photographs and measurements with their cameras and instruments and beaming them down to Earth by radio. Some satellites circle the planet from pole to pole; others circle around the equator. Most satellites orbit the Earth at a height of between 124 miles (200 km) and 496 miles (800 km) and have to travel at a speed of 5 miles (8 km) per second to stay in orbit. Communications and weather satellites are boosted to a height of 22,320 miles (36,000 km)–much higher than other satellites. At this height above the equator, a satellite circles the Earth once every 24 hours, the same time the Earth takes to spin once on its axis. This kind of orbit is called "geostationary" because the satellite seems to hover over the same spot on Earth. It takes three satellites in geostationary orbit to relay telephone calls between any two points on Earth.

Communications satellite
A communications satellite, or comsat, works a little like a mirror in the sky. It receives radio signals beamed up to it from Earth, amplifies them and sends them back to a different place on Earth.

A LIVE BROADCAST
Satellites enable events such as the Olympic Games to be watched anywhere in the world seconds after they happen.

Gas tanks
The satellite uses jets of gas from its gas tanks to stop it from drifting out of position.

Communications circuits
The satellite's communications circuits can relay tens of thousands of telephone calls at the same time.

38

PICTURING THE WEATHER
Satellite pictures can help a
weather forecaster see how
weather systems, such as
cyclones, grow and move
across the oceans. Views
such as this would be
impossible to obtain
from the ground.

Weather satellite
The world's weather constantly
changes, and the temperatures of the
sea, the land and the clouds vary all
the time. A weather satellite carries
heat-sensitive cameras that
continually monitor the weather.

STAYING IN ORBIT

If you could throw a ball hard enough, it would fly
all the way around the Earth, because the curve of
its fall would exactly match the curve of the Earth's
surface. To see this in action, make two plastic balls—
one 2 in (5 cm) across to
represent gravity, and
one ¾ in (2 cm) across to
represent a satellite.
Thread 20 in (51 cm) of
string through a thread
spool (Earth), and tie
each end to a key.
Push each key into one of
the balls. Hold the thread
spool and the large ball and start
the small ball spinning. Let the
large ball go. The satellite tries to
fly away from Earth but gravity pulls
it back. When the two forces are
balanced, the satellite orbits Earth.

Solar panels
Solar panels change
sunlight into electricity to
supply power for the
satellite's radio equipment.

Discover more in Reaching into Space

TELEPHONE
A telephone converts the sound of a caller's voice into an electric current and changes electric currents received from other telephones into sound.

Coil
An electrical signal received by the earpiece passes through a coil and creates a weak magnetic field around it.

Magnet
The magnet attracts or repels the coil and makes it vibrate.

Diaphragm
The diaphragm then vibrates to create a sound.

Diaphragm
The mouthpiece works in the opposite way to the earpiece. The caller's voice makes the diaphragm vibrate.

Coil
A coil of wire fixed to the diaphragm vibrates next to a magnet. The vibrations create an electric current that is sent on.

• COMMUNICATIONS •

Shrinking the World

Distances are usually measured in miles or kilometers, but they can also be measured in time—the time it takes to communicate over a certain distance. In past centuries, the distance to the next town might have been measured by the time it took to travel there on foot or on horseback. A more distant town might be a week away and another continent might be several months away by ship. With telephones we can now communicate with someone thousands of miles away just as quickly as we can with someone in the next room. The size of the world seems to have shrunk. Electrical communication works by changing information into electrical signals that can be sent along cables. The first telephone calls made their entire journey as electrical signals in metal cables. Today, telephone calls can also travel in the form of infrared beams along fiber-optic cables or as radio waves relayed by satellites in space.

THE PATH OF A TELEPHONE CALL

Telephones are connected to a network of exchanges that are linked together by copper cables, fiber-optic cables or radio. Every telephone is identified by a unique number and the path a telephone call takes depends on the number that is dialed. Most telephones stay in one place and are connected to the nearest exchange by cable, but some are portable. Mobile telephones can be carried anywhere, even to another country. Every few minutes, they send out coded radio signals that identify them and let the network know where they are. This enables calls to be transmitted to the correct mobile telephone from the nearest radio antenna.

Antenna
The antenna detects radio waves for the telephone's radio receiver, and also sends them out from its transmitter.

MOBILE PHONE
Mobile telephones are linked to the international telephone network by radio. Every mobile telephone contains its own radio receiver and transmitter.

STRANGE BUT TRUE
The fiber optics that carry telephone calls are made of highly refined glass. A 12-mile (19-km) block of it would be as clear to see through as an ordinary window pane. A few fibers, enough to carry 100,000 telephone calls at the same time, can pass through the eye of a needle.

Earpiece
The earpiece changes electrical signals from the radio receiver into sound.

Display
A liquid crystal display shows the number being dialed.

FIBER-OPTIC CABLES
In many parts of the world, metal telephone cables are being replaced by fiber-optic cables made of glass. Telephone calls travel along these cables as flickering infrared beams. Fiber-optic cables are much thinner, yet carry more calls than metal cables.

Battery
A mobile telephone is powered by a battery. When the battery runs out of energy, it can be recharged (filled with more electrical energy) by a charging unit.

Keypad
Telephone numbers are dialed by pressing these keys.

Microphone
The microphone changes the speaker's voice into electrical signals.

1. Local exchange
A telephone call is sent through the caller's local exchange to the nearest main exchange by cable.

2. Main exchange
The main exchange sends the call on its way to the next main exchange via cables, optical fibers or radio signals.

4. Cell base station
The call is then sent to the mobile telephone from a nearby radio antenna.

3. Mobile network
Calls made to a mobile telephone are sent through the mobile telephone network.

Computer Friendly

PERSONAL COMPUTER
Every computer, whatever its size and complexity, contains four basic elements: the input device, usually a keyboard; the memory, where information is stored; the central processing unit (CPU), which carries out the instructions; and the output device, usually a monitor and printer.

Computers have become part of our everyday lives. We use them to store a vast amount of information and process it very quickly. The processing and storage are carried out by microscopic electronic circuits called chips. The master chip, the microprocessor, controls the computer. A microprocessor may contain several hundred thousand electronic components in a space that is no bigger than your thumbnail. The chips and the rest of the equipment form only one part of a computer, the hardware, but the computer also needs instructions to tell the hardware what to do. These instructions are called computer programs, or software. Software can make a computer perform a huge range of different jobs. It may turn the computer into a word processor for writing and storing documents, a games machine for having fun, an educational tool or a very fast calculating machine.

Monitor
A computer monitor looks like a small television screen. It receives and displays information from the computer.

Ball | Wheel

Buttons
A mouse may have one, two or three buttons.

CD-ROM drive
A CD-ROM (Compact Disk Read-Only Memory) can hold all sorts of information that can be copied from the CD-ROM onto the computer, but new information cannot be recorded on it.

DRIVING THE MOUSE
A mouse is used to steer a pointer around the screen. When you move the mouse, a ball underneath it rolls and makes two slotted wheels turn. As each wheel turns, a light shines through the slots and the flashes are detected by a sensor. The number and speed of the flashes show how far, how fast and in which direction the mouse is moving. When you "click" the button or buttons on the mouse, you select different options on the screen.

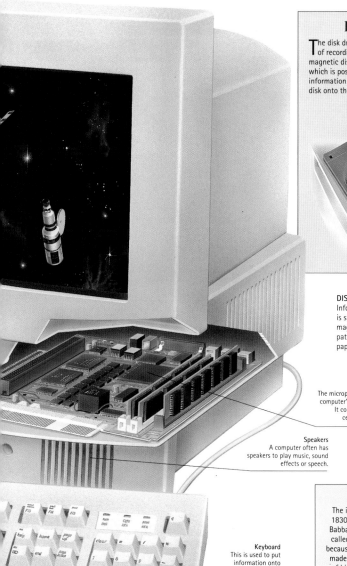

FLOPPY DISK DRIVE

The disk drive works like a tape recorder, but instead of recording information on magnetic tape it uses magnetic disks. With a disk slotted into the drive, which is positioned at the front of the computer, information can be copied onto the disk or from the disk onto the computer.

Read–write head
The read–write head records information onto the disk and reads it again when it is needed.

Case
A stiff plastic case protects the delicate disk.

DISK
Information is stored as magnetic patterns on a paper-thin disk.

Microprocessor
The microprocessor is a personal computer's master control chip. It contains the computer's central processing unit.

Speakers
A computer often has speakers to play music, sound effects or speech.

Keyboard
This is used to put information onto the computer.

DID YOU KNOW?

The idea of the computer dates back to the 1830s when English mathematician Charles Babbage tried to build a calculating machine called the Analytical Engine. Babbage failed because the parts for his machine could not be made with enough precision. However, many of his ideas were used more than 100 years later when the first computers were built.

Discover more in Recording Sound

43

View of a butterfly wing
with the naked eye

15 x magnified view
of a butterfly wing

50 x magnified view
of a butterfly wing

Seeing clearly
The image is seen
by looking into the
eyepiece. It contains
one or more lenses.

MICROSCOPE
The simplest microscope is a magnifying glass.
However, a single lens can only magnify an object up
to 15 times. For greater magnifications, a compound
microscope with several lenses is used.

Objective lenses
Three or more objective lenses with
a range of magnifying powers are
fitted to a microscope. Each can be
rotated to focus on the specimen.

Specimen
The specimen is placed
between two pieces of
glass, which must be thin
enough for light to pass
through them.

Light source
Light is reflected up to the
specimen using an angled mirror.

Focusing knob
The object can be brought into
sharp focus by turning the
focusing knob. This moves the
eyepieces closer to or farther
away from the objective lenses.

Eyepiece
Each eyepiece contains lenses that
magnify the image. One eyepiece
can be adjusted to allow for
differences between the eyes.

· LEISURE AND ENTERTAINMENT ·

A Closer Look

The human eye is an amazing organ, but there are
many things it cannot see because they are too
small or too far away. We use instruments to
magnify tiny details so they are big enough for us to
see. Microscopes, for example, can make small objects
look 2,500 times bigger than they really are. Telescopes
and binoculars produce magnified images of objects that
appear too small to see clearly because they are so far
away. These instruments are built with lenses, because
lenses can bend light rays and make our eyes think that
the light has come from a much larger object. Light
enters the instruments through a type of lens called an
objective lens. This forms an enlarged image of the
object. We look at this image through another lens, the
eyepiece, which magnifies it a little more.

44

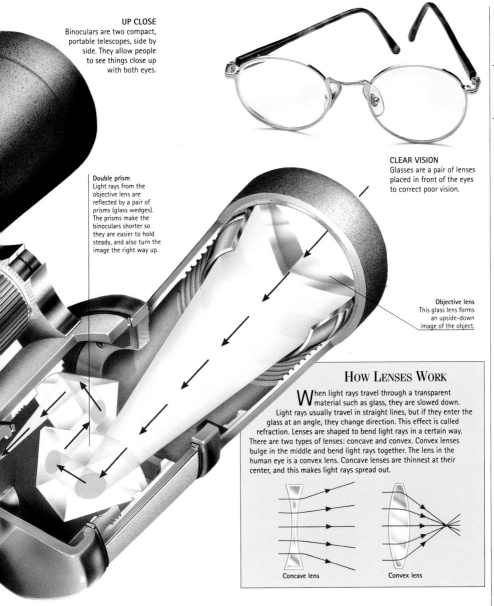

UP CLOSE
Binoculars are two compact, portable telescopes, side by side. They allow people to see things close up with both eyes.

Double prism
Light rays from the objective lens are reflected by a pair of prisms (glass wedges). The prisms make the binoculars shorter so they are easier to hold steady, and also turn the image the right way up.

CLEAR VISION
Glasses are a pair of lenses placed in front of the eyes to correct poor vision.

Objective lens
This glass lens forms an upside-down image of the object.

HOW LENSES WORK

When light rays travel through a transparent material such as glass, they are slowed down. Light rays usually travel in straight lines, but if they enter the glass at an angle, they change direction. This effect is called refraction. Lenses are shaped to bend light rays in a certain way. There are two types of lenses: concave and convex. Convex lenses bulge in the middle and bend light rays together. The lens in the human eye is a convex lens. Concave lenses are thinnest at their center, and this makes light rays spread out.

Concave lens   Convex lens

In Focus

A camera allows us to take photographs of people or scenes. Cameras vary enormously in their complexity but they all operate on the same principle–light enters the camera and falls on the film inside. Light rays enter a camera through a lens, which focuses them to form a sharp image on the film. The camera is aimed by looking through a window called the viewfinder. Many cameras automatically control the amount of light entering them, which makes them simple to use. Other cameras allow you to manually control the amount of light that falls on the film, but this makes them more complicated to use. Some cameras have separate lenses for forming the image on the film and in the viewfinder. One popular type of camera, the single lens reflex (SLR), can have both automatic and manual control. It uses the same lens for both so that the photographer always sees precisely the same scene that will be photographed.

SLR CAMERA
A single lens reflex (SLR) camera can be fitted with a variety of different lenses. A wide-angle lens is used for broad scenes, while a macro lens is added for close-ups. A zoom lens varies the magnifying power.

Winding on
After a photograph is taken, the film winder is turned to move a new piece of film behind the shutter, ready for the next photograph.

Smile!
The shutter release button is pressed to take a photograph.

Film
A springy plate presses the film into position behind the shutter, and also keeps it flat.

DID YOU KNOW?
The first permanent photograph was produced in 1827 by Frenchman Joseph-Nicéphore Niépce when he discovered that asphalt was light sensitive. However, posing for his photograph may not have been much fun. He took eight hours to take one photograph!

SINGLE-USE CAMERA
A single-use camera is sold complete with a film loaded inside it. When all the photographs have been taken, the whole camera is sent away for the film to be processed.

Plastic lens
To keep the camera simple to make and use, the plastic lens is set so that it does not have to be adjusted.

Viewfinder
The photographer looks through the viewfinder to see the picture the camera will take.

Right way up
The pentaprism is a specially shaped block of glass. It reverses the image formed by the lens so that it appears the correct way through the viewfinder.

MAKING PHOTOGRAPHS

When light falls on photographic film, it causes a chemical reaction in a light-sensitive layer called the emulsion. In the fraction of a second when a camera's shutter is open, the chemical reaction releases a tiny amount of silver from silver crystals in the emulsion. Only a few atoms of silver are released by each crystal, so the image is invisible. When the film is treated with chemicals, millions more silver atoms are released. The emulsion is then washed away in places where light did not fall on it. The picture formed on the film is a negative image—dark where light fell on it. This is turned into a photograph by shining light through the negative onto a sheet of light-sensitive paper and developing the image on the paper. Color film has three layers of light-sensitive chemicals. Each layer is sensitive to a different color. The three colors combine to form a lifelike color photograph.

Hinged mirror
The mirror reflects light entering the camera onto the viewing screen. Once the shutter release is pressed, the mirror flips out of the way to let light fall on the film.

Lens system
The lens in an SLR camera contains several separate lenses that work together to form a clear, sharp image.

HOW A CAMERA SEES

When a photographer presses a camera's shutter release button, the shutter opens and lets light stream in through the lens and fall on the film. The lens focuses the image on the film and a chemical reaction in the film captures the image.

Shutter
The shutter opens to let light fall on the film, then closes again. The time it takes to do this is either set by the photographer or calculated automatically by the camera.

Recording Sound

Recorded sound enables us to listen to music wherever and whenever we like. Recordings can be broadcast to millions of people by radio. Small, lightweight personal stereos and compact disk players allow us to enjoy music in private, even while we are on the move. Most recorded sound depends on either magnetism or light. Personal stereos and tape recorders use magnetism. Before sound can be recorded magnetically, it must first be changed into an electrical signal by a microphone. The electrical signal is then changed into a varying magnetic force that magnetizes the recording tape. Compact disk players use light. Sound is stored on the silver-colored disk as a pattern of tiny pits (holes). When light bounces off the spinning disk, the pits make the reflections vary. The varying intensity of the reflections is changed into electricity, and a speaker then converts this into sound.

COMPACT DISK PLAYER
A compact disk player is a machine that uses light, produced by a laser, to react to a spiral pattern of tiny holes in a spinning plastic disk.

PORTABLE PERSONAL STEREO
A personal stereo is a miniature tape recorder. The pattern of magnetism on the tape creates a varying current in the playback head next to the tape. The current is amplified (made larger) and then changed into sound by the earphones.

Spindles
The portable stereo contains two spindles. They wind the tape from one spool to the other, past the tape heads.

Playback head
This detects the magnetic pattern on the tape and changes it into an electric current.

HOW MAGNETIC TAPE WORKS

The record head and playback head of a tape recorder are electromagnets. The strength of the magnetic field set up by the record head varies as the current through it varies. As the tape moves past the head when recording, parts of the tape are magnetized to varying degrees. When the magnetized tape is run past the playback head, the fluctuating magnetism from the tape sets up tiny currents in a coil of wire. These are then amplified and fed to the headphones or to the speaker.

Unrecorded tape

Electromagnet

Ordered magnetic pattern

DID YOU KNOW?

The first sound recording was made by shouting at a diaphragm—a disk that vibrates in response to the sound waves of a voice. As the disk vibrated, it activated a needle attached to it. This scored a groove in a spinning cylinder covered with tinfoil. When the recording was played back, the groove made the needle and disk vibrate and recreate the voice.

MICROPHONE

Performers often wear radio microphones. The microphone changes the performer's voice into an electrical signal. Then a battery-powered radio transmitter, sometimes worn on a belt, sends it to a radio receiver, which relays it to the audience via the sound system.

Making music
On a circuit board beneath this casing, the Digital to Analog Converter (DAC) changes pulses of electricity from the photodiode (a light-sensitive conductor) into an analog (smoothly varying) electrical signal. When this is amplified, a speaker changes it into a copy of the sound recorded on the compact disk.

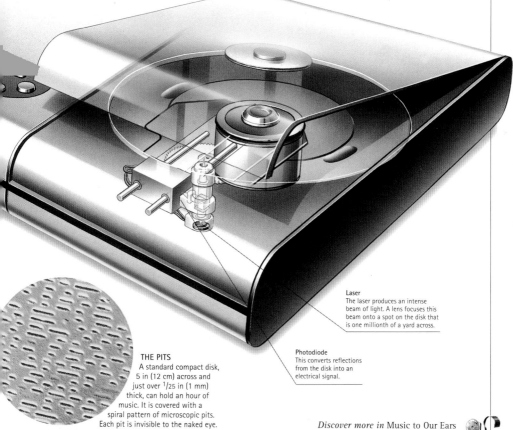

Laser
The laser produces an intense beam of light. A lens focuses this beam onto a spot on the disk that is one millionth of a yard across.

Photodiode
This converts reflections from the disk into an electrical signal.

THE PITS

A standard compact disk, 5 in (12 cm) across and just over $1/25$ in (1 mm) thick, can hold an hour of music. It is covered with a spiral pattern of microscopic pits. Each pit is invisible to the naked eye.

Discover more in Music to Our Ears

Music to Our Ears

A symphony orchestra fills a concert hall with music. From the piccolo to the double bass, all musical instruments produce sound in the same way—they make the air vibrate. Some wind instruments have woody reeds or rely on the players' lips to make the air vibrate; in others, such as the flute, the vibration comes from the players' breath passing over the sharp edge of a hole in the instrument. Stringed instruments use vibrating strings, while percussion instruments vibrate when they are struck. The first vibration from a reed or a string is usually very weak, so most instruments have a way to make more air vibrate, which in turn makes the sound louder. Stringed instruments, for example, have a sounding board that vibrates with the strings and amplifies the sound. Some percussion instruments, such as the xylophone, have hollow tubes or blocks of wood to increase the sound they make. Although musical instruments produce sound in the same way, they do not produce the same sounds. Instruments are all shapes and sizes and are made from a variety of materials. The music they make is just as varied.

PLAYING THE PIANO
A piano has wire strings that are stretched tightly in pairs or groups over an iron frame. Each set of strings has a hammer. When the player strikes the keys, strongly or gently, the hammer strikes the strings with the same force and makes them vibrate. The long strings make low notes and the short strings make high notes. When the air around the strings vibrates, a sounding board behind them amplifies the sound. The angled lid of a grand piano deflects the sound toward the audience.

PAN PIPES
When you blow across the top of a pipe, the air inside vibrates and produces a musical note. A long pipe holds more air and makes a lower note than a short pipe.

Black and white
The keyboard of a piano has 52 white keys and 36 black keys.

BOTTLED MUSIC

Tap a bottle and listen to the sound it makes. If you pour some water into the bottle, the amount of air inside is reduced and the pitch of the sound is raised. You can make your own version of a xylophone by setting out a line of bottles and filling them with different amounts of water. Tap the bottles gently with a pencil and notice the different sound each bottle makes. You can tune your xylophone by adding or pouring out water in each bottle.

Pedals
Foot pedals change the length and loudness of a note. The pedal on the right lifts all the dampers so the sound continues after the key has been released. The pedal on the left moves the hammers sideways, to make the sound softer. Some pianos have a third (middle) pedal. This sustains notes that are already sounding when the pedal is pressed.

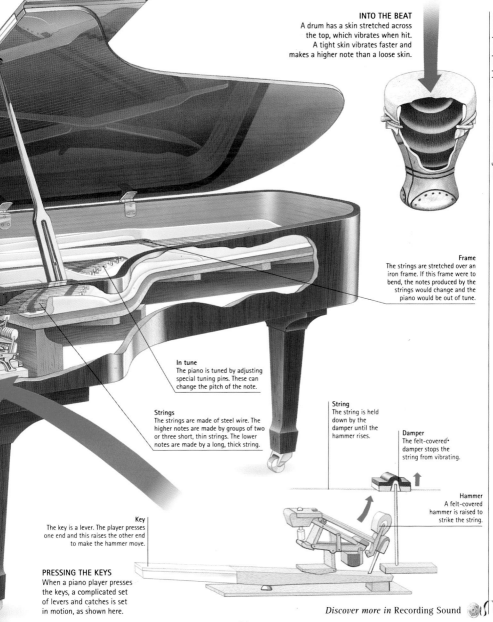

INTO THE BEAT
A drum has a skin stretched across the top, which vibrates when hit. A tight skin vibrates faster and makes a higher note than a loose skin.

Frame
The strings are stretched over an iron frame. If this frame were to bend, the notes produced by the strings would change and the piano would be out of tune.

In tune
The piano is tuned by adjusting special tuning pins. These can change the pitch of the note.

Strings
The strings are made of steel wire. The higher notes are made by groups of two or three short, thin strings. The lower notes are made by a long, thick string.

String
The string is held down by the damper until the hammer rises.

Damper
The felt-covered damper stops the string from vibrating.

Hammer
A felt-covered hammer is raised to strike the string.

Key
The key is a lever. The player presses one end and this raises the other end to make the hammer move.

PRESSING THE KEYS
When a piano player presses the keys, a complicated set of levers and catches is set in motion, as shown here.

Discover more in Recording Sound

The World in View

Television programs are usually transmitted to your home by radio waves, but they can also travel via underground cables. Your rooftop antenna detects the radio waves and converts them into an electrical signal. The television converts this signal into pictures and sound. A television picture seems to be moving, but it is really a technological trick that fools your eyes and brain. The picture on the screen is made by a glowing spot that moves from side to side and up and down so quickly that the whole screen seems to glow at the same time. Television pictures appear on the screen one after another so rapidly that they look like a single moving picture. It works because of an effect called "persistence of vision." When light forms an image (picture) on the retina (the light-sensitive layer at the back of your eye), the image stays on the retina for a fraction of a second after the light that formed it has gone. This means that when images reach your eyes very quickly, one after the other, they merge together.

TELEVISION
A television converts electrical signals from a rooftop antenna or an underground cable into pictures and sound. This is a widescreen television. Its screen is one-third wider than a normal television.

Video drum
The video drum spins at an angle to the tape. Tape heads (small electromagnets) on the surface of the video drum transfer the electrical signals into a magnetic pattern on the video tape.

Video cassette
When a standard video cassette is put into the recorder, the machine automatically opens a flap on the front edge, pulls out a loop of tape, and wraps this around the video drum.

Screen
The television screen, on which pictures are formed, is the flattened end of a large glass tube.

Phosphors
The back of the screen is coated with chemicals called "phosphors." They glow in one of three colors—red, green or blue—when electrons strike them.

Shadow mask
The shadow mask is a metal sheet with slots in it, and three electron beams pass through it on their way to the screen. The mask ensures that each electron beam strikes and lights up only one phosphor: red, green or blue.

VIDEO CASSETTE RECORDER
A video cassette recorder can record television programs by storing the electrical signals from the television antenna magnetically on video tape.

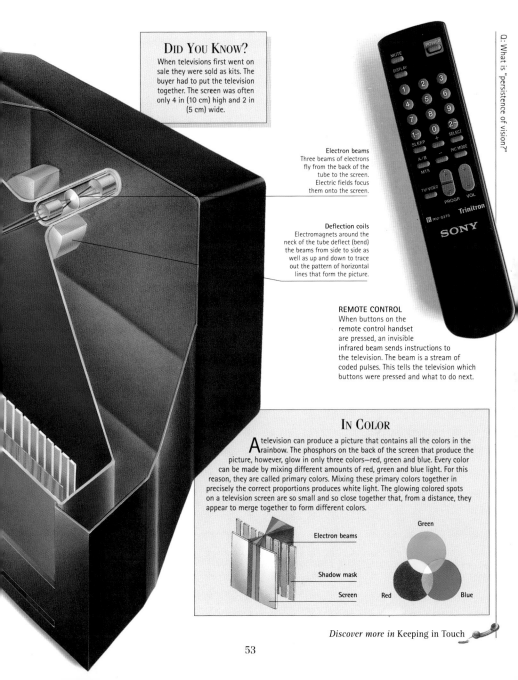

DID YOU KNOW?

When televisions first went on sale they were sold as kits. The buyer had to put the television together. The screen was often only 4 in (10 cm) high and 2 in (5 cm) wide.

Electron beams
Three beams of electrons fly from the back of the tube to the screen. Electric fields focus them onto the screen.

Deflection coils
Electromagnets around the neck of the tube deflect (bend) the beams from side to side as well as up and down to trace out the pattern of horizontal lines that form the picture.

REMOTE CONTROL
When buttons on the remote control handset are pressed, an invisible infrared beam sends instructions to the television. The beam is a stream of coded pulses. This tells the television which buttons were pressed and what to do next.

IN COLOR

A television can produce a picture that contains all the colors in the rainbow. The phosphors on the back of the screen that produce the picture, however, glow in only three colors—red, green and blue. Every color can be made by mixing different amounts of red, green and blue light. For this reason, they are called primary colors. Mixing these primary colors together in precisely the correct proportions produces white light. The glowing colored spots on a television screen are so small and so close together that, from a distance, they appear to merge together to form different colors.

Electron beams

Shadow mask

Screen

Green

Red

Blue

Discover more in Keeping in Touch

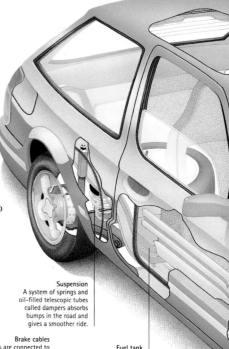

On the Road Again

When Carl Benz visualized the "horseless carriage" last century he could not have imagined how complex cars would become. Modern cars consist of several different mechanical and electrical systems all working together. The fuel system supplies fuel to the engine, and the ignition system provides electrical sparks at just the right moment to ignite the fuel. The transmission system transmits the power generated by the engine to the car's wheels. The lubrication system keeps all the moving parts in the engine covered with a film of oil so that they can slip over each other without sticking. The cooling system stops the engine from overheating and the braking system stops the car safely. The suspension system allows the wheels to follow bumps and dips in the road while the rest of the car glides along smoothly. Today there are more than 400 million passenger cars on roads around the world.

Suspension
A system of springs and oil-filled telescopic tubes called dampers absorbs bumps in the road and gives a smoother ride.

Tires
Pneumatic (air-filled) tires give a smoother ride over small bumps in the ground.

Brake cables
Brake cables are connected to levers on the handlebars. When the levers are squeezed, the cables compress the brake pads on the wheels and slow the bicycle down.

Fuel tank
The fuel that is burned in the engine is pumped from a tank at the rear of the car.

Chain
The chain passes around the gears and turns the rear wheel.

Gears
Gears make a bicycle easier to pedal. Low gears turn the wheel only a small amount and help the rider to pedal uphill.

Spokes
Thin wire spokes hold the wheels in place and let the wind blow through them instead of against the bicycle.

STRANGE BUT TRUE

In 1865, a law was passed in England to limit the speed of steam cars. They were not allowed to go faster than 2 miles (3 km) per hour in cities and they also had to travel behind a man waving a red flag!

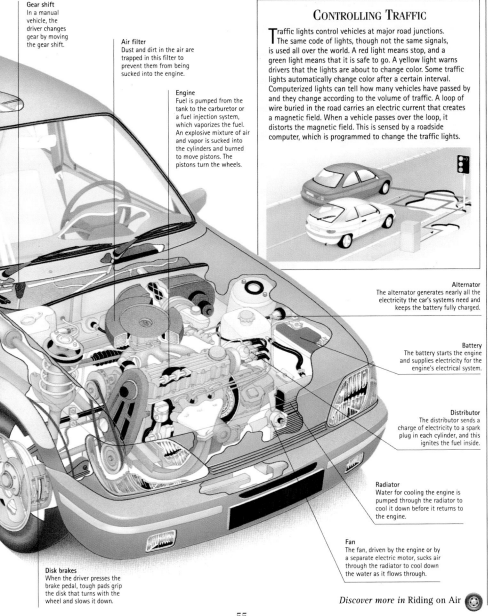

Gear shift
In a manual vehicle, the driver changes gear by moving the gear shift.

Air filter
Dust and dirt in the air are trapped in this filter to prevent them from being sucked into the engine.

Engine
Fuel is pumped from the tank to the carburetor or a fuel injection system, which vaporizes the fuel. An explosive mixture of air and vapor is sucked into the cylinders and burned to move pistons. The pistons turn the wheels.

CONTROLLING TRAFFIC

Traffic lights control vehicles at major road junctions. The same code of lights, though not the same signals, is used all over the world. A red light means stop, and a green light means that it is safe to go. A yellow light warns drivers that the lights are about to change color. Some traffic lights automatically change color after a certain interval. Computerized lights can tell how many vehicles have passed by and they change according to the volume of traffic. A loop of wire buried in the road carries an electric current that creates a magnetic field. When a vehicle passes over the loop, it distorts the magnetic field. This is sensed by a roadside computer, which is programmed to change the traffic lights.

Alternator
The alternator generates nearly all the electricity the car's systems need and keeps the battery fully charged.

Battery
The battery starts the engine and supplies electricity for the engine's electrical system.

Distributor
The distributor sends a charge of electricity to a spark plug in each cylinder, and this ignites the fuel inside.

Radiator
Water for cooling the engine is pumped through the radiator to cool it down before it returns to the engine.

Fan
The fan, driven by the engine or by a separate electric motor, sucks air through the radiator to cool down the water as it flows through.

Disk brakes
When the driver presses the brake pedal, tough pads grip the disk that turns with the wheel and slows it down.

Discover more in Riding on Air

RAILWAY SIGNALS

Colored light signals by the side of the track tell train drivers whether or not it is safe for a train to proceed.

A green signal indicates that two or more sections of track ahead of a train are clear and it is safe to proceed.

A yellow signal indicates that only one clear section of track separates two trains. As a train passes each signal, it changes to red.

A red signal shows that the track ahead is not clear and an alarm sounds in the driver's cab.

When the train has stopped, the alarm continues to sound a warning that it is not safe to proceed.

Staying on Track

I f you travel anywhere by train, your carriage will be pulled along by one of four types of locomotive. In a few places, steam locomotives are still used. Coal is burned to heat water and make steam, which pushes pistons inside cylinders to turn the wheels. Diesel or diesel-electric locomotives are now more common than steam engines. Diesel locomotives burn oil in a piston engine, which is like a giant car engine, to turn the wheels. Diesel-electric locomotives use a diesel engine to drive an electric generator. The generator powers the electric motors that turn the wheels. The world's fastest railway trains are pulled by electric locomotives that convert electrical energy directly into movement. The electricity for the motors is supplied by cables suspended over the track. The German InterCity Express (ICE) train is one of the world's fastest trains: its top speed is more than 248 miles (400 km) per hour, but it usually transports passengers at 155 miles (250 km) per hour.

ICE TRAIN
The German ICE (InterCity Express) is an example of a modern, high-speed electric train.

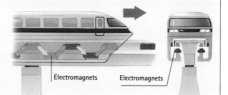

FLYING ON MAGNETISM

The maglev (magnetic levitation train) is held above its track, or guideway, by powerful forces between electromagnets (electrically powered magnets) in the track and in the train. Magnets in the sides of the train hold it steady in the guideway. Magnetic fields are also used to make the train move. Magnets ahead of the train attract it, pulling it forward, and magnets behind repel it, pushing it forward. The magnetic field ripples along the guideway, pulling the train with it. There is no contact between the train and the track and passengers have a very smooth and quiet ride.

Close behind
These carriages are divided into sections with seats in rows or facing each other.

Riding up front
The carriages at the front have fewer seats and more room than those behind.

British maglev train

Electromagnets   Electromagnets

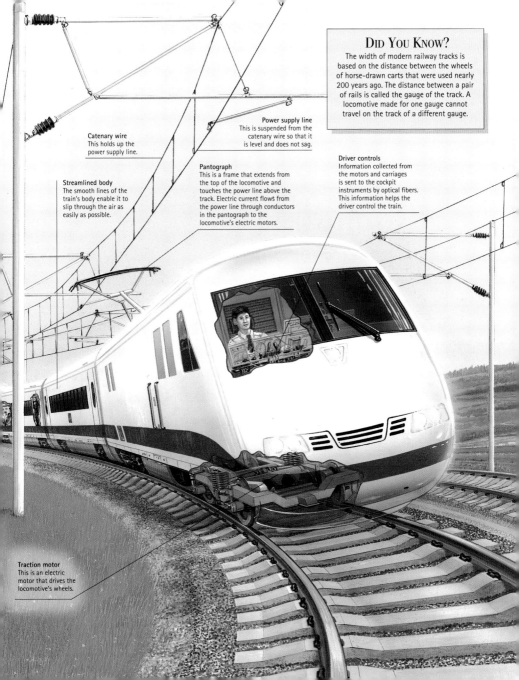

Catenary wire
This holds up the power supply line.

Power supply line
This is suspended from the catenary wire so that it is level and does not sag.

Pantograph
This is a frame that extends from the top of the locomotive and touches the power line above the track. Electric current flows from the power line through conductors in the pantograph to the locomotive's electric motors.

Driver controls
Information collected from the motors and carriages is sent to the cockpit instruments by optical fibers. This information helps the driver control the train.

Streamlined body
The smooth lines of the train's body enable it to slip through the air as easily as possible.

Traction motor
This is an electric motor that drives the locomotive's wheels.

• TRANSPORTATION •

Roaming the Oceans

When anything tries to move through water, the water resists its movement. Boat designers try to minimize water resistance, called "drag," by making boat hulls as smooth and streamlined as possible. Water underneath a boat pushes up against its hull with a force called "upthrust." If the force of the boat's weight is equal to the upthrust of the water, the boat floats. If the boat weighs more than the upthrust of the water, it sinks. A submarine, or a smaller underwater craft called a submersible, sinks under the waves by letting water into its ballast tanks to make it heavier. It rises to the surface again by forcing the water out of the tanks with compressed air, or by dropping heavy weights to make the craft lighter. Most working boats, submarines and submersibles are powered by propellers with angled blades that push against the water as they turn.

Manipulator arm
A robot arm with a mechanical claw at the end of it picks up objects from the sea bed.

SETTING SAIL
A sail is set at an angle so that wind blowing around the sail from the side reduces the air pressure in front of it, sucking the sail and the boat forward. This means that a sail can use a wind blowing in one direction to propel a yacht in a completely different direction. But a yacht can never sail directly into the wind.

BELOW THE SURFACE
Submersibles allow scientists to explore the sea bed, study living organisms in their natural surroundings and investigate shipwrecks.

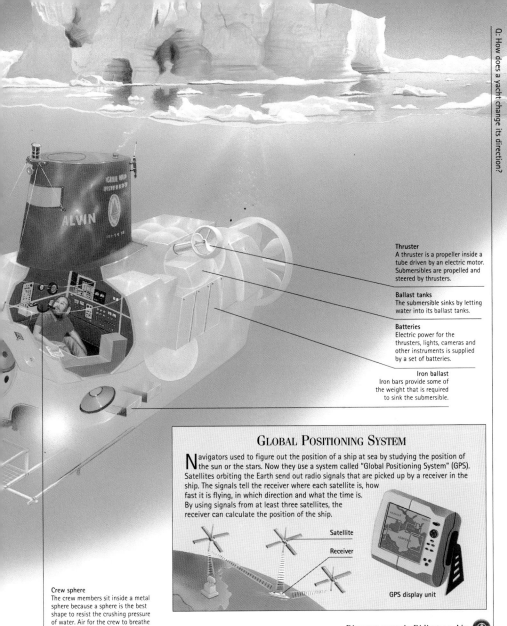

Thruster
A thruster is a propeller inside a tube driven by an electric motor. Submersibles are propelled and steered by thrusters.

Ballast tanks
The submersible sinks by letting water into its ballast tanks.

Batteries
Electric power for the thrusters, lights, cameras and other instruments is supplied by a set of batteries.

Iron ballast
Iron bars provide some of the weight that is required to sink the submersible.

GLOBAL POSITIONING SYSTEM

Navigators used to figure out the position of a ship at sea by studying the position of the sun or the stars. Now they use a system called "Global Positioning System" (GPS). Satellites orbiting the Earth send out radio signals that are picked up by a receiver in the ship. The signals tell the receiver where each satellite is, how fast it is flying, in which direction and what the time is. By using signals from at least three satellites, the receiver can calculate the position of the ship.

Satellite

Receiver

GPS display unit

Crew sphere
The crew members sit inside a metal sphere because a sphere is the best shape to resist the crushing pressure of water. Air for the crew to breathe is also stored in spherical tanks.

Discover more in Riding on Air

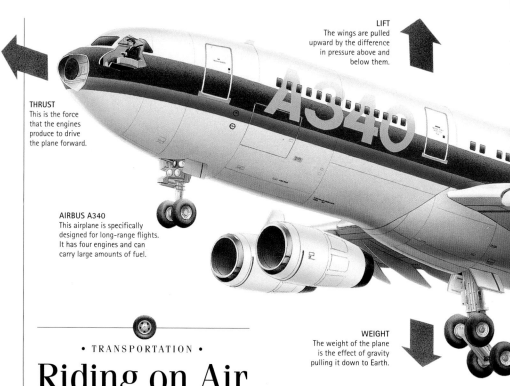

LIFT
The wings are pulled upward by the difference in pressure above and below them.

THRUST
This is the force that the engines produce to drive the plane forward.

AIRBUS A340
This airplane is specifically designed for long-range flights. It has four engines and can carry large amounts of fuel.

WEIGHT
The weight of the plane is the effect of gravity pulling it down to Earth.

• TRANSPORTATION •

Riding on Air

The Airbus A340 weighs up to 553,000 lb (251,000 kg). It looks as if it could not possibly fly, yet when it reaches a speed of up to 183 miles (295 km) per hour, its nose tips up and it soars skyward. Airplanes are able to fly because of the shape of their wings. Air is forced to travel over the curved top of the wings farther and faster than it can move past the flat underside. This has the effect of lowering the air pressure above the wings. The difference in pressure above and below the wings sucks the wings upward. When this force, called "lift," is greater than the weight of the airplane, the plane takes off. The airplane has moving panels in the wings (ailerons) and in the tail (the elevators or rudder). When these "control surfaces" are moved, the air pushing against them makes the plane turn or climb or dive.

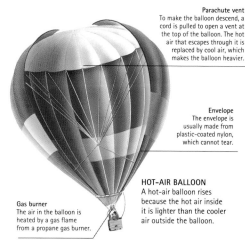

Parachute vent
To make the balloon descend, a cord is pulled to open a vent at the top of the balloon. The hot air that escapes through it is replaced by cool air, which makes the balloon heavier.

Envelope
The envelope is usually made from plastic-coated nylon, which cannot tear.

HOT-AIR BALLOON
A hot-air balloon rises because the hot air inside it is lighter than the cooler air outside the balloon.

Gas burner
The air in the balloon is heated by a gas flame from a propane gas burner.

Slats
Slats are strips that extend from the front of the wings to generate more lift and adapt the plane for flying at lower speeds.

On the inside
Spars (metal beams) running the length of the wings are linked by ribs running from front to back, forming a very strong structure.

Engine
This works when air is sucked into the engine through a large spinning fan. Fuel is mixed with air and burned in the combustion chamber. Hot gases then rush out of the engine through the exhaust nozzle, creating the force to drive the plane forward.

Rudder
Turning the rudder pushes the plane's tail out to the left or right.

Flaps
Flaps are panels that extend from the rear of the wings and function in the same way as slats.

Combustion chamber

Exhaust nozzle

Fan

Wheels
When the plane is airborne, the wheels fold up into the body.

Cabin
The passenger cabin is pressurized by air pumps to provide enough oxygen for the passengers to breathe.

DRAG
This force is caused by air resistance that tries to slow the plane down.

HOW RADAR WORKS

Pulses of radio waves are sent out from a radar transmitter. Large metal objects in the path of the waves, such as an airplane, reflect the waves. Some of the waves travel back to the dish where a receiver detects them. The time they take to be reflected from an airplane can be measured very accurately. As radio waves are known to travel at the speed of light, the distance to the airplane can be calculated. The plane's position is shown as a glowing spot on a radar screen.

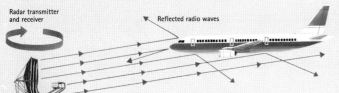

Radar transmitter and receiver

Reflected radio waves

Radar screen

Discover more in Reaching into Space

Reaching into Space

Technology has made it possible for us to reach into space. Special engines, for example, have been designed to power spacecraft in this airless environment. Fuel needs oxygen to burn, but because there is no oxygen in space, the rockets that propel spacecraft carry their own supply of oxygen, or a substance containing oxygen, which is mixed with the fuel before it is burned. When rocket fuel is burned, the hot gases produced expand rapidly and rush out through a nozzle. The force propels the spacecraft in the opposite direction. When the nozzle is turned, the jet of gases changes direction and this steers the rocket. In the early days of space travel, rockets and spacecraft could be used once only. In 1981, however, America launched a reusable space shuttle. It consists of an orbiter space-plane, two booster rockets and an external fuel tank. The orbiter takes off like a rocket, glides back from space, and lands on a runway, like an aircraft.

Flight deck
The orbiter is controlled by its commander and pilot from the flight deck.

Thermal tiles
Tiles cover the orbiter to protect it from the intense heat it encounters when it re-enters the atmosphere.

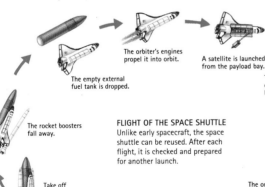

The orbiter's engines propel it into orbit.

A satellite is launched from the payload bay.

The empty external fuel tank is dropped.

The orbiter's engines fire to begin its descent.

The rocket boosters fall away.

FLIGHT OF THE SPACE SHUTTLE
Unlike early spacecraft, the space shuttle can be reused. After each flight, it is checked and prepared for another launch.

The orbiter glows red hot as it plunges through the atmosphere.

Take off

The orbiter glides down toward the runway.

The space shuttle is prepared for take off.

Touchdown

SPACE SHUTTLE

Two astronauts check a satellite before it is launched from the space shuttle orbiter's payload bay. They are linked to the orbiter by safety lines so they cannot float away into space. Small satellites are released into space by springs. Larger satellites are lifted out of the payload bay by the orbiter's robot arm. At a safe distance from the orbiter, the satellite's rocket thrusters fire to boost it into the correct orbit.

TAKE OFF!

You can make your own rocket with a balloon. Blow up a balloon and clip it closed. Attach a drinking straw to the balloon with tape. Pass a length of thread through the straw and tie it tightly to two chairs placed 7 ft (2 m) apart from each other. Launch your rocket by taking the clip off the balloon. Air rushes out of the balloon and pushes it in the opposite direction.

Robot arm
The orbiter is equipped with a robot arm for moving satellites and experiments into and out of the payload bay.

Orbital maneuvering engines
Two engines in the orbiter's tail move it to a higher or lower orbit and begin its descent at the end of the mission.

Payload bay
Satellites and even fully equipped scientific laboratories can be carried in the payload bay, which is 59 ft (18 m) long and 16 ft (5 m) across.

Thrusters
Changes to the orbiter's position are made by firing rocket thrusters contained in its nose and tail.

Main engines
The three main engines, supplied with fuel from an external tank, are fired for the first 8½ minutes of each flight.

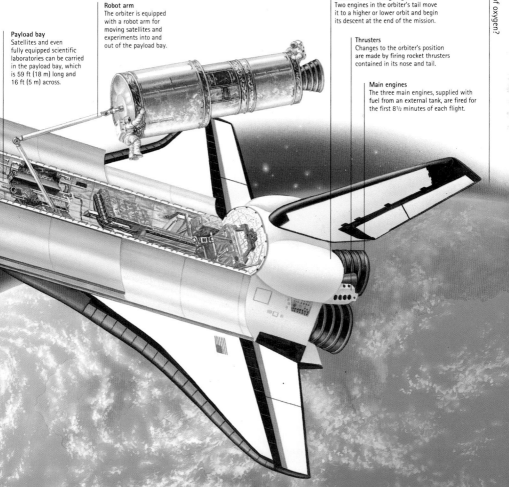

The Principles of Things

All machines, instruments and electronic devices, from a humble office stapler or a magnifying glass to a computer or the space shuttle, make use of scientific principles. Understanding a few basic scientific principles can make it easier to see why machines are built the way they are, why some machines have a certain shape and how they work.

Body that is not streamlined

Streamlined body

Aerodynamics
This is the study of how air flows around objects. An object's shape affects the way that air flows around it. Air does not flow easily around a broad, angular shape, such as a truck, but it does flow easily around a slim, smoothly curved shape, such as a car. Such an object is said to be streamlined. Streamlining is particularly important for racing cars, jet airliners, rockets or anything else designed to travel at high speed.

Equal air pressure	Lower air pressure in straw

Air pressure
Air pressure describes the way that air pushes back when it is squeezed. Air pressure is greatest near the Earth's surface because of the air above pressing down. High above the Earth, the air pressure is low. When you use a drinking straw you use air pressure. Sucking air out of the straw lowers the air pressure inside. The higher air pressure outside forces the drink up the straw to fill the lower pressure area.

Electromagnetic radiation
Light, radio and X-rays are all examples of electromagnetic radiation. They consist of waves of electric and magnetic energy vibrating through space. The only difference between them is the length of the waves. Light is the only part of the electromagnetic spectrum that our eyes can detect. The whole spectrum is shown below, with wavelengths given in meters. Each wavelength is ten times the one before it; 10^3 meters is the same as 10 x 10 x 10, or 1,000 meters.

Adequate oxygen supply	Reduced oxygen supply

Combustion
Combustion is another word for burning. It is a chemical reaction that gives out heat and light in the form of a flame. It occurs when a substance reacts quickly with oxygen, a gas in the air around us. A burning candle underneath an upturned glass soon uses up the oxygen in the air trapped under the glass. Its flame shrinks and goes out, while a candle outside the glass would continue to burn.

ELECTROMAGNETIC RADIATION

Gamma rays
Gamma rays can travel through most materials. They are stopped only by a thick sheet of steel or lead.

X-rays
X-rays can pass through some materials, and they are used to study internal structures in industry and medicine.

Ultraviolet (UV)
Invisible UV rays are responsible for producing a suntan.

Visible light
Different wavelengths of light are seen as different colors. Red light has the longest waves, while violet has the shortest.

Infrared
Television remote controls use infrared rays to send instructions to a television.

Gravity

Gravity is the force that pulls everything toward the ground, such as an apple falling from a tree. Gravity also makes the moon circle the Earth and the Earth circle the sun. The strength of an object's gravity depends on the amount of matter it is made from. Stars have a stronger gravitational pull than planets, because stars are bigger than planets.

		First-class lever
		Second-class lever
		Third-class lever

Levers

A lever is a device that transfers a force from one place to somewhere else. Every lever has a load, an effort and a fulcrum. The effort makes the lever pivot at the fulcrum, moving the load. There are three different ways of arranging the effort, load and fulcrum, called three classes of lever. A first-class lever is like a see-saw, a second-class lever is like a wheelbarrow and a third-class lever is like your own forearm.

Magnetism

Magnetism is a force produced by magnets, which can attract materials such as iron, steel, cobalt and nickel. A bar magnet has a north pole and a south pole. If a north pole is brought close to the south pole of another magnet, they attract each other. If two north poles or two south poles are brought together, they push each other away. Magnetism is also produced when electricity flows. An electric current flowing along a wire creates a magnetic field around the wire.

Reflection

When a wave strikes a surface it bounces back, like a ball bouncing off a wall. This rebounding effect is called reflection. We can see ourselves in a mirror because light waves are reflected by the mirror.

Light

Refraction

When light travels from one substance into another, such as from air to water, it changes speed and direction. This is called refraction. A wedge of glass called a prism separates light into all the colors it contains by refraction. The different colors present in sunlight are bent by different amounts, so the colors separate and form a rainbow.

Microwaves
Microwaves, used by microwave ovens and for communications, are radio waves between $^1/_{25}$ in (1 mm) and 1 in (3 cm) long.

Radio waves
Radio waves can be as short as $^1/_{25}$ in (1 mm) or as long as 62 miles (100 km).

Flight

How does a plane stay up in the air?

How can a bird fly without flapping its wings?

How can a Stealth Fighter avoid being detected by enemy radar?

Contents

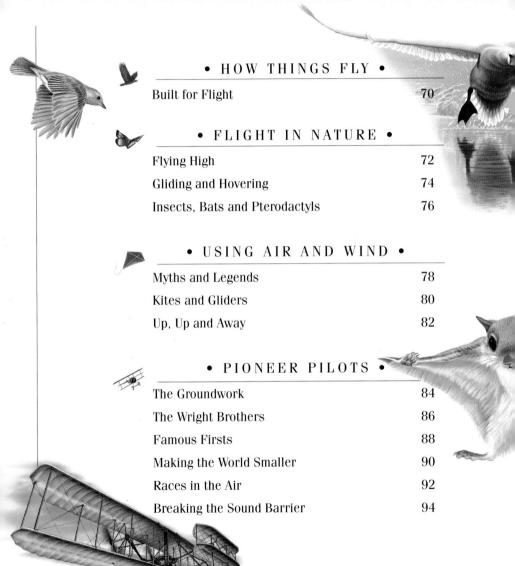

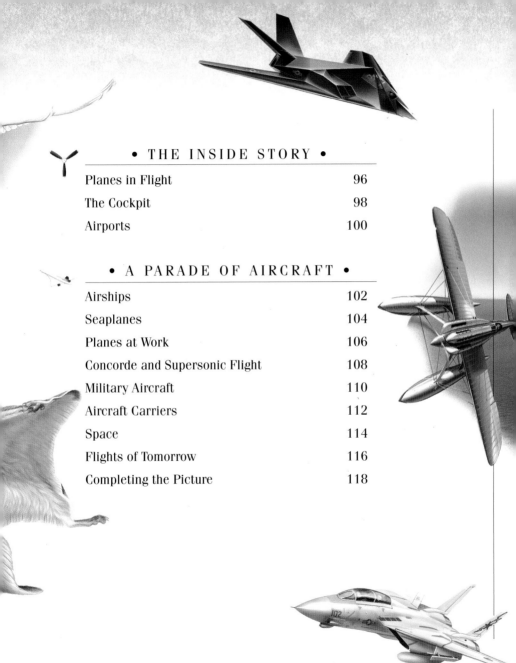

TAKE OFF
Swans are heavy birds. They need to run over the water for quite a long distance before they can build up enough speed to support their great weight in the air. For the same reason, aircraft that are laden with passengers or cargo also need a long runway to become airborne.

• HOW THINGS FLY •

Built for Flight

Have you ever watched birds in the sky and thought how easy it looks to fly? Centuries ago, people dreamed of joining birds in flight. Some went even further and flapped about, vainly, in wings made of feathers. But the human body is heavy and does not have the muscles needed for flight. The pioneers of aviation soon realized that before they could join the birds, they needed to understand how birds flew. They discovered that the wings of a bird are specially curved surfaces, called airfoils. When air flows over a bird's wings, a difference in air pressure is produced above and below the wings. This difference in pressure creates a force called "lift," which can overcome the weight of a bird or a plane. This is called heavier-than-air flying, and gliders and airplanes also fly this way. Balloons and airships are lighter-than-air fliers. They are filled with hot air (which always rises) or gases, such as helium or hydrogen, that are lighter than the air around them.

DID YOU KNOW?
Hydrofoils produce lift in water, just as airfoils give lift in air. This fast boat has hydrofoils, which lift it up and enable the boat to skim along the surface of the water.

70

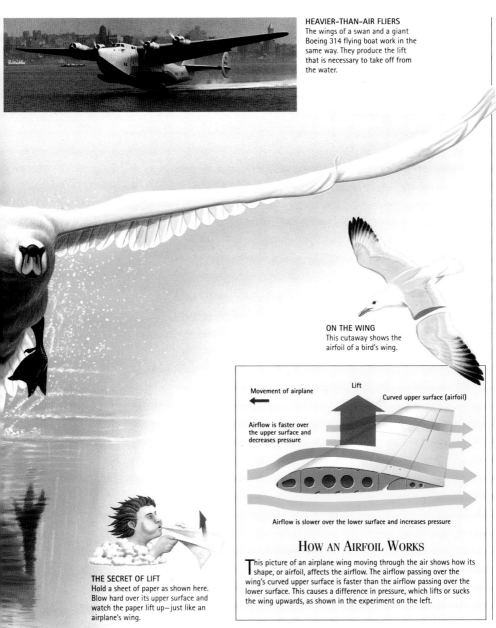

HEAVIER-THAN-AIR FLIERS
The wings of a swan and a giant Boeing 314 flying boat work in the same way. They produce the lift that is necessary to take off from the water.

ON THE WING
This cutaway shows the airfoil of a bird's wing.

Lift

Movement of airplane

Curved upper surface (airfoil)

Airflow is faster over the upper surface and decreases pressure

Airflow is slower over the lower surface and increases pressure

HOW AN AIRFOIL WORKS

This picture of an airplane wing moving through the air shows how its shape, or airfoil, affects the airflow. The airflow passing over the wing's curved upper surface is faster than the airflow passing over the lower surface. This causes a difference in pressure, which lifts or sucks the wing upwards, as shown in the experiment on the left.

THE SECRET OF LIFT
Hold a sheet of paper as shown here. Blow hard over its upper surface and watch the paper lift up—just like an airplane's wing.

Flying High

irds fly higher, farther and faster than any other flying animals. Many species can span entire oceans and continents on their migrations. Birds are also the only flying animals that regularly use the wind as a source of lift. A bird's wings are powered by two sets of muscles on the breast. In most birds, these muscles make up about one-third of their total weight. Muscle power moves the long, stiff flight feathers. Feathers at the wing tips, called primaries, propel the bird forward, while the rest of the wing generates the lift. The wings change shape as they beat up and down. They are broad and extended on the downstroke, but tucked in tight on the upstroke.

Wing shape and how a bird lives are closely related. Long wings are more efficient than short wings, but much harder to flap. Birds with long wings are usually soaring birds. Short-winged birds have less stamina, but can build up speed very quickly.

WIDE WINGS
A spotted harrier eagle has wide wings and flies slowly over open country, looking for small reptiles, birds and mammals.

IN SLOW MOTION
The wingbeat of a European robin is a smooth alternation between the wings moving upwards (upstroke or recovery stroke) and the wings moving downwards (downstroke or powerstroke). It takes place in a cycle that is almost too fast to see. The process is broken into five stages (below) to show the details.

Flight control
The long feathers of the tail help to control the flight, especially steering and braking.

FEATHER CLOSE UP
Birds are the only animals that have feathers. An electron micrograph shows the intricate structure of a feather. It is held together by minute barbs and barbules like Velcro.

TUCKING IN
The wings are tucked well into the body throughout the upstroke to reduce air resistance.

SPREADING OUT
The wings are high, fully spread and thrown forward. The feathers are overlapped and the feet are tucked against the body. The curled primary tips are like the angled blades of a propeller and pull the bird forward.

BEGINNING
At the start of the upstroke, the robin's feathers are separated. This reduces air resistance and the bird uses less energy.

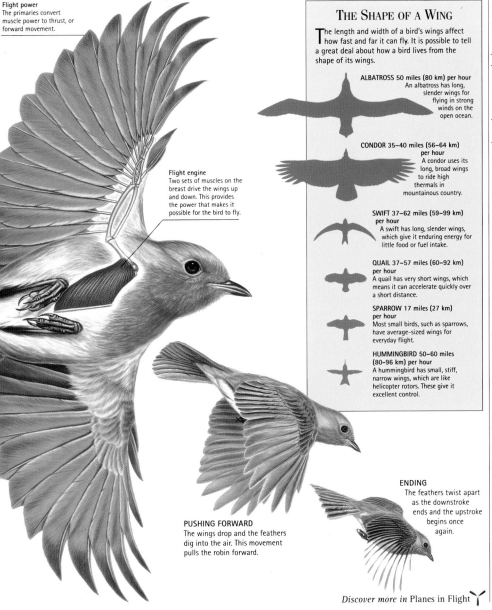

Flight power
The primaries convert muscle power to thrust, or forward movement.

Flight engine
Two sets of muscles on the breast drive the wings up and down. This provides the power that makes it possible for the bird to fly.

Q: What is the purpose of a bird's primary feathers?

THE SHAPE OF A WING

The length and width of a bird's wings affect how fast and far it can fly. It is possible to tell a great deal about how a bird lives from the shape of its wings.

ALBATROSS 50 miles (80 km) per hour
An albatross has long, slender wings for flying in strong winds on the open ocean.

CONDOR 35–40 miles (56–64 km) per hour
A condor uses its long, broad wings to ride high thermals in mountainous country.

SWIFT 37–62 miles (59–99 km) per hour
A swift has long, slender wings, which give it enduring energy for little food or fuel intake.

QUAIL 37–57 miles (60–92 km) per hour
A quail has very short wings, which means it can accelerate quickly over a short distance.

SPARROW 17 miles (27 km) per hour
Most small birds, such as sparrows, have average-sized wings for everyday flight.

HUMMINGBIRD 50–60 miles (80–96 km) per hour
A hummingbird has small, stiff, narrow wings, which are like helicopter rotors. These give it excellent control.

PUSHING FORWARD
The wings drop and the feathers dig into the air. This movement pulls the robin forward.

ENDING
The feathers twist apart as the downstroke ends and the upstroke begins once again.

Discover more in Planes in Flight

• FLIGHT IN NATURE •

Gliding and Hovering

Some forest-dwelling animals can leap safely from one tree-branch to another. The southern flying squirrel glides on a furry membrane of skin. It can fly 328 ft (100 m) when there is no wind. But this is a very basic form of flight because the squirrel cannot fly for long, and its control is very limited. True flight— for animals as well as aircraft— depends on keeping a constant flow of air over a wing's surface. Animals do this in two ways: they exploit the wind by soaring; or they rely on muscle power, through flapping wings, to maintain the airflow. Hovering is an extreme example of flapping. It gives the animal great control and maneuverability, but it also demands an enormous amount of power. A hummingbird hovers with seemingly little effort. In fact, the bird is working very hard. Imagine the enormous energy Olympic athletes use at the moment of their greatest effort. The hummingbird uses more than ten times this amount of energy to hover in the air.

TREE GLIDING
The southern flying squirrel uses the furry membrane between its outstretched limbs to glide from tree to tree. The membrane acts like a parachute and makes its fall gentle and safe. The squirrel also uses its long, furry tail as a rudder to help it maneuver.

74

LONG-DISTANCE GLIDER
Albatrosses can fly for thousands of miles without settling on the water.

Catching the Wind

Birds must keep a constant flow of air over their wings to fly. They have developed several ways of using wind rather than muscle power to do this.

Condors and vultures can ride columns of air called thermals to great heights. Thermals are caused when the sun heats open ground, such as a plowed field. This warms the air above, which then starts to rise.

When strong winds meet an obstruction such as a high cliff, they are forced upward to form updrafts. Many birds, such as kestrels and swallows, skim along the top of these updrafts for a free ride.

Strange but True

Flying fish live close to the surface in tropical seas. To escape predators, they can leave the water and glide on their broad pectoral fins. Their tails lash the water to provide thrust.

Strong winds blow over the open ocean. But these winds are weaker at the surface of the ocean (because of friction with the water) than higher in the sky. Albatrosses use these different wind strengths to soar through the air. This requires great skill but demands little energy.

Discover more in Kites and Gliders

Insects, Bats and Pterodactyls

Insects are the smallest of all flying animals. Their tiny muscles in tiny bodies are more efficient than large muscles in large bodies. Flying insects need less power than heavier animals such as bats and birds, and they are more maneuverable in the air. A housefly can somersault on touchdown to land upside down on a ceiling.

Insects, birds and bats are the only animals alive today capable of true flight. Bats fly using a membrane of skin, reinforced with muscle and tissue, which stretches between the arms and the legs (and sometimes the tail). Most bats are the size of a mouse and catch flying moths at night, but a few species weigh as much as 3 lb (1.5 kg) and feed on fruit. Long ago, at the time of the dinosaurs, another group of animals, called pterodactyls, also flew. They included the largest of all flying animals, the *Quetzalcoatlus*.

DID YOU KNOW?

Some insects, such as flies and mosquitoes, use their flight muscles to vibrate the sides of their body wall in and out. This makes the wings vibrate up and down together and creates the familiar buzzing or whining sound of an insect.

A CLOSER LOOK
A close-up of a dragonfly in flight shows some of the basic differences in the way insects, birds and bats are built for flight. Insects have two pairs of wings; birds and bats have only one. In birds and bats, the wings extend from the body, but in insects, they are entirely different structures.

CLOSER STILL
This electron micrograph shows where the wings of a dragonfly join to its body.

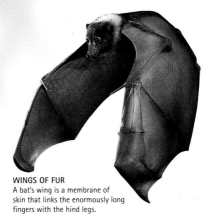

WINGS OF FUR
A bat's wing is a membrane of
skin that links the enormously long
fingers with the hind legs.

INSECTS' WING BEATS
The smaller the insect, the faster its wings
beat. This usually means its progress through
the air is also slower.

	SPEED PER HOUR	WING BEATS PER SECOND
DRAGONFLY	15 miles (24 km)	35
BUTTERFLY	14 miles (22.4 km)	10
HOUSEFLY	9 miles (14.4 km)	170
HONEYBEE	4 miles (6.4 km)	130
MOSQUITO	1 mile (1.6 km)	600

PTERODACTYLS

The first machine to copy
the wing-flapping
technique of animal flight was
built as part of a study on the
flight of extinct pterodactyls.
A team of American scientists reconstructed a
pterodactyl called *Quetzalcoatlus northropi*, which lived
about 100 million years ago. Its wing span was around
19 ft (6 m) and it was equipped with a radio receiver, an
onboard computer "autopilot" and 13 tiny electric motors
to make the wings flap. In 1986, the model flew for three
minutes over Death Valley, California.

77

A SKY BIRD
The Garuda was a giant bird that carried the Indian god Vishnu across the sky. It is also the name of Indonesia's national airline, which uses the mythical bird for its logo.

• USING AIR AND WIND •

Myths and Legends

For thousands of years, people have told stories of wondrous beings that moved through the sky with the grace and ease of the birds. The ability to fly was seen as a sign of greatness and power. The gods and the heroes of many myths and legends were set apart from ordinary people because they could fly. In Greek mythology, Icarus and Daedalus flew on wings made of feathers, twine and wax; King Kaj Kaoos of Persia harnessed eagles to his throne, while Count Twardowski of Poland flew to the moon on the back of a rooster. Many people were inspired by visions of joining their heroes in the sky. They strapped wings to their arms and jumped off towers, high buildings and even out of balloons. Some did survive their dramatic falls. In 1507, Scotsman John Damian leapt from the walls of a castle with wings made of chicken feathers and broke only his thigh. He thought he would have been more successful if he had used the feathers of a bird that could really fly. The modern hero Superman can fly "faster than a speeding bullet." It seems our desire to believe in flying heroes continues.

A CHARIOT OF WINGS
Alexander the Great was said to have flown by harnessing six griffins, mythical winged animals, to a basket. He placed meat on his spear and enticed them to fly after it.

A WINGED HORSE
According to Greek legend, Bellerophon the Valiant, son of the King of Corinth, captured a winged horse called Pegasus. He flew through the clouds to find and defeat in battle the triple-headed monster, Chimera.

REACHING FOR THE SUN
In Greek mythology, Daedalus and his son Icarus used wax-and-feather wings to escape from the island of Crete. But Icarus flew too close to the sun and the wax on his wings melted. He fell into the Aegean Sea and drowned.

TOWER JUMPERS

Through the centuries, humans have tried to copy the birds. With elaborate wings made of feathers, they jumped from towers and flapped their arms desperately as they plummeted to the ground. They did not know that humans are too heavy, and their muscles are not strong enough to fly like birds. The hearts of humans cannot pump blood fast enough to meet the demands of wing flapping, which even in a sparrow is 800 heartbeats a minute.

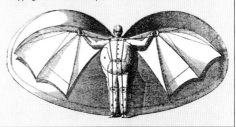

SKY BATTLE
In many legends, the forces of both good and evil had the power to fly. Here, St. Michael defends his island against a deadly dragon.

A TOUCH OF SPRING
The Egyptian goddess Queen Isis had wings like a falcon. Each year she flew over the Earth and brought spring to the land.

Discover more in The Groundwork

79

FORERUNNER OF FLIGHT
The Japanese used vibrantly
colored kites in religious
ceremonies and for
entertainment.

HANG–GLIDING
A typical hang-glider flight
begins on a high, windy ridge.
Strapped into a harness, the
pilot runs into the wind and is
lifted by the sail (wing). The
pilot holds the control bar and
shifts his or her weight around
to direct the hang-glider.

Control bar

• USING AIR AND WIND •

Kites and Gliders

The kite is the ancestor of the airplane. The
Chinese flew kites more than 2,000 years ago,
and through the years they have been used to lift
people high above battlefields on military observation,
to collect weather information and to drop supplies. Kites
inspired the English inventor George Cayley in his design
of the world's first model glider. Pioneer fliers such as
Englishman Percy Pilcher and German Otto Lilienthal
used kite designs to develop the wings of their gliders.
Lilienthal believed that aviators should learn to
glide so they could really understand the air.
The Wright brothers constructed their first glider
in 1901, and this was based on a biplane kite they
had built previously. Early gliders were launched
from hills, but modern sailplanes are towed into the
air by light planes. When they are released, they
climb and soar in the sky, using thermals of hot air.

MAORI FISHING KITE
The Maoris in New Zealand flew
birdmen kites. Some had special
significance and were used by
"tohunga" (important men) to help
them make big decisions.

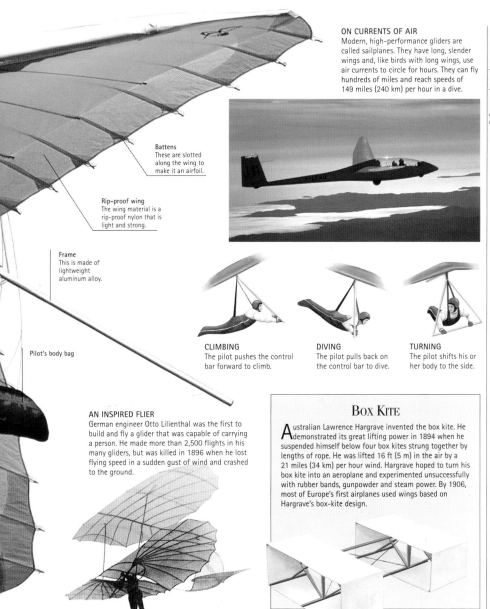

ON CURRENTS OF AIR
Modern, high-performance gliders are called sailplanes. They have long, slender wings and, like birds with long wings, use air currents to circle for hours. They can fly hundreds of miles and reach speeds of 149 miles (240 km) per hour in a dive.

Battens
These are slotted along the wing to make it an airfoil.

Rip-proof wing
The wing material is a rip-proof nylon that is light and strong.

Frame
This is made of lightweight aluminum alloy.

Pilot's body bag

CLIMBING
The pilot pushes the control bar forward to climb.

DIVING
The pilot pulls back on the control bar to dive.

TURNING
The pilot shifts his or her body to the side.

AN INSPIRED FLIER
German engineer Otto Lilienthal was the first to build and fly a glider that was capable of carrying a person. He made more than 2,500 flights in his many gliders, but was killed in 1896 when he lost flying speed in a sudden gust of wind and crashed to the ground.

BOX KITE
Australian Lawrence Hargrave invented the box kite. He demonstrated its great lifting power in 1894 when he suspended himself below four box kites strung together by lengths of rope. He was lifted 16 ft (5 m) in the air by a 21 miles (34 km) per hour wind. Hargrave hoped to turn his box kite into an aeroplane and experimented unsuccessfully with rubber bands, gunpowder and steam power. By 1906, most of Europe's first airplanes used wings based on Hargrave's box-kite design.

81

• USING AIR AND WIND •

Up, Up and Away

People created many elaborate flying machines in their attempts to see the world from the sky. Frenchmen Joseph and Etienne Montgolfier built a balloon of paper and cloth that rose in the hot air above a fire. Their next balloon had passengers: a sheep, a duck and a rooster. In 1783, before an astonished crowd in Paris, a Montgolfier balloon (below) carried two French noblemen into the skies and became the first successful flying craft. Hot-air and gas-filled balloons soon became popular. When Paris was surrounded by a Prussian army in 1870, the French smuggled people and mail out of the city in balloons. In 1897, three explorers vanished trying to reach the North Pole in a balloon. Fifty years later, scientists used balloons to study the Earth's upper atmosphere. Today, several groups of balloon enthusiasts are planning to fly nonstop around the world.

Burners
The pilot reaches up and pulls the trigger operating the burner blast valve. Two propane gas burners blast heat into the balloon.

Basket
Most baskets are woven of willow, because it is strong, light and flexible.

FULL OF HOT AIR
Hot-air ballooning is very popular and many balloonists operate joy flights for the public. A typical balloon is 59 ft (18 m) in diameter and holds 99,956 cubic ft (2,830 cubic m) of air. This is heated by two propane gas burners, each of which is powerful enough to heat 120 houses.

82

Envelope
This is made of 24 panels of polyurethane-coated, rip-proof nylon. It usually contains 10,760 sq ft (1,000 sq m) of fabric and 3 miles (5 km) of thread.

Parachute vent

Ripcord

BALLOON ADVENTURERS
In 1978, *Double Eagle*, an American gas-filled balloon, crossed the Atlantic in six days. Fifteen years later, a hot-air balloon (below) crossed Australia in 40 hours and 23 minutes.

BARRAGE BALLOONS
Clusters of gas-filled balloons were used to defend cities during the Second World War. They were tied to the ground by steel cables and obstructed enemy bombers that tried to fly low over the cities.

HOW HOT AIR BALLOONS FLY

Balloons travel with the wind, so balloonists have little control over their direction. But they can control the height to which they rise. If they want the balloon to climb, they turn the burner on. This heats the air in the balloon and produces lift. If a balloonist wants to descend, air in the balloon is allowed to cool, or the ripcord is pulled. Hot air then escapes from the parachute vent and is replaced by cooler (heavier) air. When the balloon has landed, the vent is opened to deflate the balloon.

Discover more in Airships

The Groundwork

AERIAL STEAM CARRIAGE
In 1842, William Henson designed an airplane— the first flying invention to actually look like an airplane. It was to be powered by a steam engine, which turned two propellers. Although Henson only managed to build a model, his design had many of the features used in today's airplanes.

There are many stories of gallant people and the inspired flying contraptions they constructed. Most of the machines never flew, but these inventors and engineers did much of the groundwork for the aviators who were to follow. The development of the steam engine in the nineteenth century led to serious attempts to invent steam-powered aircraft. In 1874, Felix de Temple built a monoplane that managed a short, downhill hop. Clement Ader and Hiram Maxim both built machines that lifted them, briefly, off the ground. But these aircraft were difficult to control and their coal-fired steam engines were too heavy and not powerful enough for true flight. In 1896, Dr. Samuel Langley launched an unpiloted aircraft with a steam-powered engine. It did manage to fly 1 mile (1.2 km), but then ran out of steam. Steam engines were soon replaced by light and powerful gasoline engines. They made sustained flight a reality.

LANGLEY'S LUCKLESS VENTURE
American Samuel Langley made model airplanes powered by steam. They were so successful he built a full-sized version, called the *Aerodrome*, which had a gasoline engine. Its first test flight was in 1903, when the piloted aircraft was launched from a catapult on the roof of a houseboat. But the launching mechanism failed, and *Aerodrome* plunged into the Potomac River.

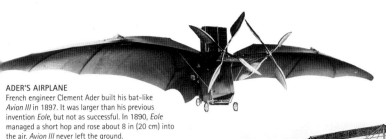

ADER'S AIRPLANE
French engineer Clement Ader built his bat-like
Avion III in 1897. It was larger than his previous
invention *Eole*, but not as successful. In 1890, *Eole*
managed a short hop and rose about 8 in (20 cm) into
the air. *Avion III* never left the ground.

MAXIM'S MONSTER
This flying giant, a triple
biplane with a wingspan of
103 ft (31.5 m), was built in
1894 by an American, Hiram
Maxim (the inventor of the
machine gun). For a few
seconds, its two steam engines
lifted Maxim and his crew from
its rail track.

> ## DID YOU KNOW?
> Artist Leonardo da Vinci (1452–1519) was
> also an engineer. He believed that
> ornithopters (wing-flapping aircraft) were
> the key to powered flight. He produced
> many plans for ornithopters, which ranged
> from strap-on wings to flying chariots.

THE SEARCH FOR AN ENGINE

In 1852, Henri Giffard built an airship,
which was powered by the first
aircraft engine— an extremely heavy,
three-horsepower steam engine. Aviators
searched for an alternative with more power and
less weight and experimented with electric motors
and engines powered by compressed air and coal
gas. In the late 1800s, Otto Daimler invented the
gasoline engine. This finally provided the light
and powerful engine needed for heavier-than-
air airplanes. The Anzani gasoline engine
shown here was invented in 1909.

Q: How did gasoline engines make true flight possible?

Wing-warping wire
This banked the plane by twisting (warping) the flexible tips of the wings.

THE FIRST FLIGHT
The 1903 *Flyer* was built of spruce, braced with wire and covered with muslin. The pilot lay on the lower wing alongside the engine. He moved the elevator lever with his left hand to climb or descend. He twisted his hips to control wires connected to the wingtips and rudders.

Propeller chain drive
Bicycle chains linked the propellers to the engine.

• PIONEER PILOTS •

The Wright Brothers

Orville and Wilbur Wright dreamed of flying. They built and sold kites to classmates at school. They opened a bicycle business when they were young men and used the profits to build aircraft. For years they experimented with, and examined the theories of, flight. What was the clue to the mystery of flight? By 1902, they had developed a glider that could carry a person. It made more than 1,000 flights. Next, they designed and built a tiny 12-horsepower gasoline engine and connected it by bicycle chains to a pair of propellers. It turned their glider into a powered airplane– the *Flyer*. In December 1903, Orville made the world's first powered flight from Kill Devil Hill, over Kitty Hawk beach in North Carolina. The first *Flyer* flew just four times, for a total of 98 seconds. Then, it was caught in a gust of wind and crashed into the sand– severely damaged.

Rudders
Two rudders helped control the direction of the plane (called yawing).

DID YOU KNOW?

Wilbur (left) and Orville Wright tossed a coin to decide who would be the world's first pilot. Wilbur won, but he stalled and crashed into the sand. Orville succeeded where his brother had failed.

AHEAD OF ITS TIME
Orville Wright was part of a team that designed this streamlined, 1920 Dayton Wright monoplane racer. It had retractable landing gear and extremely strong wings!

PROPELLER POWER

A propeller is a tiny wing that spins. As it rotates, air flows around the propeller blades and moves faster over the curved leading edge. This reduces the air pressure in front of the blade and pulls the aircraft forward. Many propellers allow pilots to adjust the blade angle for climbing, cruising and descending. This improves performance and keeps engine speed and fuel consumption low— like changing gears in a car. The propeller of the *Flyer* (shown on the right) was carved out of wood. Today, propellers are made of metal or fiberglass and carbon.

Water-filled radiator

Fuel tank

Biplane elevators
These tilted up or down to make the plane climb or descend.

Gasoline-combustion engine
The 12-horsepower engine was mounted to the side to balance the pilot's weight.

Elevator lever

Skids for landing

THAT'S MY PLANE
The Wright brothers patented their aircraft in 1906 to stop others from copying their ideas. But aviators in Europe were already designing different kinds of airplanes.

A LATER PERSPECTIVE
Orville Wright's flight covered a distance of 170 ft (51.5 m), which included the take off and landing run. The whole flight could have taken place in the passenger area of this Boeing 747-400.

Discover more in Built for Flight

Q: How did the Wright brothers turn their glider into a powered aircraft?

Famous Firsts

The progress of aviation has been marked by the achievements of many pioneers determined to dominate the skies. It all began in 1783 with the Montgolfiers' hot-air balloon. In 1853, George Cayley created the first heavier-than-air aircraft. Fifty years later, the Wright brothers introduced powered flight and five years after, carried the first plane passenger. Louis Blériot flew across the English Channel in 1909 and proved that water was no longer an obstacle. Soon, time and technology would reduce flying times, and the world would be encompassed by airplanes. In 1947, Chuck Yeager blasted through the sound barrier in his Bell X-1. Aviation sights were then set even higher, and in 1961, Russian Yuri Gagarin became the first person to fly in space. Six years later, the rocket-powered X-15A-2 reached a world-record speed of 4,497 miles (7,254 km) per hour— about 6.8 times the speed of sound.

WALKING ON THE MOON
Americans Neil Armstrong and Edwin "Buzz" Aldrin landed their lunar module *Eagle* on the moon on July 21, 1969. Their tentative steps were seen by millions of people all over the world.

PEDAL POWER
American cyclist Bryan Allen pedalled his *Gossamer Condor*, the first successful human-powered aircraft, around a 1 mile (1.6 km) figure-eight course in 1977. His aircraft cruised at 10 miles (16 km) per hour, had a wingspan of 95 ft (29 m) and weighed 72 lb (32.7 kg). It was made of cardboard and aluminum tubing covered in plastic.

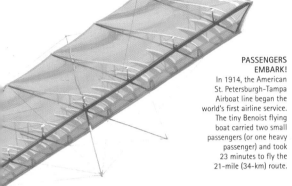

PASSENGERS EMBARK!
In 1914, the American St. Petersburgh-Tampa Airboat line began the world's first airline service. The tiny Benoist flying boat carried two small passengers (or one heavy passenger) and took 23 minutes to fly the 21-mile (34-km) route.

WOMEN OF THE AIR
Women pilots were among those setting aviation firsts. In 1910, Baroness de Laroche from France became the first female pilot. Two years later, American Harriet Quimby flew the English Channel. Another famous American Amelia Earhart (above) was the first woman to fly the Atlantic in 1932. Other solo, long-distance pilots of the 1930s were Amy Johnson of England who flew 12,162 miles (19,616 km) to Australia; Australian Lores Bonney who flew 18,054 miles (29,120 km) to South Africa; and New Zealander Jean Batten whose array of firsts included crossing the South Atlantic. In 1953, American Jacqueline Cochran was the first woman to break the sound barrier.

JETTING ABOUT
In 1939, the German Heinkel He 178 became the world's first jet-powered aircraft. It could reach speeds of 434 miles (700 km) per hour. These aircraft had a great impact on aviation. All future designs for fighter planes in the United States and Europe were jet-propelled.

A FLYING MILESTONE
The sleek 1912 Deperdussin monocoque racer was the super plane of its day. Its single-shelled (monocoque) fuselage made it streamlined and fast. Flown by Frenchman Jules Vedrines, it was the first aircraft to exceed 100 miles (161 km) per hour.

Discover more in Breaking the Sound Barrier

CROSSING THE PACIFIC
In 1928, Australians Charles Kingsford-Smith and Charles Ulm made the first aerial crossing of the Pacific Ocean. They averaged 89 miles (143 km) per hour in their Fokker VII/3m *Southern Cross* on the 7,400-mile (11,914-km) flight. They stopped for fuel in Hawaii and Fiji.

THE WHITE CLIFFS OF DOVER
Louis Blériot flew from France to England in a monoplane, powered only by a 35-horsepower engine. He crash-landed on the cliffs of Dover after his great flight.

SPIRIT OF ST. LOUIS
Lindbergh's plane, the Ryan NYP monoplane *Spirit of St. Louis*, was built especially (in just two months) for his transatlantic flight. The 3,600-mile (5,796-km) flight from New York to Paris took 33 hours and 30 minutes. The cockpit of the plane was tucked behind a huge fuel tank, and Lindbergh had to use a periscope to see in front of him.

• PIONEER PILOTS •

Making the World Smaller

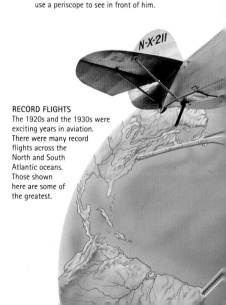

RECORD FLIGHTS
The 1920s and the 1930s were exciting years in aviation. There were many record flights across the North and South Atlantic oceans. Those shown here are some of the greatest.

As airplanes changed from rickety machines to planes that could cover huge distances, pilots dreamed of conquering large expanses of land and water. On 25 July 1909, Louis Blériot took nearly 37 minutes to become the first person to fly the 22 miles (35 km) across the English Channel. He battled wind gusts and a severely overheating engine to make a rushed downhill landing that shattered his propeller and landing gear. American Cal Rodgers survived five crashes to fly across the United States in 84 days in 1911. Ross and Keith Smith set an extraordinary record when they flew 11,426 miles (18,396 km) from England to Australia. Aviation dominated the headlines in 1927 when Charles Lindbergh flew the Atlantic, and a year later when Charles Ulm and Charles Kingsford-Smith conquered the Pacific. Jet travel eventually brought the continents less than a day's flight apart.

ALTITUDE RECORD

After flying around the world in 1934, American Wiley Post decided to break the world's altitude record. He designed the first spacesuit which, like a deep-sea diver's suit, allowed him to breathe and work in unusual pressure situations. In 1934, he reached a record 50,000 feet (15,240 m). He wore the suit when he made the first flights in jet streams— the high altitude winds used by today's airlines to increase their speed over the ground.

AROUND THE WORLD IN NINE DAYS

A long-distance aviation record was set in 1986 when Dick Rutan and Jeana Yeager flew their plane *Voyager* nonstop around the world in nine days.

FLYING FAME

Charles Lindbergh began his flying career by carrying mail across the United States. He became the world's most famous pilot after his Atlantic crossing.

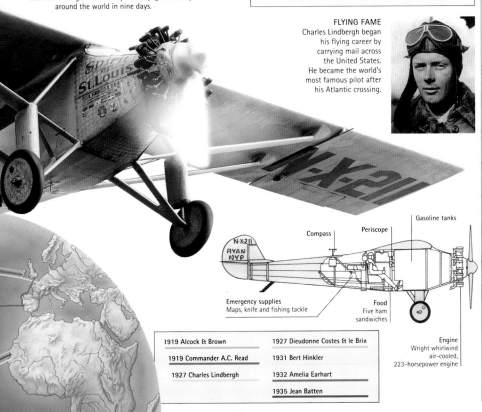

Gasoline tanks

Periscope

Compass

Emergency supplies
Maps, knife and fishing tackle

Food
Five ham sandwiches

Engine
Wright whirlwind
air-cooled,
223-horsepower engine

1919 Alcock & Brown	1927 Dieudonne Costes & le Brix
1919 Commander A.C. Read	1931 Bert Hinkler
1927 Charles Lindbergh	1932 Amelia Earhart
	1935 Jean Batten

WEDDELL-WILLIAMS
The Weddell-Williams 44 racer won the American 1933 Thompson Trophy. It averaged 238 miles (383 km) per hour around a circuit marked by pylons. The racing planes of the 1930s relied on very powerful engines, rather than streamlining, for their speed.

• PIONEER PILOTS •

Races in the Air

People have always been competitive. The Wright brothers' wondrous flying machine introduced people to powered flight– a new way to gamble their lives for glory and the rich rewards of air racing. In 1909, American Glenn Curtiss won the world's first air race, the Gordon Bennett Cup, flying a rickety, open biplane at 47 miles (76 km) per hour. Three years later, aircrafts and speeds had improved dramatically, and a Deperdussin monoplane flew at 123 miles (199 km) per hour to win the cup. In the 1930s, the National Air Races held in the United States attracted huge crowds, and pilots such as Roscoe Turner and Jimmy Doolittle became sports heroes. Twenty-three women pilots, including Amelia Earhart, entered the first women's air race. The Schneider, Thompson and Bendix races continued to thrill the public and help refine aircraft design. Many pilots today recreate the excitement of the great air-racing days by flying specially built planes at events such as the Reno Air Races held in Nevada.

SCHNEIDER TROPHY
This race for seaplanes was held across open water. Racers flew seven times around a triangular 30-mile (48-km) course.

RACING FOR GLORY
The race is on as a British Supermarine S.5 leads an Italian Macchi M.52 around a pylon during the 1927 Schneider Trophy Race for seaplanes. Both machines were very streamlined, and this made them very fast. The winning S.5 averaged 282 miles (454 km) per hour.

AIR ACE

American Jimmy Doolittle was a top air racer. In 1925, he won the Schneider Trophy in a US Army seaplane. In 1931, he won the Bendix Trophy. He set a world record of 294 miles (473 km) per hour in 1932 when he won the Thompson Trophy.

BENDIX TROPHY

This trophy was awarded to the winner of a long-distance race across the United States, held between 1931 and 1949.

AERIAL SHOW-OFFS

After the First World War, many unemployed pilots earned a living as traveling showmen, giving joyflights at city shows and country fairs. Others worked for flying circuses and entertained the crowds with spectacular aerial tricks such as walking on wings, hanging upside down from the landing gear, transferring from one plane to another in flight, or playing tennis on the wings of a Curtiss Jenny biplane.

FASTER THAN THE SPEED OF SOUND
As bullets were known to be supersonic,
Bell aircraft shaped the fuselage of the
X-1 experimental rocket plane like a .50
caliber bullet. The aircraft was powered
by a rocket motor and launched
from Boeing B-29 and
B-50 bombers

• PIONEER PILOTS •

Breaking the Sound Barrier

I magine flying through the air and striking an invisible barrier.
This happened to Spitfire and Mustang fighter pilots during
the Second World War. At speeds of around 545 miles
(880 km) per hour, their aircraft suddenly became difficult to
control. The planes shook so violently that some even broke apart.
These pilots were approaching the sound barrier, which many
experts believed no aircraft could penetrate. Breaking through that
barrier became an international challenge, and some pilots in the
early jet fighters died in their attempts. But in 1947, Chuck Yeager
blasted through the sound barrier in his specially designed Bell X-1,
powered by a rocket engine. An American F-86 Sabre jet fighter
then exceeded the speed of sound while in a dive. Today, airplanes
such as the Concorde and most military aircraft can easily fly faster
than the speed of sound.

GLAMOROUS GLENNIS
At an altitude of 43,000 ft (13;106 m), Captain Charles "Chuck"
Yeager flew through the sound barrier at 698 miles (1,126 km)
per hour in *Glamorous Glennis*, named after his wife.

THE AREA RULE
This is a method of designing an
airplane's shape to reduce drag.
The red dashes on the F-102 above
show the width of the middle
fuselage before it was trimmed
according to Area Rule
calculations.

High-speed probe
This gathers
information on air
pressure during
flight.

THE SOUND BARRIER
As it moves through the air,
an airplane makes pressure
waves that travel at the
speed of sound. They radiate
like ripples from a stone
dropped in a pond.

SUBSONIC: BELOW MACH 1
The pressure waves radiate in
front of, as well as behind, the
airplane.

Horizontal stabilizer
This moves to help control or stabilize the aircraft as it nears the sound barrier.

Cockpit
This is pressurized and has room for one pilot.

Rocket-engine plumes
The engine is a 6,000-lb (2,722-kg) thrust rocket, powered by liquid oxygen and ethyl alcohol.

SPEEDING THROUGH AIR
A Machmeter gives the speed of air as a percentage of the speed of sound, which varies with temperature. At 40,000 ft (12,192 m), where it is very cold, Mach 1.0 is 657 miles (1,060 km) per hour.

Wings
The wings are short and very thin to reduce drag at high speed.

DID YOU KNOW?
A thunderclap, rifle shot and whip-crack are tiny sonic booms. Like the boom of a supersonic airplane, they are created by shock waves— sudden increases in air pressure.

Fuselage
This is shaped like a supersonic .50 caliber machine gun bullet.

TEST PILOTS
Test flying requires skill and daring— pilots call it "the right stuff." Each new aircraft, whether it is the latest glider, jumbo jet, military fighter or space shuttle, must be tested in flight to check that it is safe and reliable. Test pilots push their machines through every imaginable flight maneuver until they are satisfied that there are no problems with the aircraft. This dangerous task has now been made easier by supercomputers that can simulate the flight performance of new designs before the actual planes are ever flown.

RANSONIC: AT MACH 1
he airplane catches up with its
wn pressure waves, which build
p into a shock wave.

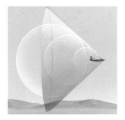

SUPERSONIC: ABOVE MACH 1
The shock waves form a cone. This causes a sonic boom when it hits the ground.

Planes in Flight

The invisible force that makes heavier-than-air aircraft fly is the flow of air around an airplane's wings. The differences in pressure above and below the wings combine to "lift" the airplane. "Lift" is one of the four forces that act upon a plane in flight, and it overcomes the plane's weight. The "thrust," or forward movement of the plane is produced by the engine, and this opposes "drag," the natural resistance of the airplane to forward motion through the air. But an airplane also needs to be stable so that it can fly smoothly and safely. Its tailplanes and the dihedral shape of its wings (which means the wings are angled upwards slightly from the fuselage) make the plane stable in the same way the tail of a kite makes it steady. The wings and tailplanes are also equipped with movable control surfaces called ailerons, elevators and rudders. These alter the airflow over the wings and tailplanes, and the pilot uses them to change the airplane's direction and height.

Rudder

Vertical tailplane (tailfin)

D-EVS

Horizontal Tailplane

Control rods
These link the pilot's controls to the elevators and rudder.

Left aileron down

PITCHING
When the elevators are up, the airplane's nose is raised above the horizon and the airplane climbs. The rudder and ailerons are in a neutral position. When the airplane descends, the elevators are down. This up or down movement of the nose is called "pitching."

D-EVSA

Elevators up

BANKING (ROLLING)
The left aileron is down and the right aileron is up, which makes the airplane bank to the right. The elevators and rudder are in a neutral position. This movement of the wings is called "rolling."

Right aileron up

ANGLE OF ATTACK
This is the angle at which the wing meets the airstream. As an airplane slows down, this angle must be increased to produce enough lift to equal its weight. When the angle reaches 14 degrees, the wing loses lift (called stalling) and the airplane descends.

LOW ANGLE
At high speed, the wing needs only a low angle of attack (about 4 degrees) to produce enough lift.

HIGH ANGLE
At low speed, a much higher angle of attack is needed to produce the same amount of lift.

STALL ANGLE
At about 14 degrees angle of attack, the airstream over the wings becomes turbulent. The plane stalls and loses height.

96

Fuselage
The body of the airplane contains the cockpit and the engine. It is streamlined to minimize the drag caused by wind resistance.

Control column
This moves backward and forward to operate the elevators, and from side to side to operate the ailerons.

DID YOU KNOW?
As an airplane flies higher, the pressure of the air outside decreases. The changes in pressure as an airplane climbs or descends can make your ears "pop." People with bad colds can experience painfully blocked ear drums.

Propeller
This has rotating blades, shaped like airfoils, which convert the engine power into forward thrust.

G115 D

Rudder pedals
These control the rudder and also operate the aircraft's brakes.

THE CONTROL SURFACES
Ailerons, elevators and the rudder are operated by the control column and rudder pedals in the cockpit. They are used to make the airplane climb or descend, roll, turn or simply fly straight and level, as shown.

Rudder right

TURNING (YAWING)
The left aileron is slightly down, the right aileron is slightly up. The rudder is moved to the right, which helps to push the airplane's nose sideways into a gentle right turn. This sideways movement of the nose is called "yawing."

D-EVSA

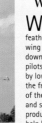

WING FLAPS
When birds land, they spread their feathers and change wing shape to touch down slowly. Airplane pilots do the same thing by lowering sections of the front and rear edges of the wing, called flaps and slats. These devices produce extra lift and help big jets to land and take off at slow speeds.

Lift

Drag

Weight

Thrust

THE FOUR FORCES
When an airplane is in straight and level flight at a steady airspeed, the four forces are in equilibrium (balanced). This means that lift is equal and opposite to weight, and thrust is equal and opposite to drag.

Discover more in Airports

The Cockpit

The cockpit is the nerve center of the airplane. It is crammed with controls, instruments and computers. Some Boeing 747 cockpits contain 971 instruments and controls. Orville Wright had none of these devices. He had to lie on the lower wing of the plane and look at the horizon to judge the plane's position in flight. Pioneer pilots navigated by comparing features on the ground with those on their maps. Pilots today do not even have to see the ground. They use navigation systems that are linked to satellites, while the computer-controlled autopilot flies the airplane far more accurately than a human ever can. The latest instrument panels have television, radar and multi-function displays that give pilots the information they need to fly their planes safely.

HAND THROTTLE
This cluster of throttle levers operates the eight engines of a giant B-52 bomber.

A PILOT'S PERSPECTIVE
The Spitfire fighter was an important player in the air battles of the Second World War. The instruments and controls shown in this Spitfire cockpit are very similar to those found in small airplanes today.

FLIGHT SIMULATORS

Jet airliners are very expensive to fly, so machines called simulators are used to train pilots and to practice emergency drills. These machines are built to resemble the cockpit of an airliner. They have all the instruments and controls of a real plane. To make the simulator even more lifelike, moving pictures of the sky and ground are projected onto the windshield— just like a giant video game.

BOEING 747
The flight crew of a modern airliner sits in a spacious cockpit using sophisticated computers and instruments that are large and easy to read.

PRIMARY FLIGHT DISPLAY
In modern airplanes, all the instruments needed for "blind" flying are combined in this single display. It has replaced the cluttered flight instrument panel that was used in airplanes such as the Spitfire.

NAVIGATIONAL DISPLAY
This display shows the airplane's position on a radar picture of the ground below. It also gives the airplane's speed, fuel flow, and the time and distance to the next position along the air route.

Airports

T he first commercial airports were built in the 1920s. They usually consisted of a large grass field with a few small buildings, a hangar and a rotating searchlight beacon to help pilots find the airport in bad weather. Airports today are miniature cities, surrounded by a web of taxiways and runways that can be more than 13,120 ft (4,000 m) long. The busiest airport in the world is O'Hare Airport in Chicago— more than 2,000 airplanes land and take off every day. Huge numbers of passengers and large amounts of baggage pass through airports. When passengers first began traveling by air, they were weighed along with their luggage! The ground staff at airports look after the equipment needed to keep the airplanes flying safely. Movement of planes through an airport is regulated by air traffic controllers, who watch and use radar to monitor the planes' flight paths.

NETWORKIN
To keep airplane traffic flowing, busy airports such a San Francisco International Airport need a network o runways, taxiways and parking bay

CONTROL TOWER
The Control Tower provides a bird's-ey view of the airport and the surroundin sky. Its staff contro the movement of a airplanes on the ground and in the air near the airport

Stacking
Aircraft circle over a radio beacon as they await their turn to land.

Outer marker
Shows that the airplane is 5 miles (8 km) from touchdown on final approach.

Middle marker
Marks the midway point of the final approach.

INSTRUMENT LANDING SYSTEM

I n bad weather, pilots use the Instrument Landing System (ILS) to land safely. Two narrow radio beams are transmitted from the touchdown point on the runway. One is called the Glide-slope, the other is called the Localizer. If pilots follow the position of these two beams on an instrument in the cockpit, they can approach the runway and land the plane without seeing the runway until the final moment.

Glide–slope beam
Shows that the airplane is descending to the runway at the right angle.

Localizer beam
Shows that the airplane is properly in line with the runway.

Inner marker
At this point, close to touchdown, pilots should be able to see the runway.

RUSH HOUR
Airport staff rush around this 375-passenger Airbus at the terminal. In just 90 minutes, the airliner must be unloaded, cleaned, restocked with food and drinks, refueled and boarded by new passengers.

DID YOU KNOW?
During takeoff and landing, the tires of large airplanes speed across the ground. Friction with the ground can make them hot enough to catch fire. To avoid this, the tires are filled with nitrogen, which does not burn, rather than air.

Toilet-waste truck
Removes waste from the aircraft toilets.

Mobile stairs
These give ground staff access to the cabin.

Cleaning service truck
This carries the cleaners and their equipment and takes away garbage from the previous flight.

Fuel-transfer vehicle
This pumps aviation fuel from underground tanks into the aircraft tanks.

Conveyor
A moving belt carries late and awkwardly shaped baggage into the aircraft hold.

Airbridge
A passenger walkway links the aircraft with the terminal.

Tractor and dollies
These bring passenger baggage to and from the terminal.

Water truck
This fills the aircraft's water tanks.

Hi-loaders
These platforms rise to load heavy containers.

Tow tractor
This pushes the aircraft from the terminal to the taxi area.

Catering truck
This stocks the aircraft with the in-flight meals and drinks.

Ground power unit

Airships

T he first airship was a sausage-shaped balloon. It was built in 1852 by French engineer Henri Giffard, who fitted his new aircraft with a small steam engine and a rudder for steering. It flew 17 miles (27 km) but did not have enough power to fly against the wind. In 1900, Count Ferdinand von Zeppelin from Germany built the first rigid airship. It was longer than a football field and had a lightweight framework that contained huge gas bags or cells, each of which was filled with hydrogen— a highly flammable gas. Between 1910 and 1913, Zeppelin airships carried more than 30,000 passengers on sightseeing flights over Germany. They were also used to bomb London in night raids during the First World War. The luxurious *Graf Zeppelin* and the *Hindenburg*, the largest rigid airship, carried thousands of passengers across the Atlantic between the two world wars. In 1937, however, the world was stunned when the *Hindenburg* exploded. The airship era came to an abrupt and tragic end.

A DRAMATIC END
The *Hindenburg* approached its mooring mast at Lakehurst, New Jersey. Suddenly, flames and smoke billowed into the sky— the airship had exploded! Amazingly, 62 of the 97 people on board escaped from the blazing airship.

TRAVELING IN STYLE
The *Graf Zeppelin* was the world's most successful airship. It was powered by five engines and had a top speed of 79 miles (128 km) per hour.

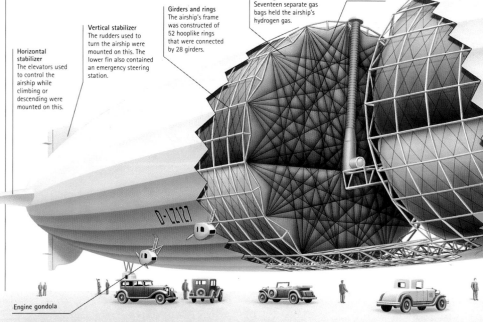

Horizontal stabilizer
The elevators used to control the airship while climbing or descending were mounted on this.

Vertical stabilizer
The rudders used to turn the airship were mounted on this. The lower fin also contained an emergency steering station.

Girders and rings
The airship's frame was constructed of 52 hooplike rings that were connected by 28 girders.

Gas bags
Seventeen separate gas bags held the airship's hydrogen gas.

Bracing wire

D-LZ127

Engine gondola

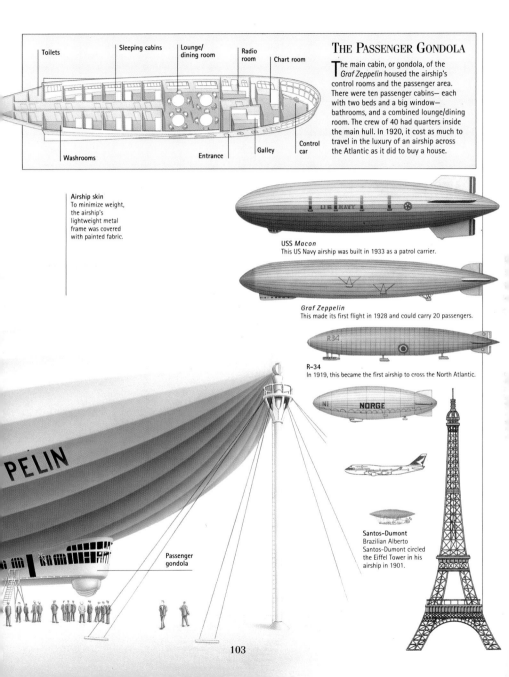

Toilets

Sleeping cabins

Lounge/ dining room

Radio room

Chart room

THE PASSENGER GONDOLA

The main cabin, or gondola, of the *Graf Zeppelin* housed the airship's control rooms and the passenger area. There were ten passenger cabins— each with two beds and a big window— bathrooms, and a combined lounge/dining room. The crew of 40 had quarters inside the main hull. In 1920, it cost as much to travel in the luxury of an airship across the Atlantic as it did to buy a house.

Control car

Washrooms

Entrance

Galley

Airship skin
To minimize weight, the airship's lightweight metal frame was covered with painted fabric.

USS *Macon*
This US Navy airship was built in 1933 as a patrol carrier.

Graf Zeppelin
This made its first flight in 1928 and could carry 20 passengers.

R–34
In 1919, this became the first airship to cross the North Atlantic.

NORGE

Santos-Dumont
Brazilian Alberto Santos-Dumont circled the Eiffel Tower in his airship in 1901.

PELIN

Passenger gondola

Seaplanes

TRANSPACIFIC

The 1930s was the age of the seaplane. People believed these aircraft were a safe way to cross stretches of water during a time when aircraft engines were thought to be unreliable. The luxurious flying boats, designed to compete with ocean liners, introduced people all over the world to the exotic reality of long-distance air travel between continents. Many airlines were now able to extend their services beyond Europe and North America. Pan American's Clipper flying boats provided the first passenger services across the Atlantic and Pacific. The Boeing 314 Clipper was the largest airplane of its day. It could carry 74 passengers and a crew of eight, and had 40 sleeping berths. Some flying boats were used to patrol the oceans for submarines during the Second World War. But the war also helped to bring about the end of the great flying boats. Land aircraft had improved enormously and airfields had been built all over the world.

Lounging and dining
At meal times, the lounge became a restaurant where diners were served by waiters.

Engines
600 hp Wright double-cyclone engines.

Wing walkway
The engineer could walk along here to make minor repairs during the flight.

Radio operators

PASSENGER COMFORT
The enormous Boeing 314 had four massive engines and could cruise at a speed of 174 miles (280 km) per hour. It could also fly 3,472 miles (5,600 km) without refueling.

Anchor
A ship's anchor was used when a mooring dock was not available.

Galley
Meals were prepared and cooked on board. On modern airliners, food is reheated.

Sponsons
These stabilizers balanced the aircraft on the water and were used to hold fuel.

AN AVIATION FAILURE
This Italian flying boat, a 9-winged Caproni Noviplano, was designed to carry 100 passengers. Instead, it crashed on its first test flight in 1921.

TYPES OF SEAPLANE

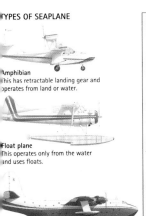

Amphibian
This has retractable landing gear and operates from land or water.

Float plane
This operates only from the water and uses floats.

Flying boat
The large hull of the flying boat is shaped like that of a boat.

GLOBAL TRAVEL

Flying boats were used on the early transoceanic airline services. Pan American Airways pioneered long-range flying boat services in the mid-1930s when its Commodores and S-42s flew to South America and Martin M-130s crossed the Pacific. In 1938, Short S-23s operated the England-to-Australia service. In 1939, a Boeing 314 made the first transatlantic airline run. Catalinas were used during the Second World War.

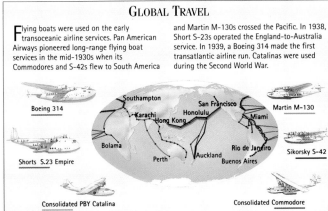

Boeing 314

Shorts S.23 Empire

Consolidated PBY Catalina

Southampton
Karachi
Hong Kong
Bolama
Perth
Auckland
San Francisco
Honolulu
Miami
Rio de Janeiro
Buenos Aires

Martin M-130

Sikorsky S-42

Consolidated Commodore

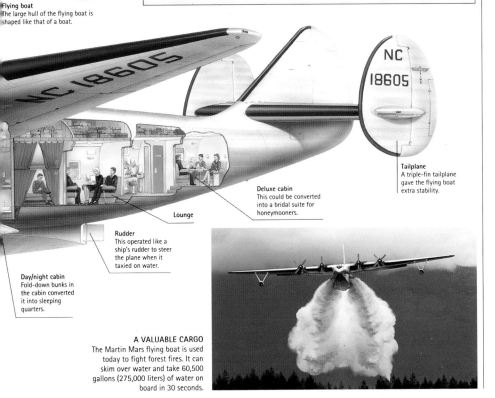

Tailplane
A triple-fin tailplane gave the flying boat extra stability.

Deluxe cabin
This could be converted into a bridal suite for honeymooners.

Lounge

Rudder
This operated like a ship's rudder to steer the plane when it taxied on water.

Day/night cabin
Fold-down bunks in the cabin converted it into sleeping quarters.

A VALUABLE CARGO
The Martin Mars flying boat is used today to fight forest fires. It can skim over water and take 60,500 gallons (275,000 liters) of water on board in 30 seconds.

105

Planes at Work

Pilots were inventive in the ways they used their wonderful new flying machines. In 1911, Frenchman Henri Pequet began an airmail service in India. Two weeks later, similar services were started in France, Italy and the United States. During the First World War, a French company built a huge biplane, which contained a portable operating theater. It landed on the battlefields carrying surgeons and nurses, who carried out emergency operations amid the chaos of war. The Huff Daland Dusters began dusting crops from airplanes in rural areas of the United States in 1924. When goldmines were built in New Guinea, all the supplies and materials were delivered by adventurous bush pilots. They landed on dangerous airstrips, carved from the jungle, in Junkers monoplanes. As the world's airplane industry grew, so too did the jobs airplanes performed. They are now used to fight fires, herd cattle, guard coasts and look for sharks, drop supplies to the victims of floods, storms and famines, and carry cargo.

A CLEAR VIEW
The slow-flying Edgely EA 7 Optica has a glass cockpit. This makes it ideal as an aerial spotter.

AIR WATCH
The coastlines in America are patrolled by specially equipped observation planes, such as this high-speed Falcon jet.

ON THE LAND
Helicopters are efficient and effective alternatives to traditional ways of rounding up horses and cattle.

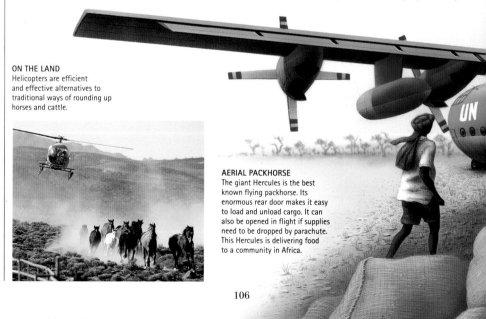

AERIAL PACKHORSE
The giant Hercules is the best known flying packhorse. Its enormous rear door makes it easy to load and unload cargo. It can also be opened in flight if supplies need to be dropped by parachute. This Hercules is delivering food to a community in Africa.

THE FLYING DOCTORS

The Royal Flying Doctor Service cares for the families who live in Australia's desolate outback. The first flying doctor flew in a slow de Havilland biplane, with an open cockpit. Today, the service uses twin-engine Beech Kingair passenger planes, which cruise at 274 miles (442 km) per hour. The doctors and nurses are on stand-by day and night for emergency calls. Their territory covers 2 million sq miles (5 million sq km), but they can reach any part of it within two hours. Every bush town and isolated homestead has a radio transmitter to call the flying doctor bases. When the service first began in 1928, these radios were run by the pedal-powered generators shown here.

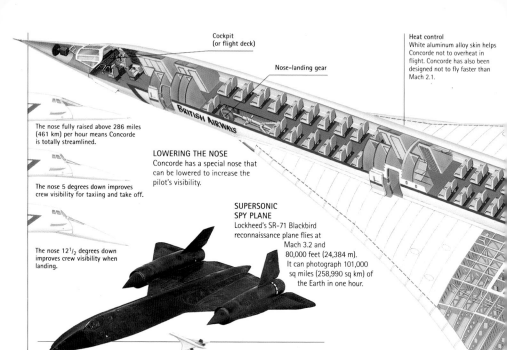

Cockpit
(or flight deck)

Nose-landing gear

Heat control
White aluminum alloy skin helps
Concorde not to overheat in
flight. Concorde has also been
designed not to fly faster than
Mach 2.1.

The nose fully raised above 286 miles
(461 km) per hour means Concorde
is totally streamlined.

LOWERING THE NOSE
Concorde has a special nose that
can be lowered to increase the
pilot's visibility.

The nose 5 degrees down improves
crew visibility for taxiing and take off.

The nose 12$^1/_2$ degrees down
improves crew visibility when
landing.

**SUPERSONIC
SPY PLANE**
Lockheed's SR-71 Blackbird
reconnaissance plane flies at
Mach 3.2 and
80,000 feet (24,384 m).
It can photograph 101,000
sq miles (258,990 sq km) of
the Earth in one hour.

• A PARADE OF AIRCRAFT •

Concorde and Supersonic Flight

In the 1960s, Britain and France joined forces to develop a passenger plane that could fly at the speed of sound—a supersonic airliner. The Americans also began to design such a plane, but abandoned it when costs skyrocketed and concentrated instead on building large subsonic jumbo jets. In 1969, Concorde, the world's first commercial supersonic plane, made its debut flight. The 14-plane Concorde fleet started service with British Airways and Air France in 1976. By then, the Soviet Union had built a supersonic airliner called the Tu-114. It crashed tragically at an airshow and never flew passenger services. Cruising at twice the speed of sound, Concorde can fly from New York to London in three-and-a-half hours. It is a technological triumph, but has not been a great commercial success. It is expensive to operate, seats only 100 passengers, and is banned from many cities due to its sonic boom. Passenger services were suspended after a fatal crash in July 2000, but resumed in November 2001.

WIND TUNNELS
Airplanes are tested in wind tunnels, which imitate
the airflow they will experience in flight. This
model is covered with fluorescent paint, which will
highlight the airflow and any problem areas.

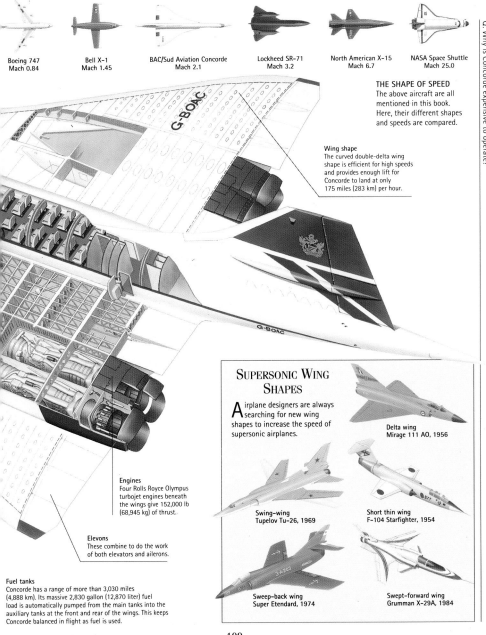

Boeing 747
Mach 0.84

Bell X-1
Mach 1.45

BAC/Sud Aviation Concorde
Mach 2.1

Lockheed SR-71
Mach 3.2

North American X-15
Mach 6.7

NASA Space Shuttle
Mach 25.0

THE SHAPE OF SPEED
The above aircraft are all mentioned in this book. Here, their different shapes and speeds are compared.

Wing shape
The curved double-delta wing shape is efficient for high speeds and provides enough lift for Concorde to land at only 175 miles (283 km) per hour.

SUPERSONIC WING SHAPES

Airplane designers are always searching for new wing shapes to increase the speed of supersonic airplanes.

Delta wing
Mirage 111 AO, 1956

Swing-wing
Tupelov Tu-26, 1969

Short thin wing
F-104 Starfighter, 1954

Sweep-back wing
Super Etendard, 1974

Swept-forward wing
Grumman X-29A, 1984

Engines
Four Rolls Royce Olympus turbojet engines beneath the wings give 152,000 lb (68,945 kg) of thrust.

Elevons
These combine to do the work of both elevators and ailerons.

Fuel tanks
Concorde has a range of more than 3,030 miles (4,888 km). Its massive 2,830 gallon (12,870 liter) fuel load is automatically pumped from the main tanks into the auxiliary tanks at the front and rear of the wings. This keeps Concorde balanced in flight as fuel is used.

Military Aircraft

Wars have accelerated the development of aircraft. At the beginning of the First World War, most planes could not fly further than 100 miles (161 km) and had a maximum speed of around 62 miles (100 km) per hour. The first military planes were used only for observing enemy activity from the air, but they were soon adapted for fighting. By the end of the war, bombers could travel almost 2,000 miles (3,220 km) and reach speeds of 150 miles (241 km) per hour. The most famous war planes of the Second World War were the piston-engine Spitfire, Mustang and Messerschmitt 109 fighters, the Flying Fortress and Lancaster bombers. Rocket-powered planes and jet fighters were also introduced, and the first all-jet air battle took place during the 1950–53 Korean War. Military planes today are controlled by onboard computers and can fly at supersonic speeds.

Jet nozzles
Wide, flat jet nozzles reduce and disperse exhaust to make the aircraft less visible to infrared missiles at night.

AIR–TO–AIR MISSILES
Modern fighter pilots do not even need to see their enemy. They can destroy other aircraft with deadly missiles guided by electronic or infrared devices.

UNDETECTABLE
Lockheed's F-117A Stealth Fighter is designed to be invisible on enemy radar screens. Most airplanes are rounded, but the surfaces of this plane are faceted (like a diamond) and highly polished to deflect and disperse radar signals.

A TIMELY EXIT
A pilot escapes in a rocket-powered ejection seat as his fighter explodes. Moments later, a parachute will open and bring him safely to the ground.

1915 German Fokker E.1 Eindecker
This could fly at 79 miles (128 km) per hour and was the first fighter equipped with a real forward-firing machine gun.

1917 French Spad X11
This great biplane fighter was flown by French, American and British airmen. It could fly at 129 miles (208 km) per hour.

1917 English Handley Page 0/400
This giant bomber could carry a 2,002-lb (906-kg) load of bombs and travel at 79 miles (128 km) per hour.

1938 English Supermarine Spitfir
More than 20,000 Supermarine Spitfires were built during the Second World War. This model cou fly at 357 miles (576 km) per hour.

JUMP JETS

This McDonnell Douglas AV-8B version of the British Harrier jump jet is flown by the United States Marine Corps. It is called a V/STOL (Vertical/Short Take Off and Landing) airplane because it can hover, land and take off like a helicopter. Its jet exhausts shoot out horizontally to the rear like an ordinary fighter plane when it is flying normally. The jet nozzles direct the exhaust vertically downwards for landing and taking off. The Harrier takes a short run to help it "jump" into the air when it has a heavy load.

V–tail
This V–tail replaces the usual vertical and horizontal tailplanes. It is slanted back to deflect or inhibit enemy radar and infrared sensors.

VIEWING ON SCREEN
When fighter pilots are in combat, they need to be able to look around and react quickly. Important instrument information is displayed on the visor of this special helmet, which means the pilot does not have to look down at the instrument panel.

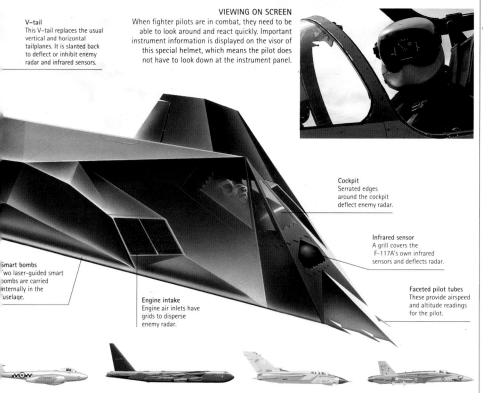

Cockpit
Serrated edges around the cockpit deflect enemy radar.

Infrared sensor
A grill covers the F-117A's own infrared sensors and deflects radar.

Smart bombs
Two laser-guided smart bombs are carried internally in the fuselage.

Engine intake
Engine air inlets have grids to disperse enemy radar.

Faceted pilot tubes
These provide airspeed and altitude readings for the pilot.

1943 English Gloster Meteor
This was powered by two engines and was the first British jet fighter. It could fly at 600 miles (969 km) per hour.

1952 United States Boeing B-52 Stratofortress
This giant bomber was powered by eight jets and flew a record 12,421 miles (20,034 km). It could fly at 595 miles (960 km) per hour.

1974 German, English and Italian Panavia Tornado
This strike aircraft has variable sweep wings. It can travel at 1,446 miles (2,333 km) per hour.

1978 United States McDonnell Douglas F/A 18C Hornet
This is also used by Australia, Canada, Spain and Kuwait. It can fly at 1,317 miles (2,124 km) per hour.

Aircraft Carriers

A coal barge was the first, and perhaps most unlikely, aircraft carrier. It towed observation balloons during the Civil War. In 1910, American stunt pilot Eugene Ely flew his Curtiss biplane from a platform on the cruiser USS *Birmingham*. The first true carrier, however, was built by the British during the First World War. Its narrow landing deck was very dangerous and returning pilots were forbidden to land. They had to ditch their planes in the sea. Aircraft carriers did improve, and by the Second World War they had replaced battleships as the most important naval ships. The bombers from a Japanese carrier force made a devastating attack on Pearl Harbor in 1941, and the major sea battles in the Pacific were fought by squadrons of carrier-based planes. Today, huge nuclear-powered carriers are the most powerful ships in the world.

COUNTING DOWN
The plane is in launch position on the carrier. The catapult crew, wearing green jackets, are in place. The catapult officer (the "shooter"), wearing a yellow jacket, gives the signal to launch the plane.

THE LANDING PATTERN
Carrier pilots fly 5-mile (8-km) wide circles at different heights while "hawking" (watching) the carrier, waiting to land. When the last aircraft are ready to launch, the pilots take turns to join the approach pattern. They time their descent to land the moment the deck is clear.

LAUNCHING
A holdback device on the catapult's shuttle (launcher) stops a plane from rolling forward, even when it is under full power, until the catapult is fired.

Jet-blast deflector
Retractable steel walls deflect the jet exhaust away from the deck.

Catapult track

A FLOATING AIRDROME
This carrier has four launching catapults and a landing deck. Flight operations are controlled by an officer, called the "air boss," at primary flight control. The captain of the carrier runs the ship from the navigation bridge.

Anti-aircraft gun

112

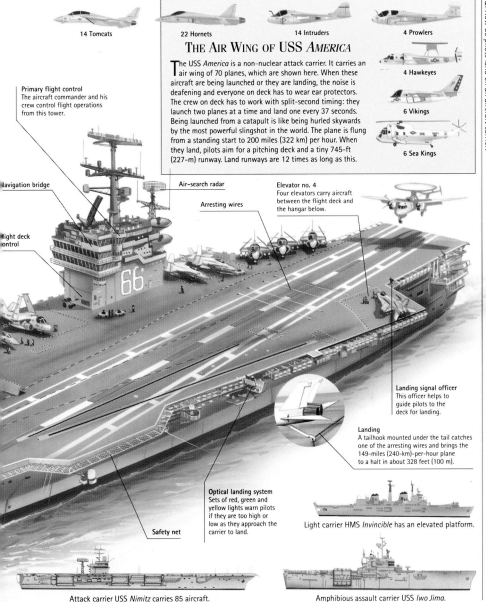

14 Tomcats

22 Hornets

14 Intruders

4 Prowlers

The Air Wing of USS *America*

The USS *America* is a non-nuclear attack carrier. It carries an air wing of 70 planes, which are shown here. When these aircraft are being launched or they are landing, the noise is deafening and everyone on deck has to wear ear protectors. The crew on deck has to work with split-second timing: they launch two planes at a time and land one every 37 seconds. Being launched from a catapult is like being hurled skywards by the most powerful slingshot in the world. The plane is flung from a standing start to 200 miles (322 km) per hour. When they land, pilots aim for a pitching deck and a tiny 745-ft (227-m) runway. Land runways are 12 times as long as this.

4 Hawkeyes

6 Vikings

6 Sea Kings

Primary flight control
The aircraft commander and his crew control flight operations from this tower.

Navigation bridge

Air-search radar

Elevator no. 4
Four elevators carry aircraft between the flight deck and the hangar below.

Arresting wires

Flight deck control

Landing signal officer
This officer helps to guide pilots to the deck for landing.

Landing
A tailhook mounted under the tail catches one of the arresting wires and brings the 149-miles (240-km)-per-hour plane to a halt in about 328 feet (100 m).

Optical landing system
Sets of red, green and yellow lights warn pilots if they are too high or low as they approach the carrier to land.

Safety net

Light carrier HMS *Invincible* has an elevated platform.

Attack carrier USS *Nimitz* carries 85 aircraft.

Amphibious assault carrier USS *Iwo Jima*.

113

Goddard 1926 (USA)
First successful
liquid-propellant rocket
traveled 184 feet
(56 m).

V-2 1942
(Germany)
First successful military rocket
reached an altitude of
53 miles (85 km).

Sputnik I 1957 (Russia)
First satellite to go into
orbit. It carried a radio
transmitter.

Gagarin 1961
(Russia)
First person in space. His
spaceship was called *Vostok 1.*

Hubble telescope 1990 (USA)
Space telescope was launched
into orbit.

Columbia 1981 (USA)
The world's first reusable
space shuttle was
launched into space.

Viking 1976 (USA)
Viking I and *II* set down
on Mars and discovered
no signs of life.

Skylab 1973 (USA)
Allowed people to
live in space for
several weeks.

Armstrong 1969 (USA)
Neil Armstrong
was the first
person to set foot
on the moon.

• A PARADE OF AIRCRAFT •

Space

S pace is an unknown but challenging territory. Since the thirteenth century, when the Chinese first used rocket power, people have been slowly acquiring the knowledge and technology to make travel in space possible. In 1903, Russian Konstantin Tsiolkovsky proposed using liquid-fuel rockets for space travel. American Robert Goddard successfully launched the first liquid-fueled rocket in 1926. Forty-three years later, *Apollo 11* was launched using the biggest rocket ever built. It carried astronauts Neil Armstrong, Edwin "Buzz" Aldrin and Michael Collins and their lunar lander *Eagle* to the moon. Millions of people all over the world listened to Armstrong's voice, crackling with static, announce that this was a "giant leap for mankind." In 1981, the first reusable space shuttle, *Columbia*, made 37 orbits of the world. This machine lifts off like a rocket, circles the Earth like a satellite, then uses its wings to glide back to Earth.

WORKING IN SPACE
During a 1994 shuttle mission, astronaut
Kathryn Thornton made repairs to the
Hubble space telescope.

WEATHER REPORT
This cloudless picture of Europe and North Africa is made up of several photographs. They were taken by cameras aboard weather satellites orbiting the Earth in space.

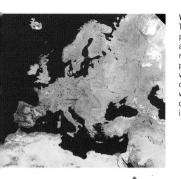

FLIGHT OF THE SHUTTLE

Three rocket engines and two booster rockets, with power equal to 140 jumbo jets, lift the shuttle off its launch pad and place it into orbit at 17,468 miles (28,175 km) per hour.

Two minutes after launch and 28 miles (45 km) up into the sky, the rocket boosters are released and parachute back to Earth.

At 69 miles (112 km) the main 686-ton (700-tonne) fuel tank falls away and burns up. It re-enters the Earth's atmosphere as the shuttle heads into orbit.

RE-ENTRY
Coming out of orbit, the shuttle slows down to re-enter the Earth's atmosphere. It glides to a landing and touches down at 215 miles (346 km) per hour.

DID YOU KNOW?

Each booster rocket contains 2 million lb (907,184 kg) of aluminum powder that steadily burns over two minutes. The tremendous heat and pressure generated is blasted from the rocket nozzle and causes the shuttle to lift.

USA
United States
NASA
Atlantis

Flights of Tomorrow

C an you imagine the airplanes of the future? Will passenger jumbo-jets be equipped with stores, restaurants, cabins and space for 1,000 passengers? How fast will the airplanes of tomorrow be able to travel? Supersonic planes such as Concorde are the fastest passenger planes today. They travel at Mach 2.1, and to exceed this, airliners would need to fly at about 60,000 feet (18,288 m). But at this high altitude, jet engines can harm the ozone layer. Supersonic planes are also very costly to build and this makes it expensive to travel on them. Special-purpose airplanes such as the Bell/Boeing Tiltrotor, which is part-helicopter and part-airplane, will definitely be part of aviation's future. This aircraft can fly between cities, land in tiny downtown airports and on the roofs of buildings. But what will be the ultimate flight of tomorrow? Perhaps flying to space in a space-shuttle airliner.

PICTURE THIS!
Boeing has imagined a supersonic airliner of the 21st century. It could carry about 300 people nonstop for 6,000 miles (9,600 km) at the same speed as Concorde today.

FLYING INTO THE FUTURE
A tiltrotor airliner takes off from a landing pad, or vertiport, on a Hong Kong rooftop. These 200-passenger planes of the future will hover like helicopters, whisk between cities at 341 miles (550 km) per hour, and help to reduce congestion at the crowded airports of tomorrow.

UPSTAIRS, DOWNSTAIRS
This two-storey jumbo of the future will fly at about the same speed as a Boeing 747, but it will be able to carry more than 500 passengers.

THE FLIGHT STAGES OF A TILTROTOR
It uses its rotors to take off (or land) vertically, like a helicopter.

It tilts its rotors and starts to move forwards.

The tilt rotors operate as normal propellers and it flies like an airplane.

TOGETHER TO MARS

The United States and Russia are working on an exciting program called "Mars Together." In the late 1990s, American and Russian rockets will launch a series of robot spacecraft, which will orbit, land and explore the surface of the Red Planet. In another program called "Fire and Ice," the two nations will explore the sun and the planet Pluto.

Completing the Picture

It is difficult to get a really good view of an airplane when it is on the ground or flying noisily overhead. This page, however, shows three-way views of 12 of the airplanes that have appeared as main images in this book. A three-way view is a standard aviation drawing that allows you to inspect and identify the plane's vital statistics: its shape; the number, type and position of the engines; and its wingspan.

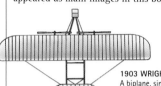

1903 WRIGHT *FLYER*
A biplane, single-piston engine.
Max speed: approx 30 miles (48 km) per hour.
Wingspan: 40 feet (12.3 m).

SUPERMARINE SPITFIRE
A monoplane fighter, single-piston engine.
Max speed: 407 miles (656 km) per hour.
Wingspan: 36 feet (11 m).

BOEING 314 CLIPPER
A monoplane flying boat, four piston engines.
Max speed: 192 miles (309 km) per hour.
Wingspan: 141 feet (43 m).

GRAF ZEPPELIN
An airship, five piston engines.
Max speed: 79 miles (128 km) per hour.
Length: 774 feet (236 m).

SUPERMARINE S.5
A monoplane float plane, single-piston engine.
Max speed: 282 miles (454 km) per hour.
Wingspan: 27$^1/_2$ feet (8.4 m).

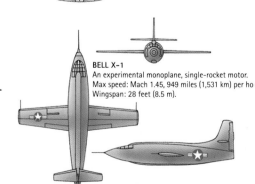

BELL X-1
An experimental monoplane, single-rocket motor.
Max speed: Mach 1.45, 949 miles (1,531 km) per ho
Wingspan: 28 feet (8.5 m).

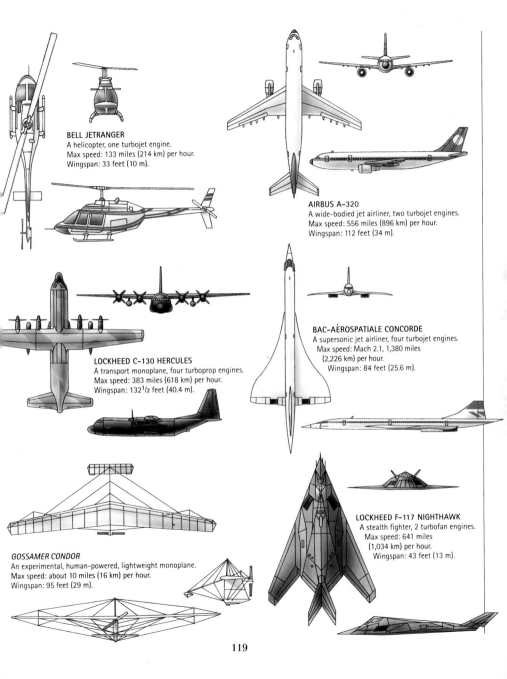

BELL JETRANGER
A helicopter, one turbojet engine.
Max speed: 133 miles (214 km) per hour.
Wingspan: 33 feet (10 m).

AIRBUS A-320
A wide-bodied jet airliner, two turbojet engines.
Max speed: 556 miles (896 km) per hour.
Wingspan: 112 feet (34 m).

LOCKHEED C-130 HERCULES
A transport monoplane, four turboprop engines.
Max speed: 383 miles (618 km) per hour.
Wingspan: 132^1/$_2$ feet (40.4 m).

BAC-AÉROSPATIALE CONCORDE
A supersonic jet airliner, four turbojet engines.
Max speed: Mach 2.1, 1,380 miles
(2,226 km) per hour.
Wingspan: 84 feet (25.6 m).

GOSSAMER CONDOR
An experimental, human-powered, lightweight monoplane.
Max speed: about 10 miles (16 km) per hour.
Wingspan: 95 feet (29 m).

LOCKHEED F-117 NIGHTHAWK
A stealth fighter, 2 turbofan engines.
Max speed: 641 miles
(1,034 km) per hour.
Wingspan: 43 feet (13 m).

Great

Inventions

How can a human use the heart of a pig?

How do inventors use natural energy?

What are bio-inventions?

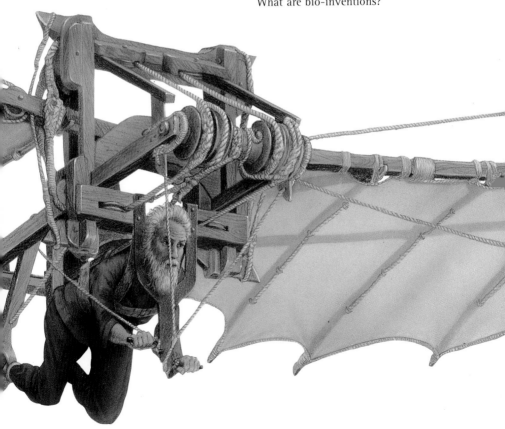

Contents

All About Invention

Inventions have shaped our world. We benefit from the work of great inventors every day: when we switch on a light, use a computer, call a friend or watch a video. Simple inventions, from buttons and zippers to Coca-Cola™ and cornflakes, make our lives easier and more enjoyable. Some inventors toil for years to perfect an idea, while others work together, sharing their discoveries and insights. The famous inventor Thomas Edison said that "genius is 1 per cent inspiration and 99 per cent perspiration." He developed the electric light bulb, after painstakingly completing 9,000 experiments! Every invention creates new knowledge, and this knowledge is used to make other inventions. Inventors protect their inventions by taking out patents—agreements with the government where inventors reveal their secrets in exchange for the right to make, use or sell their inventions. Patents can last up to 21 years, and then the invention becomes public property. During the past 500 years, a staggering 25 million products, processes and devices have been invented and patented.

NOBEL INTENTIONS

It is hard to predict all the effects an invention might have. In 1867, after three years of work, Swedish inventor Alfred Nobel made nitroglycerine safe by mixing it with a stabilizing mineral. He called this new, doughlike substance dynamite. Nobel intended that his invention would prevent accidental deaths in engineering, but when the world went to war, dynamite was used to kill and destroy. To make amends for the devastation his invention caused, Nobel put his fortune into founding the Nobel prizes. These are awarded to people who make outstanding contributions to chemistry, physics, medicine, literature, economics and peace.

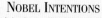

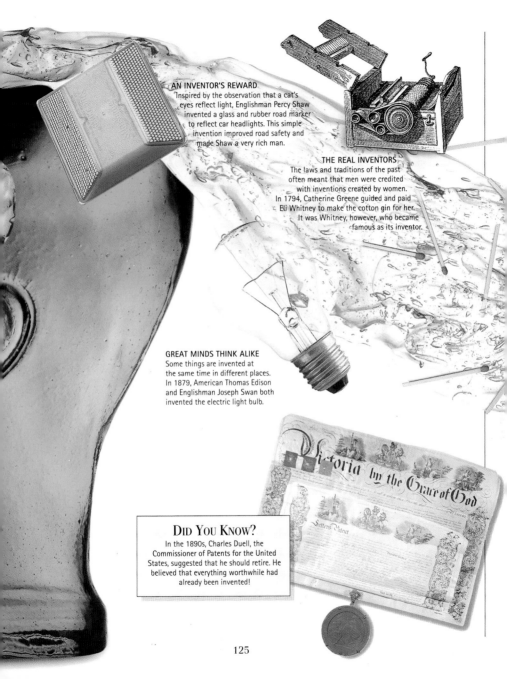

AN INVENTOR'S REWARD

Inspired by the observation that a cat's eyes reflect light, Englishman Percy Shaw invented a glass and rubber road marker to reflect car headlights. This simple invention improved road safety and made Shaw a very rich man.

THE REAL INVENTORS

The laws and traditions of the past often meant that men were credited with inventions created by women. In 1794, Catherine Greene guided and paid Eli Whitney to make the cotton gin for her. It was Whitney, however, who became famous as its inventor.

GREAT MINDS THINK ALIKE

Some things are invented at the same time in different places. In 1879, American Thomas Edison and Englishman Joseph Swan both invented the electric light bulb.

DID YOU KNOW?

In the 1890s, Charles Duell, the Commissioner of Patents for the United States, suggested that he should retire. He believed that everything worthwhile had already been invented!

LIGHT UP

Gustave Pasch of Sweden patented safety matches in 1845, but they were not manufactured until 1855. Earlier types ignited without warning, or gave off dangerous gases.

A QUICK SHAVE

In 1901, a traveling salesman named King Camp Gillette invented the disposable razor blade in the United States.

AN EXPENSIVE WASH

Soap was invented in Sumeria 4,000 years ago, but it was not until the 1820s that it became cheap enough for most people to buy.

SAFETY PIN

In 1849, Englishman Charles Rowley and American Walter Hunt both invented a clever device—the safety pin.

BUTTON UP

Buttons have been used since ancient times. The modern two-holed button is thought to have been invented in Scotland about 4,000 years ago.

• EVERYDAY LIFE •

Simple Things

We use buttons, bottles, knives, nails, safety pins, combs, coat hangers and other simple inventions every day. What would we do without them? Imagine a supermarket without canned foods, bread, ice cream, cartons of milk or packets of crackers! People are constantly re-inventing simple things, such as toothbrushes and bottle caps, by using new ideas, materials and technologies. Other inventions, such as pins and coat hangers, were perfect when they were invented and have changed very little since. A special word for simple, yet ingenious, inventions was coined in New York in 1886. Frenchman Monsieur Gaget sold thousands of miniature models of the Statue of Liberty to American sightseers. The New Yorkers who bought these statues called them "gadgets," and the word has been used ever since to describe clever but simple devices.

LOCK AND KEY

In 2000 BC, the Egyptians made wooden locks and keys to secure royal treasures. These locks and keys were decorated with gold to show their importance. In 1865, Linus Yale Jr. patented a drum-and-pin lock that could be mass-produced.

IP UP

his ingenious device was invented by merican Whitcomb Judson in 1891. ut it was 19 years before the zipper became faster and easier to use than buttons.

CHOPSTICKS

Chopsticks were used by the Chinese about 4,000 years ago.

PINS

The Egyptians made pins from copper, thorns and fishbones about 4,000 years ago. In 1625, John Tilsby started the first pin factory in England.

LEAN TEETH

ncient civilizations ad various ways of leaning teeth, but he toothbrush as we now it was invented by William Addis in 1780.

MONEY, MONEY, MONEY

The first coins were probably made in ancient China. They were cast in bronze and designed to look like everyday tools, such as knives and spades. In the ancient kingdom of Lydia, round coins were used in 600 BC. They were stamped with a lion and bull imprint on one side in honour of the king. Special marks on the other side showed the weight and quality of the coins. As time went on, trade flourished and so did money. Paper money probably originated in ancient China, but paper bank notes were first issued in 1661 by the Bank of Stockholm in Sweden. In 1988, scientists in Australia invented plastic bank notes that last four times as long as paper notes and can be recycled.

COAT HANGER

Thomas Sheraton, a famous furniture maker, built permanent coat hangers into wardrobes in the 1790s. Loose coat hangers, called "shoulders," were invented 100 years later.

CANNED FOOD

Englishman Peter Durand first thought of preserving food in tin canisters. He sold his idea to John Hall and Bryan Donkin who set up a canning factory, or "preservatory," in 1811.

SUNSHADE, RAINSHADE

The umbrella was first used as a sunshade in ancient China. It was not used as a shield against the rain until much later. The term umbrella comes from a Latin word meaning "little shadow."

TABLE MANNERS

Cutlery was introduced to the world's dinner tables more than 400 years ago. The fork became part of Italian tableware in the 1500s and was used in England, France and America in the 1600s.

4000 BC	REED ROPES Mesopotamia
1800 BC	BATHTUB Babylonia
808	BANK Italy
1648	SEWING THIMBLE Netherlands
1875	SUGAR CUBES Eugen Langen Germany
1896	ICE-CREAM CONES Italo Marcioni USA
1924	PAPER HANDKERCHIEFS (tissues) USA
1991	RECYCLED VINYL BOTTLES USA

THE LOOKING GLASS
Glass mirrors were invented in Venice,
Italy, perhaps 600 years ago. They were
made by sticking a very thin layer of
tinfoil onto glass
using mercury.

A royal treat
Ice cream was probably
first enjoyed in ancient
China. It was re-invented
in Europe in the 1300s
and became a popular
dessert among royalty.

• EVERYDAY LIFE •

Around the House

We use household inventions so frequently, it seems as if we have always had them. But these everyday items and devices were once new and exciting. Inventions for the house were designed to take the effort out of household chores. Before the vacuum cleaner, people would laboriously beat their carpets to remove the dust; before the lawn mower, cutting grass was a back-breaking task done by hand with a scythe. Before the electric refrigerator, food needed to be bought daily and any leftovers thrown away. When they were first produced, most household inventions were handmade, unreliable and expensive. Mass production made many goods cheaper, more reliable and available to everyone. Household inventions are constantly being improved and updated. What labor-saving inventions will we use in houses of the future?

FRIDGES FOR THE FUTURE
The first electric refrigerators were cooled by poisonous
substances, such as ammonia. After the 1930s,
manufacturers used a non-toxic substance called Freon.
But Freon was found to cause the breakdown of ozone
in the Earth's atmosphere. This environmentally
friendly fridge was invented in Germany in 1991.

Frozen food
After observing how the Inuit (Eskimos) preserved their fish, Clarence Birdseye saw the potential for quick-freezing many foods. Birdseye introduced frozen foods to American stores in about 1930.

"IT BEATS AS IT SWEEPS AS IT CLEANS"

In 1901, Hubert Cecil Booth was convinced he could build a machine that picked up and filtered dust using suction. He did just that, but the cleaner was so large it had to be set on a trolley and needed two people to operate it. James Murray Spangler invented a more portable model and in 1908 patented the electric Suction Sweeper. Spangler sold his idea to a man named William Hoover. His surname and the above slogan became household words in many countries.

Margarine
A shortage of butter in France in the 1860s led Emperor Napoleon III to encourage the development of a substitute. In 1869, Hippolyte Mège-Mouriès invented a paste made from animal fats. He called it margarine.

Chocolate
Chocolate is made from the seeds of a tropical tree called *cacao*. Early Central and South Americans crushed the seeds to make a drink. The first chocolate bar was made in Switzerland in 1819 by François-Louis Cailler.

SEWING MACHINE
In the early 1800s, the first sewing machines were not welcomed by tailors, who thought their jobs were under threat. In 1851, Isaac Singer invented the first efficient domestic sewing machine.

DID YOU KNOW?
During the Second World War, two scientists developed a device to produce microwaves. Little did they know their work to improve radar would later be used to develop a new cooking appliance—the microwave oven.

Perfect packaging
Convenient "waxed cardboard" cartons with pull-out spouts were invented in Sweden by Ruben Rausing in 1951.

RESTROOM
The flush toilet was invented by Sir John Harington in 1589. He installed one in his own house and one in the house of his godmother, Queen Elizabeth I of England. But they did not become common until nearly 200 years later.

100 BC
SAUNA BATH
Finland

1795
CORKSCREW
Samuel Hershaw
England

1857
TOILET PAPER
USA

1886
DISHWASHER
Josephine Cochrane
USA

1901
ELECTRIC WASHING MACHINE
Alva Fisher
USA

1909
ELECTRIC TOASTER
USA

1917
ELECTRIC HAND DRILL
Duncan Black and Alonso Decker
USA

1926
AEROSOL SPRAY
Erik Rotheim
Norway

1945
MICROWAVE OVEN
Percy Le Baron Spencer
USA

1985
SOAPLESS ULTRASONIC WASHER
Nihon University
Japan

At a Factory

FORKLIFT TRUCK
Forklift trucks, developed in America and Australia during the Second World War, are the "worker bees" of a factory. They scurry from place to place carrying heavy loads on wood or plastic pallets.

Long ago, members of a family or tribe would make everything they needed by hand. As the population grew, people worked together in factories. In 1798, American Eli Whitney invented mass production. He received an order for 10,000 guns, but realized it was impossible for craftspeople to produce this quantity in the time available. He decided to divide the work up into separate jobs, so that different people could make different parts of the gun, which was then put together later. At the same time in England, James Watt's new steam engines were installed in factories, and soon the Industrial Revolution began. The new steam factories employed many people, but the work was dangerous, cramped, time-consuming and often boring. Many factories today offer good lighting, filtered air, protective clothing and rest breaks. Robots are used for the worst jobs, and people work fewer hours in safer conditions.

PROTECTIVE CLOTHING
The heat, the cold and dangerous chemicals used in today's factories mean that suits such as this are needed to protect the worker from danger.

KEEPING IT ALL TOGETHER
Handmade metal nails were first used by the Egyptians to hold coffins together 5,000 years ago. Tiny metal screws for joining wood date back to 1760. Before welding, metals were joined with pins or rivets.

WORK TIME

The time clock was invented by American W. H. Bundy in 1885. Workers were given their own specially numbered key, which they inserted into the clock when they arrived at work. This key activated the clock to print their key number and their arrival time on a strip of paper. When the workers were ready to leave, they repeated this process, so their employer could check the number of hours they had worked. The Australian phrases to "bundy on," meaning to arrive at work, and to "bundy off," meaning to leave work, have helped to keep the inventor's name alive.

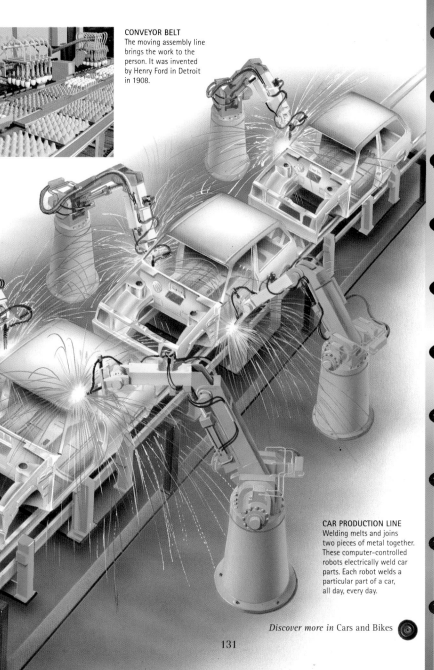

CONVEYOR BELT
The moving assembly line brings the work to the person. It was invented by Henry Ford in Detroit in 1908.

CAR PRODUCTION LINE
Welding melts and joins two pieces of metal together. These computer-controlled robots electrically weld car parts. Each robot welds a particular part of a car, all day, every day.

Discover more in Cars and Bikes

1500 BC
IRON SMELTING
Asia Minor

870 BC
PULLEY WHEEL
Assyria

200 BC
CRANK HANDLE
China

100 BC
BAROULKOS
(CRANE)
Hero of Alexandria

1550
NUTS, BOLTS AND
WRENCHES
France

1804
PATTERN-WEAVING
LOOM
Joseph-Marie
Jacquard
France

1856
STEEL MAKING
Henry Bessemer
England

1903
OXYACETYLENE
WELDING TORCH
Edmond Fouche
and Jean Picard
France

1946
ROBOT
AUTOMATION
Delmar Harder
USA

1983
ROBOT-MAKING
ROBOT
Japan

131

On a Farm

The plow and irrigation have tamed more farmland than any other farming inventions. People first grew crops in the Middle East about 10,000 years ago, but planting, harvesting and watering them by hand was a slow process. In Egypt and India, nearly 4,500 years later, farmers prepared the ground for planting with wooden plows pulled by oxen. The Egyptians invented a machine called a shaduf, which helped them take water from the River Nile to irrigate or water their crops. Barbed wire was another great farming invention. Farmers used it to divide huge areas of Africa, North America and Australia into separate wheat, cattle and sheep farms in the 1800s. These enormous new farms revolutionized farming. Farmers now needed faster ways of harvesting grain, wool, meat and milk. The old methods were soon replaced by machines that did the work of hundreds of people.

FORK
Iron forks were invented by the Romans. "Pitchforks" were used to "pitch" or stack hay in the field.

RAKE
Wooden rakes were invented in Europe in about AD 500 to gather grain that had been threshed or beaten off its stalks.

SPADE
Wooden spades with iron blades were invented by the Romans about 2,000 years ago.

BARBED WIRE
In 1867, American Lucien Smith invented barbed wire and made it possible for farmers to fence off their lands.

OUR JOBS IN ONE
In 1884, Australian Hugh McKay invented the horse-drawn harvester. It combined cutting, threshing, winnowing and bagging wheat grain into one operation. Combine harvesters with gasoline or diesel engines are now used 24 hours a day, with lights at night, to harvest the crops.

WATERING THE CROPS
The Egyptian shaduf is a little like a seesaw. A long wooden pole, balanced on a crossbeam, has a rope and bucket at one end, and a heavy stone weight to counterbalance it at the other. The weight of the rock makes it easier to lift a heavy bucket of water.

PLOW
Plows made from wood and stag antlers were invented in Egypt and India about 5,500 years ago. Simple ox-drawn plows are still used on family farms in many countries.

TRACTOR
Three-wheeled steam tractors, built by the Case company of America in 1829, were very heavy and often became stuck in the soft soil. Modern tractors were pioneered by Henry Ford in 1907.

DID YOU KNOW?
Superphosphates—artificial chemicals that enrich the soil—were invented by Sir John Bennett Lawes in England in 1842. But fertilizers often run into the rivers and oceans, killing fish and making algae grow.

1794
COTTON GIN (SEPARATOR)
Catherine Greene and Eli Whitney
USA

1831
GRAIN REAPER (CUTTER)
Cyrus McCormick
USA

1833
STEEL PLOW
John Lane
USA

1860
NUTRICULTURE
Julius von Sachs
Germany

1868
GRANNY SMITH APPLE
Maria Smith
Australia

1889
MODERN MILKING MACHINE
William Murchland
Scotland

1924
AERIAL CROP DUSTING
USA

1939
DDT PESTICIDE
Paul Müller
Switzerland

1975
AXIAL COMBINE HARVESTER
International Harvester
New Holland, USA

At School and the Office

During some stage of their lives, people in many countries spend time in schools and offices. Schools were first set up in Greece between 800 BC and 400 BC to teach boys subjects such as mathematics and astronomy. In the 1800s, people were employed in offices to monitor the staff and accounts of the first factories. Today, on a desk at school or the office you will find inventions large and small. It is hard to imagine our lives without them. Desks would be strewn with papers because we would not have paperclips or staplers to keep them together. How would we write without pens and pencils and make straight lines without rulers? How would you cut paper without scissors? In fact, if the Chinese had not invented paper, there would be nothing to write on, cut or keep in an orderly pile.

BLACKBOARD
James Pillans, a Scottish teacher, invented the blackboard in 1814 so that all his students would be able to see the maps he drew.

Stapler and staples
Englishman Charles Gould invented these in 1868.

Correction fluid
Bette Graham of the United States invented this white fluid in 1959.

Glue
In 3000 BC, the Egyptians used glue to stick furniture together.

Rubber bands
Native South Americans used the white sap of rubber trees to make rubber bands.

Sticky tape
In 1939, ten years after inventing masking tape, Richard Drew invented clear sticky tape.

Scissors
Two-bladed shears were invented in 1500 BC by the Chinese. These became our modern scissors.

Eraser
This was invented in 1752 by Magellan from Portugal.

Felt pens
These were invented in 1960 by the Japanese company Pentel.

Paperclips
These were patented in 1900 by Norwegian Johann Waaler.

HITTING THE KEYS
Carlos Gliddens and Christopher Sholes named their typewriter the "literary piano". In 1873, the Remington Fire Arms Company undertook to manufacture it, and in 1876, it was displayed at the Centennial Exposition in the United States.

MOVING LIGHT
The desk lamp is a little like an arm. It can be moved about for close work or kept in the same position. An adjustable desk lamp was designed by George Cardardine in 1934.

PHOTOCOPIER
The first photocopiers used messy chemicals and sensitive papers to photograph documents. In 1938, Chester Carlson invented a dry copying process that used plain paper. The first photocopiers of this kind were sold in 1959.

Pencils
Soft graphite was used for pencils in England from 1564.

FAX MACHINE
A round 1900, German scientist Arthur Korn invented an electric cell that could detect dark and light areas on paper. He used it to send a photograph by telephone line from Germany to England in 1907. Almost 70 years passed before people realized how useful this invention would be in the office. The fax (which is short for facsimile, meaning an exact copy or a reproduction) now plays an important part in offices. This machine makes it possible to communicate instantly with people all over the world.

Sticky notes
These were invented in 1980 by American Arthur Fry.

Discover more in From Quill to Press

800 BC
SCHOOL
Greece

1806
CARBON-COPY PAPER
Ralph Wedgewood
England

1837
SHORTHAND WRITING
Isaac Pitman
England

1858
PENCIL WITH ERASER ATTACHED
Hyman Lipman
USA

1901
ELECTRIC TYPEWRITER
Thaddeus Cahill
USA

1903
WET PHOTOCOPIER
George Beidler
USA

1959
CORRECTION FLUID
Bette Graham
USA

1990
NO-LICK STAMPS
Australia Post
Australia

At Play

P eople are always inventing ways to have fun. The Egyptians threw stone balls at upright pins in a game similar to bowling about 5,000 years ago. The Greeks played "soccer" with inflated animal bladders about 2,500 years ago. Some games seem timeless—hopscotch, marbles, tick-tack-toe and rope skipping are as popular today as when they were first played. Dolls have delighted young and old for centuries. They have been made of many different materials, from apples and animal skins to china and plastic. In 1823, baby dolls were made to cry. Soon, they were talking as well. Today, the games industry is booming as inventors create new and exciting games that challenge all who play them.

CHECKMATE
Chess was invented in about AD 500 in India. The moves we play today were first used in Europe in the mid-1500s. The winning position "checkmate" comes from *shah mat,* Arabic for "the king is dead."

BARBIE DOLL
In 1958, Ruth Handler invented Barbie, a dress-up doll complete with a wardrobe of clothes and a way of life. More than one billion Barbie dolls have been sold in 140 countries.

NINE OR TEN PINS?
In 1845, nine-pin bowling had become so popular in the state of Connecticut that it encouraged heavy gambling. A law was passed that banned the game of "bowling at nine pins." The eager bowlers added a tenth pin and kept on bowling!

DID YOU KNOW?

The very first roller skates, invented by a Belgian musician Joseph Merlin in 1760, had wheels in one line—similar to today's rollerblades.

LEGO™

The Danish word *leg-godt* means to play well. Ole Kirk Christiansen chose the name "Lego" for his line of toys. By 1955, his toy plastic bricks that can be joined to construct things such as buildings, machines, people and animals were known as Lego all over the world.

2450 BC
DOMINOES
Mesopotamia

1200 BC
CHECKERS
Egypt, Sri Lanka

1450
GOLF
Scotland

1823
CRYING DOLLS
Johann Maelzel
Belgium

1850
MAH–JONG
China

1882
JUDO
Jigoro Kano
Japan

1891
BASKETBALL
James Naismith
USA

1929
YO-YO
Donald Dwean
USA

1931
MONOPOLY
Charles Darrow
USA

1992
VIDEO BOARD GAME
Brett Clements
and Phillip Tanner
Australia

GAMES, GAMES, GAMES

In 1972, American Nolan Bushnell invented the first successful computer game. It was like table tennis, and was called *Pong*. In 1978, *Space Invaders* was introduced and became a big success. Today's electronic games, such as *Where in the World is Carmen Sandiego?*, use full color animation, speed and constantly changing tactics to outwit even the best human players. The computer game *Lunicus* (below) pits players against a giant bee in the year 2000.

137

People Movers

Mass transportation was invented to carry large numbers of people at one time. The first omnibus (a Latin word that means "for everyone") was built by Englishman George Shillibeer in 1829. It had 22 seats and was pulled by three horses. Robert Stephenson's steam locomotive, the *Rocket*, traveled so fast in 1829 that Dr Dionysys Lardner was moved to predict: "passengers, unable to breathe, will suffocate". People loved train travel, and railway tracks spread across the countryside like giant spider webs. In 1863, railways went underground in London, and five years later they were built overhead in New York. Streetcars soon appeared in city streets everywhere, and in huge stores people rode on escalators invented by Jesse Reno in 1894 as rides in a New York amusement park.

A SMALL BEGINNING
Walter Hancock's steam-powered motor bus of 1831 (known as the *infant*) could carry only ten passengers. Buses today, such as this English double-decker, have gasoline or diesel engines, and can carry more than 70 people.

Pantograph
Overhead cables transfer power to many electric trains. The pantograph has springs that keep it in constant contact with these cables.

Safety doors
Modern train doors that are opened and closed by compressed air are electronically controlled by the train operator. A sensor in the doorframe detects anything stuck in the doorway.

Railway tracks
William Jessop of England made the first raised metal rails for a railway in 1789, but the metal wheels and tremendous weight of the new steam locomotives broke them.

AROUND THE WHEEL

The invention of the wheel made an enormous difference to people in many countries. The first wheels are thought to have been developed in about 3500 BC in southwest Asia. They were made from planks of wood cut into a circle. These solid, heavy wheels were replaced eventually by lighter spoked wheels, a design which was perfected by Leonardo da Vinci in the 1400s. Wheels with wire spokes were developed in about 1800, and in 1895 André and Edouard Michelin introduced air-filled tires on cars.

CABLE TRAM
Cable trams such as this one have been running up and down the steep hills of San Francisco, California since 1873. They are pulled along by a cable that is set inside a groove in the road.

ON TRACK
Monorails glide along a single track above crowded streets. They look very modern, but a cable-powered monorail took passengers around the Lyon Exposition in France as early as 1872.

ELECTRIC TRAIN
This is one of the Eurostar trains, a fleet of electric trains in Europe. The French TGV—one of the fastest electric trains—travels at an average speed of 161 miles (260 km) per hour between Paris and Lyon.

1640
TAXI FLEET
Nicholas Sauvage
France

1802
STEAM LOCOMOTIVE
Richard Trevithick
Wales

1857
STEAM SAFETY ELEVATOR
Elisha Otis
USA

1879
ELECTRIC LOCOMOTIVE
Werner von Siemens and Johann Malske
Germany

1888
ELECTRIC TRAMWAY
Frank J. Sprague
USA

1908
CABLE CAR
Switzerland

1964
VERY FAST TRAIN
Japanese National Railways
Japan

1981
SUPER FAST TRAIN (TGV)
France

Cars and Bikes

Cars and bikes revolutionized transportation for people everywhere. Bikes gave everyone the freedom to travel where they wished, over long distances and at speeds of up to 43 miles (70 km) per hour—downhill, anyway. The design of bikes changed considerably, from the dandy horse to the penny-farthing, before the modern bike was invented in 1879. The first motor car was a three-wheeled road steamer, invented by Nicolas-Joseph Cugnot in 1769. It traveled at 3 miles (5 km) per hour and could be overtaken by most people walking at a brisk pace! Handmade, gasoline-engined cars were invented in 1885 by German Karl Benz, but they were very expensive. When cars such as the Volkswagen Beetle were mass-produced in the 1940s, they became more affordable. Millions of people could now enjoy the pleasures of driving—and the horrors of traffic jams.

A CLASSIC HARLEY
In Germany in 1894, brothers Heinrich and Wilhelm Hildebrand and Alois Wolfmüller built the first motorbikes to have two-stroke engines and pneumatic tires. This motorbike is a 1917 model Harley Davidson.

DID YOU KNOW?

In 1983, Richard Noble traveled faster by car than anyone else ever has. In his jet-powered "Thrust 2," he sped across the Black Rock Desert in Nevada at 633.089 miles (1,019.467 km) per hour.

Dandy horse
1790

Penny-farthing
1870

Safety bike
1879

Aerodynamic bike
Early 1980s

A POPULAR BEETLE
Commonly known as the Beetle, this Volkswagen (VW) is one of the most popular cars in the world. About 23 million have been produced since the first model hit the road in 1936.

Back-to-front car
The VW has a most unusual design element: the engine is at the back and the spare tire is at the front.

Headlights
Driving at night was very difficult until 1925, when headlamps that had both short and long beams of light were invented.

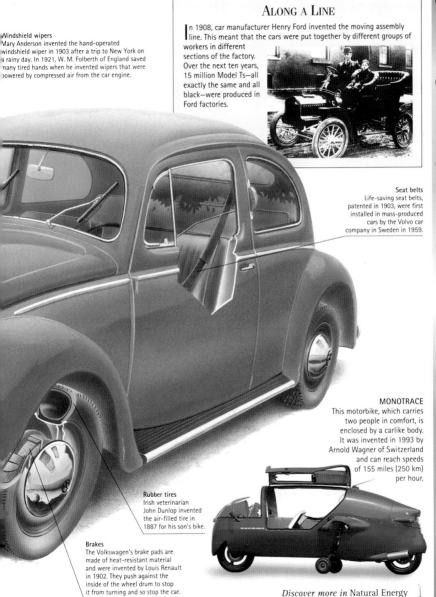

Windshield wipers
Mary Anderson invented the hand-operated windshield wiper in 1903 after a trip to New York on a rainy day. In 1921, W. M. Folberth of England saved many tired hands when he invented wipers that were powered by compressed air from the car engine.

ALONG A LINE

In 1908, car manufacturer Henry Ford invented the moving assembly line. This meant that the cars were put together by different groups of workers in different sections of the factory. Over the next ten years, 15 million Model Ts—all exactly the same and all black—were produced in Ford factories.

Seat belts
Life-saving seat belts, patented in 1903, were first installed in mass-produced cars by the Volvo car company in Sweden in 1959.

MONOTRACE
This motorbike, which carries two people in comfort, is enclosed by a carlike body. It was invented in 1993 by Arnold Wagner of Switzerland and can reach speeds of 155 miles (250 km) per hour.

Rubber tires
Irish veterinarian John Dunlop invented the air-filled tire in 1887 for his son's bike.

Brakes
The Volkswagen's brake pads are made of heat-resistant material and were invented by Louis Renault in 1902. They push against the inside of the wheel drum to stop it from turning and so stop the car.

Discover more in Natural Energy

3200 BC
WHEELED CHARIOT
Mesopotamia

231
WHEELBARROW
China

1861
VELOCIPEDE
Pierre and Ernest Michaux
France

1885
GAS PUMP
Sylvanus Bowser
USA

1885
SPARK PLUGS
Etienne Lenoir
France

1914
ELECTRIC TRAFFIC LIGHT (RED ONLY)
Alfred Benesch
USA

1974
SAFETY AIRBAG
General Motors Corporation
USA

1993
COLLISION AVOIDANCE RADAR
Japan

FINDING THE WAY

Lighthouses can guide
ships that are close to
shore, but navigation in
the open ocean is more
difficult. Early sailors relied
on the sun, the moon and
the stars. Between 850 and
1050, the Chinese invented
the magnetic compass to help
guide their ships. The mariner's
astrolabe (left), and later the
sextant and chronometer, made more accurate
readings of the heavens possible. Today, ships use
satellite signals to navigate.

Safety lines
Ships throughout the
world are marked with a
Plimsoll line, invented
by English politician
Samuel Plimsoll in 1876.
When a ship is being
loaded and the water
level reaches the line, it
means that no more
cargo can be added.

• TRANSPORTATION •

On and Under Water

Boats have a long history. They probably predate the wheel. As early as 40,000 years ago, dugout canoes were paddled across shallow seas. Sails were added by the Egyptians about 5,000 years ago, and the Chinese attached a rudder to the stern for steering about 2,000 years ago. The invention of the steam engine led to the development of bigger, faster and safer ships. Robert Fulton enthralled Americans when he steered his paddle steamer down the Hudson River in 1807. When Isambard Brunel launched his giant steamships in the 1840s, he inspired people everywhere with thoughts of speedy Atlantic crossings. Transport under the sea, however, progressed slowly until the First World War. When the Germans launched their U-boats, submarine warfare was born. Today, undersea vessels are also used for deep-sea salvage and exploration.

CATCHING THE WIND
An exciting new wind sport
was created in 1958 when
Peter Chilvers of England invented
the sailboard. Jim Drake of the
United States modified the
design and patented it in 1968.

Propeller power
Francis Smith in England and
John Ericsson in America invented
the screw propeller in the mid-1830s.
Propellers were more reliable and could
drive ships faster than paddlewheels.

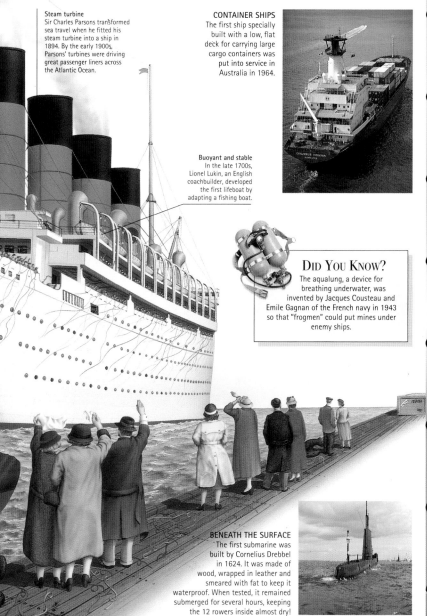

Steam turbine
Sir Charles Parsons transformed sea travel when he fitted his steam turbine into a ship in 1894. By the early 1900s, Parsons' turbines were driving great passenger liners across the Atlantic Ocean.

CONTAINER SHIPS
The first ship specially built with a low, flat deck for carrying large cargo containers was put into service in Australia in 1964.

Buoyant and stable
In the late 1700s, Lionel Lukin, an English coachbuilder, developed the first lifeboat by adapting a fishing boat.

DID YOU KNOW?
The aqualung, a device for breathing underwater, was invented by Jacques Cousteau and Emile Gagnan of the French navy in 1943 so that "frogmen" could put mines under enemy ships.

BENEATH THE SURFACE
The first submarine was built by Cornelius Drebbel in 1624. It was made of wood, wrapped in leather and smeared with fat to keep it waterproof. When tested, it remained submerged for several hours, keeping the 12 rowers inside almost dry!

7500 BC
REED BOATS
Middle East

880 BC
INFLATED LIFE PRESERVER
King Assur-nasur-apli II
Syria

285 BC
LIGHTHOUSE
Egypt

1735
MARINE CHRONOMETER
John Harrison
England

1915
SONAR
Paul Langevin
France

1955
HOVERCRAFT
Christopher Cockerell
England

1963
JETSKI
Clayton Jacobson
USA

1984
WAVE-PIERCING CATAMARAN
Phillip Hercus
New Zealand

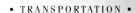

To Outer Space

Space is a dangerous place. There is no air, no food, no sound and no gravity. Without a spaceship or a spacesuit to provide air pressure, your body would explode into billions of tiny pieces. The Italian mathematician and astronomer Galileo Galilei saw the surface of the moon through a telescope in 1609. People soon began to dream of reaching such a wondrous place. In 1965, American astronauts "walked" in space wearing shiny, airtight, water-cooled spacesuits, helmets and oxygen tanks. They were attached to their spaceship by ropes. Four years later, *Apollo 11* landed on the moon and Neil Armstrong became the most famous moonwalker in history. *Apollo 11* astronauts had personal jetpacks, invented by NASA, to help them travel quickly over the moon's surface. But in space, they had to tie themselves to their spacecraft—a jetpack could send them spinning into deep space forever.

ROCKET MAN
In 1926, an American professor, Robert Goddard, invented a liquid-fueled rocket. It flew to a height of only 44 ft (13.5 m) before it crash-landed. Ten years passed before scientists paid any attention to his invention.

LIFE IN SPACE

Without gravity, life in space must be well planned. Astronauts have to drink through straws and strap themselves into their bunks to sleep. They eat packaged food that has to be sticky so they can keep it on their forks or spoons. A sneeze inside a spacecraft will send an astronaut flying backwards and turn the sneeze into a ball of floating liquid. Astronauts have to vacuum their skin after a shower—a towel would simply brush water droplets into the air. Toilet waste is collected, shredded, dried and stored until the spaceship returns to Earth.

Space toilet and shower

MOON BUGGY
The Lunar Rover, invented by the Boeing company in 1971, was the first car in space. It could speed across the moon at 37 miles (60 km) per hour, and could be completely folded up and stored in the spaceship.

Space food

SPACE SHUTTLE
...uttles were first launched in
...81. The crews of these
...usable craft collect satellites
...r return to Earth, repair
...isting satellites or carry new
...tellites into Earth's orbit.

DID YOU KNOW?
Many materials we use every day were
invented for space travel. Sunglasses
designed to reduce glare in space are
worn on Earth. A material called Kevlar,
invented to build lighter and stronger
spacecraft, is used in bicycles.

A GIANT LEAP
The *Sputnik 2* satellite, launched in
1957, carried the first traveler in
space—a Russian dog named Laika.
In 1961, Russian Yuri Gagarin
became the first human to travel
in space. He circled the Earth in
Vostok 1 for 108 minutes.

A HOME AWAY
Space stations are like
homes in space. The
Russians launched space
station Salyut in 1971,
and the Americans followed
two years later with Skylab.

SATELLITES
Scientists have invented
satellites that orbit the Earth. They
are used to send radio, television and
telephone signals around the Earth, or
to transmit scientific information.

1608
TELESCOPE
Hans Lippershey
Netherlands

1668
REFLECTING
TELESCOPE
Isaac Newton
England

1926
LIQUID-FUELED
ROCKET
Robert Goddard
USA

1932
RADIO TELESCOPE
Karl Jansky
USA

1942
A4 ROCKET
Wernher von Braun
and Walter
Dornberger
Germany

1957
SPUTNIK 1
Dr Sergei Korolyov
USSR

1969
APOLLO 11
NASA
USA

1981
SPACE SHUTTLE
NASA
USA

1989
GALILEO SPACE
PROBE
NASA
USA

1990
HUBBLE SPACE
TELESCOPE
NASA
USA

REED PEN
A hollow reed can be cut to a point and used as a pen with ink or paint.

PENCIL
In 1792, Jacques Nicolas Conté invented a hard, long-wearing pencil made from clay mixed with powdered graphite and covered in cedar wood.

HIEROGLYPHS
The Egyptians developed a type of picture writing called hieroglyphs about 3000 BC. They either carved these pictures into stone, or painted them onto walls or papyrus with a pen cut from a reed from the River Nile.

FOUNTAIN PEN
This pen, which stores ink in its handle, was invented in 1884 by American Lewis Waterman.

CALLIGRAPHY BRUSH
This is used by trained writers, called calligraphers, to paint words onto rice paper and silk.

BALLPOINT PEN
In 1938, Laszlo Biro from Hungary invented a pen that used a rolling ball instead of a nib.

SEAL OF APPROVAL
A seal is a device used to imprint design or symbol that represents family or a company. The first sea were used in Sumeria and India fo signing documents.

• COMMUNICATIONS •

From Quill to Press

Technology has made it possible for us to communicate in different ways. People all over the world write with pens and pencils; many key their words into computers. About 30,000 years ago, however, people drew pictures on cave walls to tell stories or record news. The ancient Sumerians replaced this form of picture writing with shapes that they pressed into wet clay. The Chinese invented 47,000 characters that were written onto paper or silk with a brush. The first books in Europe were written with ink and a quill—a feather plucked from a goose and cut to a sharp point. The Chinese printed books by using wooden blocks. Later, they created movable, fire-baked characters that could be dipped into ink and pressed onto paper, but they wore out quickly. This problem was solved by the invention of reusable type made of hard metal. The printing press was soon in operation.

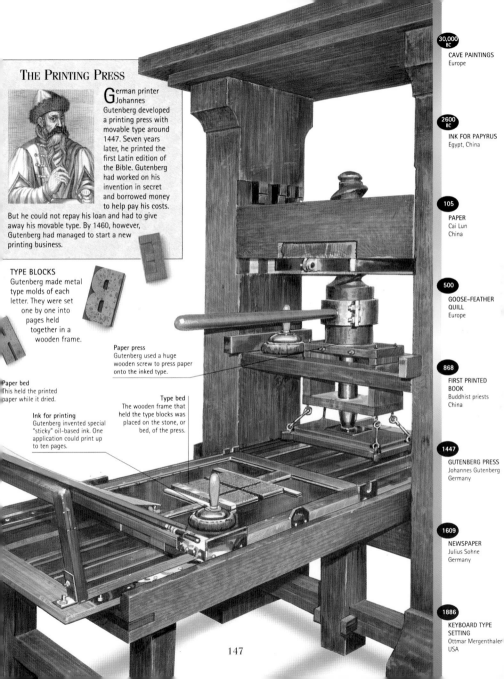

THE PRINTING PRESS

German printer Johannes Gutenberg developed a printing press with movable type around 1447. Seven years later, he printed the first Latin edition of the Bible. Gutenberg had worked on his invention in secret and borrowed money to help pay his costs. But he could not repay his loan and had to give away his movable type. By 1460, however, Gutenberg had managed to start a new printing business.

TYPE BLOCKS
Gutenberg made metal type molds of each letter. They were set one by one into pages held together in a wooden frame.

Paper press
Gutenberg used a huge wooden screw to press paper onto the inked type.

Paper bed
This held the printed paper while it dried.

Type bed
The wooden frame that held the type blocks was placed on the stone, or bed, of the press.

Ink for printing
Gutenberg invented special "sticky" oil-based ink. One application could print up to ten pages.

30,000 BC
CAVE PAINTINGS
Europe

2600 BC
INK FOR PAPYRUS
Egypt, China

105
PAPER
Cai Lun
China

500
GOOSE-FEATHER QUILL
Europe

868
FIRST PRINTED BOOK
Buddhist priests
China

1447
GUTENBERG PRESS
Johannes Gutenberg
Germany

1609
NEWSPAPER
Julius Sohne
Germany

1886
KEYBOARD TYPE SETTING
Ottmar Mergenthaler
USA

FIBER OPTICS

In 1976, the Western Electric Company of Atlanta, Georgia used pulses of laser light to send voice, video and computer messages through fibers made of glass, which were called optical fibers.

EXCHANGING WORDS

As a telephone wire could carry only one message at a time, Bell and Gray invented the telephone exchange system in 1878. Within ten years, hundreds of wires connected houses to buildings called exchanges, where operators joined the wires so that people could speak to each other.

MAKING CONNECTIONS

On February 14, 1876, Scottish-born Alexander Graham Bell and American Elisha Gray both applied to patent the telephone in the United States, although neither of them had made a telephone that really worked. Gray was two hours later than Bell and lost the race to claim the telephone as his invention. People were very eager to invest in this exciting new technology, and by the end of 1877, Bell was a millionaire.

ELECTRIC TELEGRAPH

In 1837, the American Samuel Morse used a magnet to interrupt the flow of electricity in a wire. This could be heard at the other end of the wire as a "beep." The beeps were formed into a code (Morse code) that operators learned to understand and translate back into words.

CANDLESTICK TELEPHONE

This was the most common telephone in the world for many years. The early models did not have dials, and you had to rattle the side hook to alert the operator. Frenchman Antoine Barnay invented the dial in 1923.

• COMMUNICATIONS •

Along a Wire

Messages can be passed from person to person in many ways. Voices, horns and beating drums carry messages through the air; written messages are sent by mail. All early methods of communication relied on how far people could run, see, hear or shout. The first messages to travel further and faster were sent by telegraph in the 1830s. The telegraph consisted of a special code of electric "beeps" that traveled huge distances along wires in a few minutes. At the other end, the receiver sorted the beeps into words and delivered the message in person. In 1851, the first telegraph cable was laid under the English Channel between Dover and Calais. By 1866, undersea cables provided the first transatlantic telecommunications link. Today, wires still connect millions of people by telephone, fax and computer.

DID YOU KNOW?

You can make a string telephone by
connecting two containers, or cans, by a
long length of string. Pull the string tight
and speak into one container. The person at
the other end should be able to hear what
you say. Children have made string
telephones since the 1600s.

WIRED FOR SOUND
The electric microphone, which amplifies, or increases, sound, was invented in 1916 and first tested at Madison Square Gardens, New York.

HEADPHONES
These were developed from hearing-aid technology invented in 1901 by American Miller Hutchinson.

• COMMUNICATIONS •

Pictures and Sound

People told stories and cast shadow shapes on the walls of firelit caves thousands of years ago. In the 1640s they used early projectors called "magic lanterns" to shine candlelight through hand-painted glass slides. Frenchman Gaspard Robert invented the marvellous Fantasmagoria in 1798. This special kind of magic lantern projected moving pictures and shadows of ghosts and monsters onto sheets in a dark room. Audiences screamed and fainted at the sight, and quickly lined up to see it all over again. In 1891, the brilliant American inventor Thomas Edison developed a moving-picture camera called a Kinetograph and opened the way for silent movies and stars of the screen such as Charlie Chaplin. Within the next 40 years, inventions such as radio, television, film and video brought new dimensions to the quality of pictures and sound.

PUMPING THE PIANO
The Fotoplayer Company of Berkeley, California invented the Fotoplayer in 1915. This huge piano was powered by an air pump and played music programmed by a roll of punched paper. It provided sound effects for silent movies.

Light-activated
sound strip

RADIO
In 1906, Reginald Fessenden broadcast voice and music on radio waves for the first time.

TIMING SOUND

Phonograph records added lifelike sound to movies. But it was difficult to match the record with the action on the movie; voices often finished after the actors had stopped mouthing the words. In the 1929 movie *Hallelujah*, sound was recorded as a pattern on the film. This pattern was "read" by a light-sensitive cell that synchronized sound with the moving pictures.

PICTURES TO EUROPE

In 1962, the first television pictures were sent from the United States to Europe. They bounced off the *Telstar* satellite in space and were collected by a dish-shaped antenna like this.

BOX "BROWNIE" CAMERA

In 1888, George Eastman invented the roll-film camera. In 1900, he developed the box "brownie" model. It sold for US 50 cents, which included the film and the developing.

COMPACT DISC

Japanese and Dutch scientists invented the CD in 1981. It records sounds as microscopic changes in the surface of a plastic disc. The changes are "read" by a laser in the CD player and changed back into sound electronically.

JOHN LOGIE BAIRD

Mechanical television was invented by the Scottish engineer John Logie Baird in 1923. His television was a combination of inventions and discoveries by other people. Baird's cameras and receivers contained a spinning disc invented by Nipkow of Germany in 1884. The disc translated pictures into dots of light in eight lines that were focused onto a tiny television screen.

TWO FOR ONE

The video camera was invented in America in 1931, but it was larger than a human and could only send pictures, not record them. In 1981, the Sony Corporation of Japan invented a hand-held video camera that records as well.

1939 German television   John Logie Baird and the first TV transmitter

1640
MAGIC LANTERN
Athanase Kircher
Germany

1827
PHOTOGRAPH
Joseph Nicéphore Niepce
France

1877
WHEEL OF LIFE ANIMATOR
Emile Raynaud
France

1895
RADIO TELEGRAPH
Guglielmo Marconi
Italy

1906
FEATURE FILM
Charles and John Tait
Australia

1931
ELECTRONIC TELEVISION
Vladimir Zworykin
USA

1962
SATELLITE TELEVISION
Telstar
USA

1980
WALKMAN™ TAPE PLAYER
Akio Morita
Japan

2000+
3-D TELEVISION

SAXOPHONE
Adolphe Sax of Belgium invented the saxophone in 1841. He patented it in 1846 and spent the next 11 years practicing and studying before teaching students at the Paris Conservatoire.

Reed
The reed is a thin piece of wood or metal inside the mouthpiece.

• COMMUNICATIONS •

Musical Instruments

S ounds are made by waves or vibrations of air. The first musical instrument was the human voice. When we sing, the air in our throat vibrates. It echoes around our mouth and nose and, hopefully, comes out as music. The first musicians learned to use objects such as shells to make their voices louder, and soon realized that musical sounds could be made by plucking the string of a hunting bow or blowing into an animal bone. Most musical instruments have evolved through the tiny improvements made by instrument makers year after year. The first wind instruments were simple animal-bone flutes; stringed instruments, such as the harp, developed into the violin and the piano. Computers now enable us to create a wide variety of sounds electronically without blowing, hitting, plucking or strumming anything at all. Music can be made by everyone.

Keys
Mechanical keys for wind instruments were invented in the 1800s by the German instrument maker Theobald Boëhm.

GUITAR
A type of guitar was played in the Middle East from about 1000 BC. The modern acoustic guitar was invented in 1850 by Antonio de Torres, a Spanish instrument maker.

INDONESIAN GAMELAN ORCHESTRA
These players use mainly percussion instruments, including the saron and bonang, to play the melody by heart. They use string instruments such as the rebab and chelempung to enhance the sound.

PAN FLUTE
The first flutes were made from the hollow bones of sheep. Leg bones were punched with holes that were covered by the fingers to alter the pitch of the notes.

PLAYING THE PIANO

In 1710, Italian Bartolommeo Christofori invented keys attached to small hammers to strike strings. He called his invention the "piano-forte," which means soft and loud.

PLAYING PERCUSSION

This Chinese instrument dates back to about 1000 BC. Different-sized sheets of metal were hung on a frame and struck with a wooden mallet.

BEATING TIME

No one knows who made the first drum. These African drums have animal skin stretched over one end, which the player hits to produce a sound.

NOTATING MUSIC

Greek scholars first wrote down music in 500 BC as a line of alphabetical signs. The signs showed musicians if a note should be played high or low. Symbols called "neumes," which stood for notes or groups of notes, were introduced in AD 650. In 1026, Italian Guido d'Arezzo introduced a system whereby the neumes were placed high or low on a line to show the pitch of the note. Rhythm and timing were added in about 1500.

15th-century music score

Original Handel score

5000 BC
DIDGERIDOO
Australia

3000 BC
HARP
Sumeria, Egypt

220 BC
PIPE ORGAN
Ctesibius
Alexandria

1300
HARPSICORD
Europe

1816
METRONOME
Johann Maelzel
Germany

1821
HARMONICA
C. F. L. Buschmann
Germany

1935
ELECTRIC GUITAR
Adolph Rickenbacher
USA

1965
SYNTHESIZER
Robert Moog
USA

153

THE WATER CLOCK
The Egyptians developed a water clock about 3,500 years ago to tell the time at night. Shaped like a bucket, the clock had a scale marked on the inside to mark the water level and a hole near the bottom through which water trickled.

The time that had passed could be measured by reading the scale.

• INSTRUMENTS AND MACHINES •

Clocks and Calculators

Humans have always been fascinated by time. The first clocks used natural rhythms such as the movement of the sun to measure time. Later, the desire to divide up the day more precisely led to the invention of mechanical clocks. Powered by the energy stored in a metal spring, or weights on a chain, these clocks relied on an important device called the escapement, which turned the energy into a regular movement. By the mid-1600s, the accuracy of clockwork cogs and gears had caught the attention of mathematicians, and counting machines were invented to take the hard work out of sums. Inventors everywhere were inspired by the mechanical clock. They imagined that clockwork could be used to power all their wonderful ideas for the future.

154

THE MECHANICAL CALCULATOR

In 1642, 19-year-old Blaise Pascal built a simple arithmetic machine for his father, whose job involved counting money. The machine used clockwork gears to automatically add (up to eight-digit figures) or subtract. Some years later, a great mathematician, Gottfried Leibniz, developed Pascal's machine into a new model that could add, subtract, multiply, divide and find the square root of numbers. This was the starting point for all true calculators, and eventually, computers.

MARKING TIME
Inspired by the action of a church lamp swinging steadily during an earth tremor, Italian Galileo Galilei invented the pendulum in 1581. The first pendulum clock was made in 1656 by a Dutch scientist, Christiaan Huygens.

KEEPING THE CHANGE
In 1879, American James Ritty invented the cash register to discourage his bar staff from stealing the profits. The register used a clockwork mechanism to add, total and print transactions.

PLOTTING THE HEAVENS
The orrery is a clockwork model that shows the movements of planets around the sun. It was named after the English Earl of Orrery who had the first one built in about 1720.

ANCIENT ADDITION
Invented in Babylonia about 3000 BC, the abacus is still used throughout Asia to add, subtract, divide and multiply numbers.

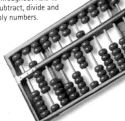

3400 BC
NUMBERS
Middle East

3000 BC
ABACUS
Babylonia

400
CANDLE AND FUSE CLOCK
Byzantium

725?
MECHANICAL WATER CLOCK
Yi Xing
China

1335
CHIMING CLOCK
Italy

1624
CLOCKWORK CALCULATOR
Wilhelm Schickard
Germany

1840
ELECTRIC CLOCK
Alexander Bain
England

1847
ALARM CLOCK
Antoine Redier
France

1907
MODERN WRISTWATCH
Louis Cartier and Hans Wilsdorf
France, Switzerland

1948
ATOMIC CLOCK
Frank Libby
USA

Computers and Robots

Two hundred years ago, people who figured out, or computed, difficult mathematical problems were called "computers." Today, computers are machines that use electronic circuits to store information such as numbers, words, pictures, sounds, shapes and calculations in code. Computers are used to control the most complex tools that have been invented—robots. These sophisticated machines are faster, more accurate and stronger than people. They can work in places and conditions where people could not survive, and they do not get bored doing the same thing every day! Virtual reality is a new invention that uses technology in a unique way. Special helmets, gloves and sensors connect a person's sight, hearing and touch to a computer. In the future, virtual reality will enable surgeons in one country to perform an operation in another country.

A SLAVE TO THE JOB
The word robot comes from the Czech "robotnik," which means "work slave." A robot can do many things faster and better than humans can.

LAPTOP COMPUTER
In 1987, Clive Sinclair of England invented a portable, or laptop, computer that weighed less than 2 pounds (1 kg).

SILICON CHIPS
Microscopic electrical circuits etched into chips of silicon were invented in 1959 by American Jack Kilby. These wafers hold hundreds of tiny silicon chips—each one powerful enough to run a small computer.

A SHEARING BREAKTHROUGH
In 1986, a robot invented by Australian farmer Lance Lines sheared the fleece of a sheep in about 90 seconds. The robot was programmed to be an efficient and safe shearer.

Robotic arm
Electric motors and hydraulic fluid move the robot's arm.

Moving robot
The robot slides along a rail.

Held tight
The sheep is gently clamped onto a platform.

CHARLES BABBAGE

In 1834, Charles Babbage invented a huge, mechanical "analytical engine"—the first mechanical computer. This machine was as big as a bus and could store and retrieve calculations from its memory. Babbage spent 40 years trying to build the machine, but he never completed it—the tools and materials of the time were not as advanced as his visionary invention.

375 BC
AUTOMATON
"FLYING DOVE"
Archytas of Tarentum
Italy

1834
ANALYTICAL ENGINE
Charles Babbage
England

1859
BINARY LOGIC
George Boole
England

1907
AUTOMATIC TOTALIZATOR
George Julius
Australia

1941
COLOSSUS COMPUTER
Max Neuman and Alan Turing
England

1954
COMMERCIAL MAGNETIC MEMORY COMPUTER
IBM
USA

1962
COMMERCIAL ROBOTICS
Unimation
USA

1975
PERSONAL COMPUTER
H. Edward Roberts
USA

1985
CD-ROM
Philips/Hitachi
Netherlands, Japan

2000+
VIRTUAL SURGERY

157

Early Power

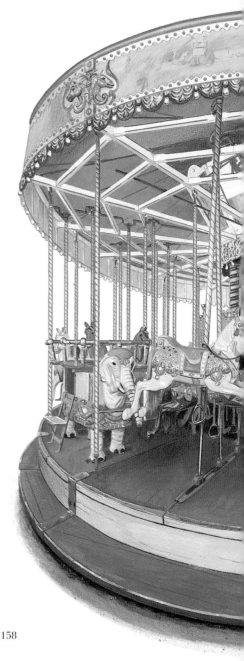

Hero, a mathematician in ancient Alexandria, first used steam power to make a metal ball spin. But he considered his invention a toy. More than 1,600 years later, inventors experimented with steam power again, but this time, the results of their efforts revolutionized people's lives. In the 1700s and 1800s, the steam-powered engine was adapted to do almost everything: pump water, drive factory machinery, propel ships, plow fields and even drive fairground rides. Some of the early steamships made so much smoke and noise that people were very reluctant to travel on them! The age of steam lasted for almost 200 years until the internal-combustion engine and electricity took over. Steam, however, is not as old-fashioned as you might think. Most of the electricity that gives us power today is produced by huge, steam-driven machines called turbines.

STRANGE BUT TRUE

Henry Seely of New York was ahead of his time. He invented the electric iron in 1882, but he could not sell it because nobody had electricity in their houses!

SMOOTHING OUT THE BUMPS
Without the steamroller, invented by Frenchman Louis Lemoine in 1859, roads would never have been smooth enough for the first fragile cars.

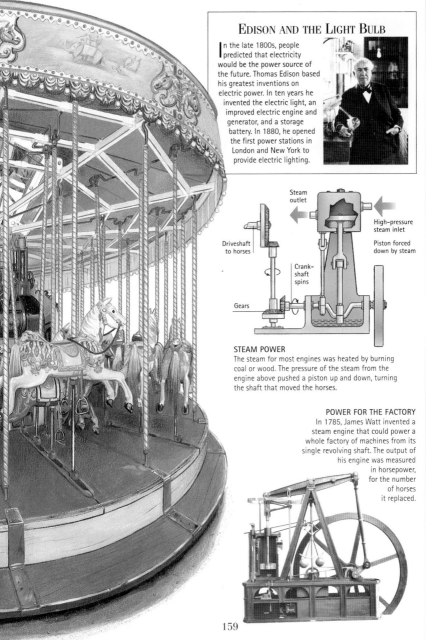

EDISON AND THE LIGHT BULB

In the late 1800s, people predicted that electricity would be the power source of the future. Thomas Edison based his greatest inventions on electric power. In ten years he invented the electric light, an improved electric engine and generator, and a storage battery. In 1880, he opened the first power stations in London and New York to provide electric lighting.

STEAM CYLINDER ENGINE
Denis Papin
France

1698

STEAM PUMP
Thomas Savery
England

1712

ENGINE WITH BOILER
Thomas Newcomen
England

1730

STEAMBOAT
Jonathan Hulls
England

Steam outlet

High-pressure steam inlet

Driveshaft to horses

Piston forced down by steam

Crank-shaft spins

Gears

1800

VOLTAIC CELL
Allesandro Volta
Italy

STEAM POWER
The steam for most engines was heated by burning coal or wood. The pressure of the steam from the engine above pushed a piston up and down, turning the shaft that moved the horses.

1821

ELECTRIC MOTOR
Michael Faraday
England

POWER FOR THE FACTORY
In 1785, James Watt invented a steam engine that could power a whole factory of machines from its single revolving shaft. The output of his engine was measured in horsepower, for the number of horses it replaced.

1829

STEAM PLOW
USA

1878

ELECTRIC DC GENERATOR
Thomas Edison
USA

1882

ALTERNATING CURRENT GENERATOR
Nikola Tesla
USA

1884

STEAM-POWERED TURBINE
Sir Charles Parsons
England

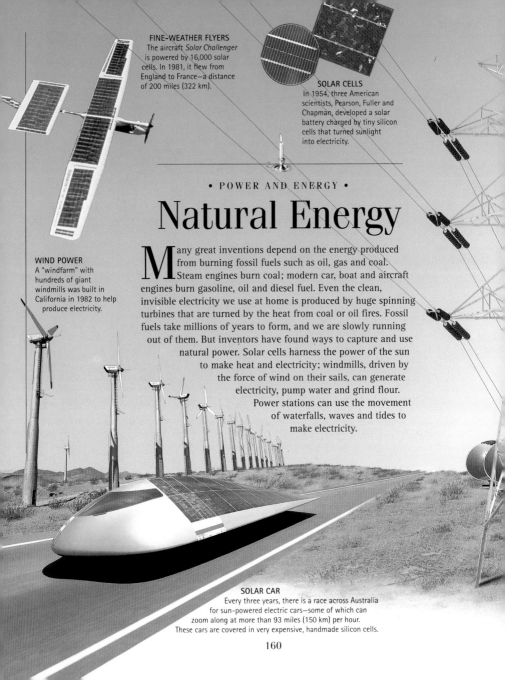

FINE-WEATHER FLYERS
The aircraft *Solar Challenger* is powered by 16,000 solar cells. In 1981, it flew from England to France—a distance of 200 miles (322 km).

SOLAR CELLS
In 1954, three American scientists, Pearson, Fuller and Chapman, developed a solar battery charged by tiny silicon cells that turned sunlight into electricity.

• POWER AND ENERGY •

Natural Energy

Many great inventions depend on the energy produced from burning fossil fuels such as oil, gas and coal. Steam engines burn coal; modern car, boat and aircraft engines burn gasoline, oil and diesel fuel. Even the clean, invisible electricity we use at home is produced by huge spinning turbines that are turned by the heat from coal or oil fires. Fossil fuels take millions of years to form, and we are slowly running out of them. But inventors have found ways to capture and use natural power. Solar cells harness the power of the sun to make heat and electricity; windmills, driven by the force of wind on their sails, can generate electricity, pump water and grind flour. Power stations can use the movement of waterfalls, waves and tides to make electricity.

WIND POWER
A "windfarm" with hundreds of giant windmills was built in California in 1982 to help produce electricity.

SOLAR CAR
Every three years, there is a race across Australia for sun-powered electric cars—some of which can zoom along at more than 93 miles (150 km) per hour. These cars are covered in very expensive, handmade silicon cells.

160

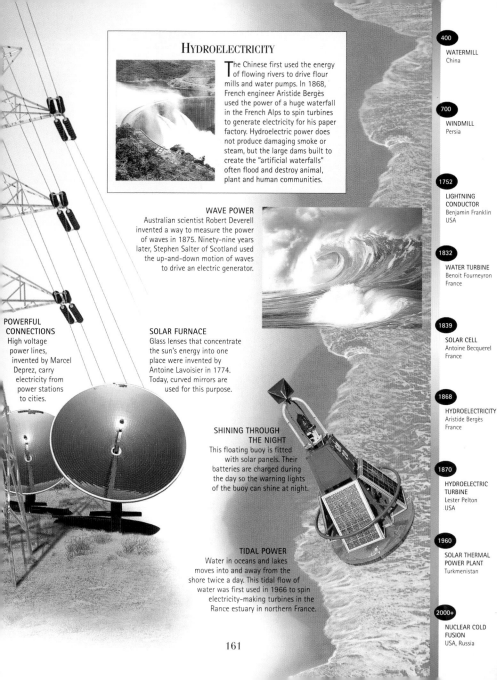

HYDROELECTRICITY

The Chinese first used the energy of flowing rivers to drive flour mills and water pumps. In 1868, French engineer Aristide Bergès used the power of a huge waterfall in the French Alps to spin turbines to generate electricity for his paper factory. Hydroelectric power does not produce damaging smoke or steam, but the large dams built to create the "artificial waterfalls" often flood and destroy animal, plant and human communities.

WAVE POWER
Australian scientist Robert Deverell invented a way to measure the power of waves in 1875. Ninety-nine years later, Stephen Salter of Scotland used the up-and-down motion of waves to drive an electric generator.

POWERFUL CONNECTIONS
High voltage power lines, invented by Marcel Deprez, carry electricity from power stations to cities.

SOLAR FURNACE
Glass lenses that concentrate the sun's energy into one place were invented by Antoine Lavoisier in 1774. Today, curved mirrors are used for this purpose.

SHINING THROUGH THE NIGHT
This floating buoy is fitted with solar panels. Their batteries are charged during the day so the warning lights of the buoy can shine at night.

TIDAL POWER
Water in oceans and lakes moves into and away from the shore twice a day. This tidal flow of water was first used in 1966 to spin electricity-making turbines in the Rance estuary in northern France.

400
WATERMILL
China

700
WINDMILL
Persia

1752
LIGHTNING CONDUCTOR
Benjamin Franklin
USA

1832
WATER TURBINE
Benoit Fourneyron
France

1839
SOLAR CELL
Antoine Becquerel
France

1868
HYDROELECTRICITY
Aristide Bergès
France

1870
HYDROELECTRIC TURBINE
Lester Pelton
USA

1960
SOLAR THERMAL POWER PLANT
Turkmenistan

2000+
NUCLEAR COLD FUSION
USA, Russia

161

DEADLY BEAUTY
This razor-sharp sword was used by Japanese Samurai warriors to kill their enemy in one blow. Invented around 1200, a Samurai sword is made from up to 20 layers of steel beaten together. It is then sharpened until it can split a human hair in half.

BOW AND ARROW
The English longbow, invented in about 1330, was very accurate and could kill an enemy 1,640 ft (500 m) away.

• WAR AND PEACE •

Into Battle

The history of the world is full of stories of war. Many inventions were developed especially as weapons to be used in battle. Clubs, axes and swords were wielded in the hand-to-hand combat of early wars. In the third century BC, the Greek scholar Archimedes gave soldiers curved and polished shields to reflect the sun into the eyes of Romans invading the city of Syracuse. People still fight with wooden shields in New Guinea. A European knight in the 1500s rode into battle covered from top to toe in metal armor that protected him from the swords, arrows, axes and spears of his enemies. His horse was also covered in menacing armor. But if the knight fell or was knocked from his horse, the weight of the armor made it difficult for him to stand and defend himself. He became an easy target for enemy knights.

BATTLE-AX
Copper axes were first invented in Mesopotamia in 4000 BC, probably as a hand tool, but they soon became weapons. This is a Viking battle-ax from the thirteenth century.

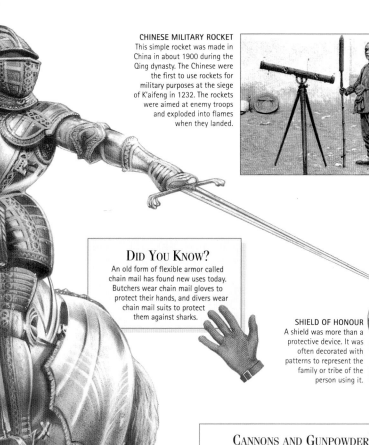

CHINESE MILITARY ROCKET
This simple rocket was made in China in about 1900 during the Qing dynasty. The Chinese were the first to use rockets for military purposes at the siege of K'aifeng in 1232. The rockets were aimed at enemy troops and exploded into flames when they landed.

DID YOU KNOW?
An old form of flexible armor called chain mail has found new uses today. Butchers wear chain mail gloves to protect their hands, and divers wear chain mail suits to protect them against sharks.

SHIELD OF HONOUR
A shield was more than a protective device. It was often decorated with patterns to represent the family or tribe of the person using it.

CANNONS AND GUNPOWDER
Gunpowder was first invented around AD 850 by chemists in China. They used it in fireworks, but later it became popular as an explosive in China, India and Europe. The first gun was a metal firing tube made in Europe in about 1300 by a monk. Cannons were invented in Italy at about the same time, but they produced a spray of iron bullets and did not always hit their target. In the 1500s, the first accurate cannons were invented in France. Warfare took a new turn.

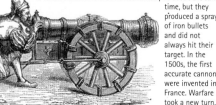

Discover more in Today and Tomorrow

163

200,000 BC
SPEAR
Europe

230,000 BC
STONE AX
Kenya

30,000 BC
BOW AND ARROW
Africa, USA, Europe

3000 BC
SWORD
Mesopotamia

500 BC
CATAPULT
Greece

400 BC
CHAIN MAIL ARMOR
Italy

300 BC
HAND-HELD CROSSBOW
China

850
BLACK GUNPOWDER
China

Today and Tomorrow

GUIDED MISSILES
All modern strategic missiles and space rockets were developed from the work of a team of Second World War German scientists, who created more than 20 types of missiles. Air-to-air guided missiles, such as this one, are used for aerial combat.

Technological progress is faster in times of war. Each side tries to make weapons and machines that are bigger and better than those of the enemy. At the beginning of the First World War, for example, the typical flying speed of an airplane was 70 miles (113 km) per hour. By the end of the war this speed had doubled. In the Second World War, the Germans introduced two inventions that later transformed flight—the turbojet, which became the basis of modern aircraft, and the ballistic missile, which took aviation from the skies into space. Some wartime inventions, such as the tank, are only suited to war, but many have other uses. The antibiotic penicillin, which saves many lives, was invented in 1941 to cure the infected wounds of soldiers. It is impossible to say whether more good has come from wartime inventions than bad.
But one thing is certain, many things were invented *because* of war.

ARMORED TANK
The tank lurched onto First World War battlefields in 1916, thanks to the combined efforts of a number of inventors and British army officers. This modern United Nations tank has a swiveling gun turret and lookout post.

Modern armor
Heavy steel armor plating was first used in America in 1862 to strengthen warships. Today's tanks are encased in lightweight but strong metal alloys, plastics and even ceramics.

RADIO DETECTION AND RANGING
In 1935, the scientist Robert Watson-Watt was asked by the British army to invent a "radio death ray" for warfare. Instead, he invented radar, which detects enemy aircraft using radio waves.

STEALTH FIGHTERS
Difficult to detect because of their shape and a radar absorbent coating, F-117 fighter bombers are designed for precision attack. They were used by the United States in the Gulf War in 1991.

LETHAL WEAPONS
Grenades have been around for more than 500 years. In the 1600s, French soldiers, called *grenadiers*, were trained specially as throwers. Plastic explosives, once unwrapped from their sausagelike skins, can be molded into position. They were used in military operations to shatter parts of bridges and buildings.

NIGHT VISION
Since the 1950s, scientists have been working on devices to make it possible for soldiers to fight in the dark. The night-vision goggles shown below are sensitive to low levels of light, such as reflected starlight or moonlight. The goggles intensify this light and allow soldiers to see, move and shoot at night as well as they can during the day. In theory, 24-hour war is now possible.

Caterpillar tracks
In 1904, Benjamin Holt built a tractor that laid down its own track under the rear wheels to travel over mud. A continuous track belt under all the wheels of a tank enables it to break through fences and go over deep gullies.

Discover more in Into Battle

165

1861
SEA MINE
USA

1862
MACHINE GUN
Richard Gatling
USA

1883
AUTOMATIC MACHINE GUN
Hiram Maxim
USA

1902
EXPLODING BULLET
John Pomeroy
New Zealand

1915
SONAR
Paul Langevin
France

1944
V2 ROCKET BOMB
Wernher von Braun and Walter Dornberger
Germany

1945
ATOMIC FISSION BOMB
Project Manhattan scientists
USA

1952
HYDROGEN BOMB
USA

1984
STUN GUN
USA

1985
FLASHBALL GUN
François Richet
France

Healers and Healing

Two hundred years ago, visiting the doctor was a risky business. Operations were performed without proper anesthetic, open wounds often became infected, and many deadly diseases could not be treated. Today, doctors can vaccinate, anesthetize, sterilize and treat with antibiotics. Dramatic discoveries and ingenious inventions led to these life-saving procedures. In 1928, for example, Alexander Fleming discovered a mold that could fight germs. Twelve years later, Howard Florey and Ernst Chain developed this substance and invented the first antibiotic—penicillin. Many of the tools now used by doctors were invented in the 1800s: the stethoscope, which listens to the heartbeat; the endoscope, which allows doctors to peer inside the body; and the sphygmomanometer, which measures blood pressure.

THE POINT OF IT
A syringe is a piston in a tube that can suck up liquids and then squirt them out. The medical syringe attached to a hypodermic (beneath the skin) needle was perfected in 1853 by Scotsman Alexander Wood.

UNDER ANESTHETIC
Only 200 years ago, patients stayed awake during operations. Many had to be tied or held down. In 1846, American dentist William Morton used the chemical ether to anesthetise a patient while a tumor was removed from the man's neck.

TRADITIONAL MEDICINE

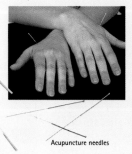

Many cultures treat illnesses with ancient medical inventions. Aboriginal people in Australia make poultices and medicines from bush plants. Tribal shamans in South America combine plants and rituals to make people better. Chinese doctors invented acupuncture more than 4,000 years ago. Acupuncturists insert very fine needles into special points on the body to stimulate the nerves and help the body to heal itself.

Acupuncture needles

166

GERM FREE
Doctors once operated with their hands and instruments covered with blood from the previous operation. In 1865, Joseph Lister used carbolic acid to sterilize his hands, tools and the air during an operation.

STRANGE BUT TRUE

In 1667, a blood transfusion was carried out using a lamb as the donor. The patient, a 15-year-old boy who was bleeding to death, survived!

THE BARE BONES
In 1895, German Wilhelm Röntgen discovered a ray that passed through flesh but not through bone. As it was such a mystery ray, he called it X-ray. This marvellous ray was used to take pictures of the human skeleton such as this—the first full-length X-ray of a person, complete with sock suspenders and keys in the pocket.

2600 BC
ACUPUNCTURE
Emperor Huang Ti
China

1270
EYEGLASSES
Court of Kublai
Khan
China

1626
MEDICAL
THERMOMETER
Santorio
Italy

1796
VACCINATION
Edward Jenner
England

1816
STETHOSCOPE
René Laennec
France

1854
NURSING CORPS
Florence
Nightingale
England

1896
SPHYGMOMANO-
METER
Scipione Riva-
Rocci
Italy

1899
ASPIRIN
Felix Hoffman
Germany

1928
FLYING DOCTOR
SERVICE
John Flynn
Australia

AN UNLIKELY PAIR
Embryos of twins can be frozen and then implanted separately, years apart. The result: twins who are not the same age!

Freezing cells
Embryos can be frozen when they consist of only a few cells.

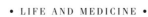

FREEZE–THAW IVF
In 1983, an Australian team led by Carl Wood invented a way to fertilize and grow human embryos in glass tubes and then freeze and store them. The embryos can be thawed and implanted into the womb up to ten years later.

• LIFE AND MEDICINE •

Marvels of Medicine

Medicine has entered an exciting new stage. With today's technology, doctors can now observe the human body working on the inside without cutting it open. Vaccines and genetically engineered viruses can help the body to repair itself, and some inventions can actually replace broken and damaged organs. Artificial parts include electronic ears, plastic stomachs, mechanical hearts, heart pacemakers and ceramic hips. Doctors today use carpentry techniques and stainless steel or plastic nuts, screws and bolts to hold broken bones together. Fifty years ago, these bones never would have healed. Skin, kidneys, heart, liver, ova, sperm, lungs, corneas and bone marrow can be transplanted from person to person. Microsurgery rejoins the smallest blood vessels and nerves that have been cut in accidents.

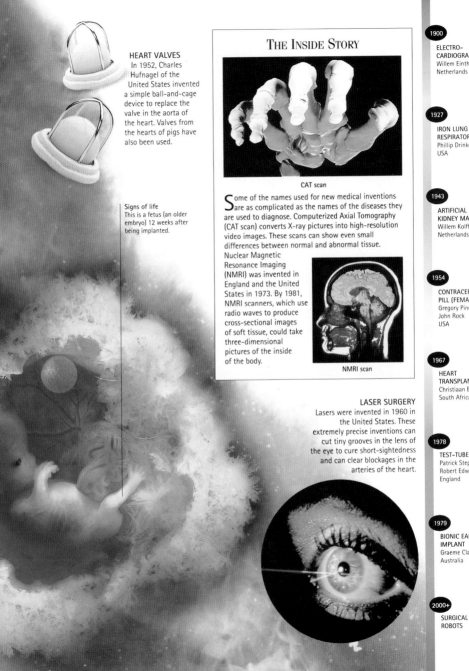

HEART VALVES

In 1952, Charles Hufnagel of the United States invented a simple ball-and-cage device to replace the valve in the aorta of the heart. Valves from the hearts of pigs have also been used.

Signs of life
This is a fetus (an older embryo) 12 weeks after being implanted.

THE INSIDE STORY

CAT scan

Some of the names used for new medical inventions are as complicated as the names of the diseases they are used to diagnose. Computerized Axial Tomography (CAT scan) converts X-ray pictures into high-resolution video images. These scans can show even small differences between normal and abnormal tissue. Nuclear Magnetic Resonance Imaging (NMRI) was invented in England and the United States in 1973. By 1981, NMRI scanners, which use radio waves to produce cross-sectional images of soft tissue, could take three-dimensional pictures of the inside of the body.

NMRI scan

LASER SURGERY

Lasers were invented in 1960 in the United States. These extremely precise inventions can cut tiny grooves in the lens of the eye to cure short-sightedness and can clear blockages in the arteries of the heart.

1900
ELECTRO-CARDIOGRAPH
Willem Einthoven
Netherlands

1927
IRON LUNG RESPIRATOR
Phillip Drinker
USA

1943
ARTIFICIAL KIDNEY MACHINE
Willem Kolff
Netherlands

1954
CONTRACEPTIVE PILL (FEMALE)
Gregory Pincus and John Rock
USA

1967
HEART TRANSPLANT
Christiaan Barnard
South Africa

1978
TEST-TUBE BABY
Patrick Steptoe and Robert Edwards
England

1979
BIONIC EAR IMPLANT
Graeme Clarke
Australia

2000+
SURGICAL ROBOTS

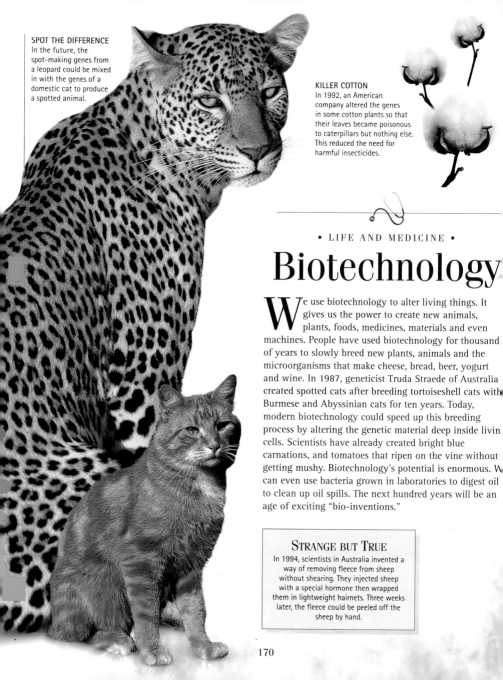

SPOT THE DIFFERENCE
In the future, the spot-making genes from a leopard could be mixed in with the genes of a domestic cat to produce a spotted animal.

KILLER COTTON
In 1992, an American company altered the genes in some cotton plants so that their leaves became poisonous to caterpillars but nothing else. This reduced the need for harmful insecticides.

• LIFE AND MEDICINE •

Biotechnology

We use biotechnology to alter living things. It gives us the power to create new animals, plants, foods, medicines, materials and even machines. People have used biotechnology for thousands of years to slowly breed new plants, animals and the microorganisms that make cheese, bread, beer, yogurt and wine. In 1987, geneticist Truda Straede of Australia created spotted cats after breeding tortoiseshell cats with Burmese and Abyssinian cats for ten years. Today, modern biotechnology could speed up this breeding process by altering the genetic material deep inside living cells. Scientists have already created bright blue carnations, and tomatoes that ripen on the vine without getting mushy. Biotechnology's potential is enormous. We can even use bacteria grown in laboratories to digest oil to clean up oil spills. The next hundred years will be an age of exciting "bio-inventions."

> **STRANGE BUT TRUE**
> In 1994, scientists in Australia invented a way of removing fleece from sheep without shearing. They injected sheep with a special hormone then wrapped them in lightweight hairnets. Three weeks later, the fleece could be peeled off the sheep by hand.

TRANSGENIC PIGS

The heart of a pig is similar in shape and size to the human heart. People and pigs, however, have very different genes. Scientists in England have developed a virus that carries human genes into pigs. This makes it possible for the human body to accept the heart of a pig in a transplant.

CHANGING GENES
Every living cell has spiral-shaped deoxyribonucleic acid (DNA), discovered in 1953 by Francis Crick and James Watson at Cambridge in England. The DNA is made up of genes that control how the cell works. Biotechnologists have learned how to alter the genes and change living cells.

6000 BC
BEER
Mesopotamia

1000 BC
CHEESE
Nomad tribes
Middle East

1972
OIL-DIGESTING
MICROBES
Dr Ananda M.
Chakrabarty
USA

1975
MONOCLONAL
ANTIBODY
George Kohler and
Cesar Milstein
England

1984
TRANSGENIC
PLANT
University of Ghent
Belgium

1986
BLACK TULIP
Geert Hageman
Netherlands

1989
GENE SHEARS
James Haseloff and
Wayne Gerlach
Australia

1990
CROWN GALL
BACTERICIDE
Dr Alan Kerr
Australia

1991
LONG-LIFE
TOMATO
USA

It Might Have Been!

The history of invention is full of brilliant, yet crazy ideas. Can you imagine what life might be like today if Thomas Edison had developed his idea of anti-gravity underwear, or if Alexander Bell had persisted in trying to invent a talking fire alarm rather than a telephone? Inventions inspire people to think of the future and its possibilities, but trying to predict the success of inventions and how people will react to them has never been easy. The first motor cars had to travel through towns at walking pace behind a man carrying a red flag because they frightened people and horses. People thought that motor cars would never replace the horse and buggy. In the 1950s, the president of IBM predicted that, at most, no more than one computer per country would ever be built! The computer endured, but other inventions had a short life. This collection from the past 200 years shows you just what might have been.

Up and over

In the early 1900s, many railway systems used single tracks. This patent from 1904 shows how an express train could drive over the top of an all-stations train and double the number of passengers on the line. But imagine being a passenger!

Watch out!

In the early days of motor cars, pedestrians and cars were always at odds in the battle for the roadways. The car above was a fanciful invention that vacuumed up "jaywalkers" who strayed onto the road.

Toothbrush

This simple invention, which originated in China, has been the subject of continual "improvements." This design by Luis Reinold in 1941 may have been good for brushing the back of your teeth!

The umbrella cap

The umbrella, invented thousands of years ago in China, has been the subject of hundreds of improvements. These include a gutter, ventilation holes, folding mechanisms, see-through plastic, and this capbrella from about 1904.

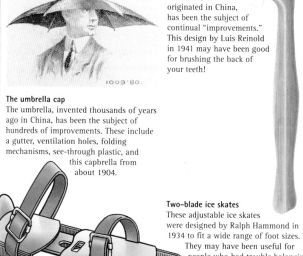

Two-blade ice skates

These adjustable ice skates were designed by Ralph Hammond in 1934 to fit a wide range of foot sizes. They may have been useful for people who had trouble balancing on one blade, but how would you be able to steer?

Anti-gravity underwear

This 1878 cartoon illustrates the effect of Thomas Edison's anti-gravity underwear. If this ingenious idea had become a reality, wearers would have beaten the Wright brothers in the race to become airborne.

Helicycle

In 1936, Igor Sikorsky's helicopter became the first practical machine to hover in the air. It caught the imagination of Daniel Gumb in 1945, who invented the personal helicopter based on the design of a motorcycle.

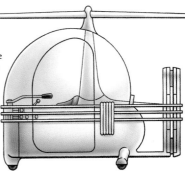

‡bot farmers

the age of steam, some people predicted at farmers would be able to put their et up; steam-powered robots would do the heavy farming work. vo hundred years later, actors and harvesters are st and efficient, but these achines still need to be perated by farmers.

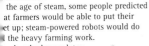

Folding baby carriage

This is a drawing of a folding carriage patented in the 1890s. The wheels came off and the whole thing folded up into a handbag. A very successful baby carriage that folded up into a neat package was patented by Harold Cornish of Australia in 1942.

Pedestrian scoop

In the 1920s, road fatalities involving pedestrians rose dramatically, as cars became faster while their brakes remained primitive. This led to quite a few inventions that would scoop up pedestrians or bounce them out of the way.

‡g power

his design for a pedal-driven propeller n the back of a surfboard was drawn by ert Lee in 1942. Today's surfskiers use paddle to power their crafts through le waves.

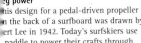

A wheelbarrow?

In 1884, this trunk that becomes a luggage cart was patented—about a hundred years before travel cases with pop-out caster wheels and retracting handles became popular!

Great Buildings

What is the oldest wooden building in the world?

Which building does not have right angles?

How do architects design buildings
to withstand earthquakes?

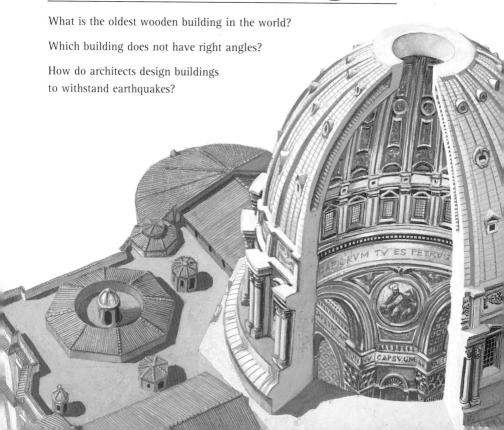

Contents

A Place to Live

People must have shelter to survive. They will die without protection from the sun, rain, wind and cold. Today, people can live in almost every part of the world because they have learned to build walls and to put a roof over their heads. For centuries, people had no tools to cut or move trees and large stones, so the first houses were built from materials that were easy to handle, such as grasses, vines and small stones. They discovered that hard rocks with sharp edges could cut trees and other rocks, and these became the first building tools. Many centuries later, people melted metals from rocks to make stronger, sharper tools. In places with little stone or wood, people made sun-dried bricks out of mud to build their houses. Some of the earliest cultures in history were the first to discover and use many of the basic building materials still used today.

The r
A waterproof roo
made from grass
thatching. Bundles
swamp grass are t
to a wooden frame
that each bun
overlaps the ones r
to it and below

178

STONE HUT

Walls of stone shaped with tools surround clusters of houses in the village of Haaran on the Turkish–Syrian border. Each house has several rooms and each room has its own dome. Some even have a second story. Smoke from fires used for cooking escapes through holes in the roof.

MAKING BRICKS

Sun-dried mud bricks were perhaps the first synthetic building material ever made. A mixture of mud and straw is pressed into molds then laid out in the sun to dry, as seen here. The straw holds the bricks together so they do not crumble. As rain will dissolve sun-dried bricks, a coating of lime is added or a wide roof is built to protect the walls.

BEEHIVE HUT

This hut on Dingle Peninsula in Ireland looks like a beehive. It was built centuries ago by a monk who piled up small flat stones cleared from his fields. He stacked each circle of stones on top of the circle below and made each stone slope downwards slightly towards the outside, so rain could not get in.

The walls
Here a man is weaving mats from palm fronds or leaves, which will become the walls of his hut. Weaving stiffens the fronds.

SOUTH PACIFIC WOVEN HUTS

On the Trobriand Islands of Papua New Guinea, houses are still built from small trees cut with stone tools. The pieces of wood are tied together with vines to form the frame of each house. The island people use plant materials to complete the house. Grass and leaves bend easily and people thought they seemed too weak to use for building until they discovered how hard it was to pull them apart.

Discover more in Games and Entertainment

179

Early American Empires

PALACE OF THE GOVERNORS
This palace in Uxmal, Mexico, is decorated with carved serpents and the Mayan rain god Chac. Religious leaders lived in its cool corbel-vaulted rooms.

The oldest architectural monuments in the Americas are found in present-day Mexico and along the west coast of South America. Early civilizations there had neither iron tools nor animals that could be trained to pull carts, yet the people constructed enormous stone buildings. The Olmecs and later civilizations in Mexico such as the Toltecs and Aztecs lived in scattered farm villages. These peoples had one religion and their religious centers were cities of stone such as Teotihuacán, where temples stood on top of tall pyramids. The peace-loving Mayan people lived in the rainforests of the Yucatan Peninsula in Mexico and they also built their religious centers of stone. In the fifteenth century, the Incas ruled an empire 2,480 miles (4,000 km) long in the Andes mountains of Peru. Their many towns were united by paved roads and a fast mail system. Incan stonemasons cut, polished and fitted stones together so tightly that a knife blade will not slide between them even today.

Stairway of gods
Two sides of the pyramid have steep stairs. A row of carved masks of Chac, the god of rain, line both sides of the staircase.

PYRAMID OF THE SUN
This pyramid, built in the third century in Teotihuacán, Mexico, stands on a high platform and is surrounded by volcanoes. Stone covers a core of dirt and lava carried to the site by thousands of workers over a period of 30 years. Aztecs lived there centuries after its Teotihuacán builders had disappeared. They believed this pyramid had been built by the gods themselves.

PYRAMID OF THE MAGICIAN
The Mayans built this pyramid in Uxmal, Mexico, in the ninth century. It has an unusual oval shape and two temples at the top. The peoples of Mexico built high platforms, or pyramids, for their temples so they would be closer to the gods in the heavens.

At the top
The temples on the pyramid are stone replicas of Mayan thatched huts. Gifts were offered before statues of gods inside the corbel-vaulted rooms.

THE CITADEL
Offerings were placed on this Chac Mool, a god sculpted as a man lying on his back, which sits near an eleventh-century Toltec pyramid in Chichen Itza, Mexico. The pyramid has a steep staircase on each side and a temple at the top.

CORBELED ROOFS
A building constructed of stone posts and horizontal beams will collapse if the beams have to support heavy walls or if the posts are not set close enough together. Stone doorways and stone roofs or vaults, such as the one shown here, can be built with small stones called corbels. Each stone lies on top of the last stone and has one end sticking out over the opening. Once the stones or corbels from both sides of the opening meet at the top, stones placed on top of the roof will hold it in place.

DID YOU KNOW?
The Pyramid of the Magician encloses three older temples. In Mexico, a new pyramid and temple often encased an earlier one. A completely furnished temple ready for use was discovered within the Pyramid of the Sun.

INCAN RUINS
Important religious ceremonies took place in Machu Picchu, an Incan town high in the Andes mountains of Peru. The plain stone walls of important Incan buildings were covered with plates of pure gold.

Early Civilizations

THE PYRAMIDS OF GIZA
These three pyramids were built more than 4,500 years ago as tombs for Egyptian pharaohs. The largest of the three, the Great Pyramid of Pharaoh Khufu, contains nearly two-and-a-half million stone blocks.

TEMPLE FOR A GOD
The huge columns of Egyptian temples still stand like stone forests in the desert above the banks of the Nile. This complex at Karnak was built over a period of 1,200 years. Here a statue of a pharaoh and his daughter stands outside the temple.

DID YOU KNOW?
Some Egyptian architects today are also building vaulted structures out of sun-dried bricks. The buildings stay cool and the materials do not damage the environment.

More than 5,000 years ago, a great civilization developed in Mesopotamia, the land between the Tigris and Euphrates rivers, then spread eastward along the north coast of the Indian Ocean. The Egyptian civilization developed beside the River Nile soon after. People traveled between the two areas and brought new ideas and inventions with them. Egypt had many workers and plenty of stone, and the Egyptians built huge pyramids and temples using simple tools and techniques. Because they did not have the wheel, 20 men pulled each stone to the pyramid on a wooden sled. Both stone and wood were scarce in Mesopotamia. The people there invented new materials such as bricks molded from clay and baked in an oven or dried by the sun. They then built wheeled carts to transport the bricks.

Steers and dragons
The symbols of the Babylonian weather god Adad and of the city's protector, the god Marduk, decorate the Ishtar Gate.

ISHTAR GATE

In the sixth century BC, King Nebuchadnezzar built a road called the Processional Way. This road led from his palace in the city of Babylon, the main city of Mesopotamia, to a ceremonial hall for New Year's celebrations. The Processional Way passed through the city's double walls at the Ishtar Gate.

PARADE OF LIONS

Every animal lining the walls of the Processional Way was brick, cast from special molds so that the bodies curved out from the wall. Each of the lions was made up of 46 specially molded and glazed bricks.

INVENTING THE ARCH

A stone laid across an open space like a doorway is brittle and will break if a heavy weight is placed on it. To avoid this, the supports of ancient stone buildings were set close together. Mesopotamians invented the arch so they could build wide, open rooms. Bricks or small stones set in a curve form an arch. The weight of each stone pushes it against the next until one pushes against a thick wall, called a buttress. The buttress presses the stones together and holds the arch in place. A vault is a ceiling built with arches.

Arched vault

Buttress

Supports weight

Glazed bricks
The bricks on the walls were painted with a glasslike mixture then baked to produce glowing colors.

An arched vault
The passage through the gate was 13 ft (4 m) wide, which was only possible because it was covered by an arch.

GREEK ORDERS
The Greeks built in three styles
called orders. You can recognize
the different orders by the style of
the wide section at the top of each
column, which is called a capital.

Doric order
This style has thick columns
and plain capitals.

Ionic order
The thinner columns of this style
are topped by a capital with two
wide spirals called volutes.

Corinthian order
This order is more elaborate,
and the capital is decorated
with acanthus leaves.

Frieze
A narrow band of
carving encircles the
top of the temple
wall and shows the
procession on Athena's
festival day.

The goddess Athena
The tall wooden statue
of Athena had an ivory
face, arms and feet. She
wore clothing made of
gold plates that weighed
2,500 lb (1,134 kg).

• THE CLASSICAL AGE •

Monuments to the Gods

I n the fifth century BC, most Greeks lived in small
city-states on islands in the Aegean Sea and in
mountain valleys near its coast. The Greeks built
temples as homes for their gods so the gods would live
among them and defend their cities. The first temples were
built of timber and sun-dried brick and looked like the
Greeks own huts. Later temples were built on top of a
three-stepped platform and surrounded by columns. When
the wooden temples decayed they were replaced by stone
temples, which looked exactly the same. The main goal of
the Greeks was to make their temples look perfect. They
built with the purest white marble and architects used
geometry to design the temples so that all the
proportions fit together in harmony.

TEMPLE OF ATHENA NIKE
The design of this small
temple, dedicated to the
goddess Athena, is based on
a typical Greek hut. It was
built in the Ionic style.

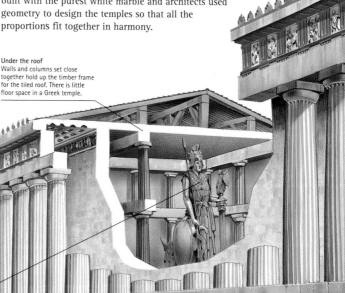

Under the roof
Walls and columns set close
together hold up the timber frame
for the tiled roof. There is little
floor space in a Greek temple.

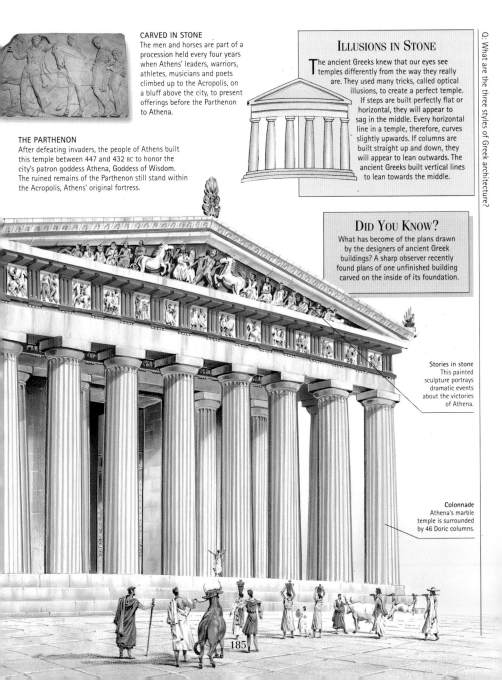

CARVED IN STONE
The men and horses are part of a procession held every four years when Athens' leaders, warriors, athletes, musicians and poets climbed up to the Acropolis, on a bluff above the city, to present offerings before the Parthenon to Athena.

THE PARTHENON
After defeating invaders, the people of Athens built this temple between 447 and 432 BC to honor the city's patron goddess Athena, Goddess of Wisdom. The ruined remains of the Parthenon still stand within the Acropolis, Athens' original fortress.

ILLUSIONS IN STONE

The ancient Greeks knew that our eyes see temples differently from the way they really are. They used many tricks, called optical illusions, to create a perfect temple. If steps are built perfectly flat or horizontal, they will appear to sag in the middle. Every horizontal line in a temple, therefore, curves slightly upwards. If columns are built straight up and down, they will appear to lean outwards. The ancient Greeks built vertical lines to lean towards the middle.

DID YOU KNOW?

What has become of the plans drawn by the designers of ancient Greek buildings? A sharp observer recently found plans of one unfinished building carved on the inside of its foundation.

Stories in stone
This painted sculpture portrays dramatic events about the victories of Athena.

Colonnade
Athena's marble temple is surrounded by 46 Doric columns.

Take-out food shop and viewing gallery

Swimming pool
Every Roman boy was expected to be able to read and to swim. Baths in colder parts of the empire had indoor, heated swimming pools.

COLOSSEUM
The 50,000 seats at the Colosseum in Rome stood on rings of concrete-vaulted passages, which were reached by stairs. Every spectator could leave the Colosseum in five minutes through exits called vomitoria. The Colosseum was used for many activities. It was flooded for mock sea battles, and gladiators tested their skills against lions that leapt into the arena when hidden doors snapped open.

Frigidarium
The Frigidarium was at the center of the baths and was a popular place to meet friends. Four baths filled with cold water gave the room its name.

WORKING OUT
A mosaic on the floor of the baths in the Villa Casale, a private country house in Piazza Armerina, Sicily, shows women exercising. Many public baths had a separate bathing area for women.

• THE CLASSICAL AGE •

Roman Recreation

B y the first century AD, Rome was a great empire. It reached from the Caspian Sea in the east and the British Isles in the north, to North Africa in the south. The Romans built roads with hard surfaces to connect their many cities. Aqueducts brought water to the cities from mountain springs. Luxury goods arrived in Rome's large harbors from every part of the known world. Romans in the cities bought food in take-out restaurants to eat in apartments with glass windows. They spent their free time watching plays or sporting events such as chariot races. They gathered at public baths to exercise and relax. Roman emperors ordered the construction of lavish buildings for public recreation to make themselves popular with the citizens. Roman engineers used synthetic materials such as concrete to construct these buildings, which were decorated with statues, mosaics and imported marble.

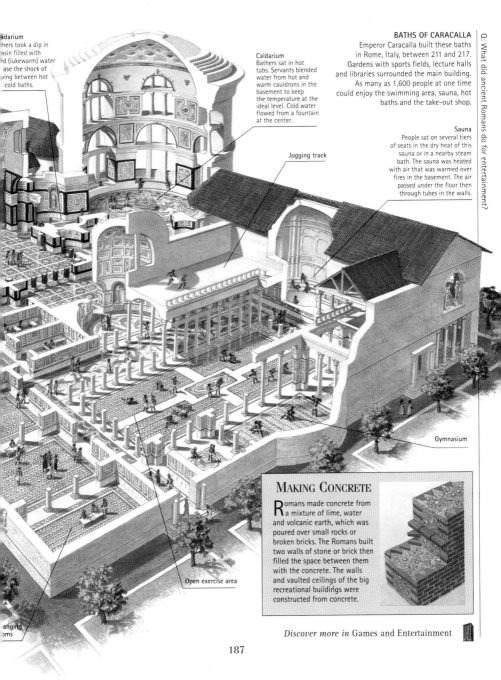

idarium
hers took a dip in
asin filled with
d (lukewarm) water
ase the shock of
ving between hot
cold baths.

Caldarium
Bathers sat in hot tubs. Servants blended water from hot and warm cauldrons in the basement to keep the temperature at the ideal level. Cold water flowed from a fountain at the center.

BATHS OF CARACALLA
Emperor Caracalla built these baths in Rome, Italy, between 211 and 217. Gardens with sports fields, lecture halls and libraries surrounded the main building. As many as 1,600 people at one time could enjoy the swimming area, sauna, hot baths and the take-out shop.

Sauna
People sat on several tiers of seats in the dry heat of this sauna or in a nearby steam bath. The sauna was heated with air that was warmed over fires in the basement. The air passed under the floor then through tubes in the walls.

Jogging track

Gymnasium

MAKING CONCRETE

Romans made concrete from a mixture of lime, water and volcanic earth, which was poured over small rocks or broken bricks. The Romans built two walls of stone or brick then filled the space between them with the concrete. The walls and vaulted ceilings of the big recreational buildings were constructed from concrete.

Open exercise area

anging
oms

Discover more in Games and Entertainment

SRI RANGANATHA
This tower in Mysore, India is
one of 15 giant gateways through
the five walls that enclose a
Hindu shrine. The gateways were
built between the eleventh and
seventeenth centuries. The shrine
itself is quite small and crowded
by the priests' houses and the
assembly rooms for pilgrims.

DID YOU KNOW?
Even a small Hindu shrine can be
seen from anywhere in a village
because of the tall, carved shikhara
above it. The shikhara represents a
holy mountain that is thought of as
a staircase to the heavenly world.

RANAKPUR TEMPLE
The Ranakpur temple honors Mahavira,
the founder of Jainism. Jains believe
that a person lives many lives, including
those of animals. Jains try not to hurt
any living creature. One of Ranakpur's
large corbeled domes rises above the
courtyard. The dome rests on two
stories of columns and is surrounded by
smaller domes.

• EMPIRES OF THE EAST •

Foundations of Religions

As early as 2500 BC, great civilizations flourished south of the Himalayan mountains, in what is now India. Three world religions began there—Hinduism, Buddhism and Jainism. All three teach that life, like a circle, has no end. It returns again and again as do the seasons. They believe that a person's soul comes back to live another life in a new body. This is called reincarnation. Hinduism began about 1500 BC. Hindus worship alone on most occasions, and many make pilgrimages to temples to pay homage to their gods. Hindu temples have richly decorated exteriors and pilgrims worship outside. The most important part of a temple is a small shrine with no windows, which is the home of the god. A tall, curved shikhara, or tower, rises above the shrine, and a series of open porches are used for assemblies and religious dancing.

BUILDING IN ROCK

In the second century BC, Buddhist monks built a monastery at Ajanta by cutting artificial caves into the cliffs above the river (left). Carvers chipped off unwanted rock and carried it away leaving a building behind. The columned entrance of the vihara (right), where the monks lived, led to a rectangular room surrounded by galleries. Each monk had a square cave that opened onto a gallery. Stone walls and ceilings were rubbed smooth then covered with paintings or carved with sculpture. The monastery also had a chaitya, or meeting hall, where people gathered to worship and study.

Shrine

Assembly hall

MYTHS IN STONE
The lively sculptures on the outside of Kandariya Mahadeo represent many of the figures in stories from Hindu mythology.

KANDARIYA MAHADEO TEMPLE
More than 1,000 carved figures cover this eleventh-century temple in Khajuraho. At first glance it looks like a mountain of rock covered with rows of sculpture. The temple stands on a high platform with the shrine under the tall shikhara at one end and a deep entrance at the other. Processions move through a passageway, which wraps around the halls and shrine.

EMPLE FLOOR PLAN
Mathematical rules control the design of Hindu temples. Many small squares make up the floor plan of the temple. A square, which never changes, symbolizes the heavenly world.

Spiritual Journeys

Many different peoples live on the islands and peninsulas of Southeast Asia and they all have unique lifestyles. From early times, traders from all parts of Asia sailed along these coastlines and seaways. They traded goods and spread new ideas. Hinduism and Buddhism arrived from India, and Islam and Christianity came from further west to join the many local religions. Some of the greatest buildings in the area were built for Buddhist worship. Siddhartha Gautama, called the Buddha or the Enlightened One, founded Buddhism in India in the sixth century BC. He taught that every person could hope to achieve nirvana—a peaceful life beyond death where there is no suffering. Buddhists build stupas over relics of their spiritual leaders. A stupa is usually shaped like a dome and often stands on a square platform. Pilgrims walk along a path on the platform and meditate on the spiritual journey they will have to make to achieve nirvana.

BOROBUDUR
This Buddhist shrine has stood in a jungle on the island of Java in Indonesia since the beginning of the ninth century. It was built to look like a mountain. The stupa has eight stories or terraces. Pilgrims walk around each one on their way to the top.

ENTRY PAVILION
This magnificently carved gatehouse at Angkor in Cambodia leads to Angkor Wat, a twelfth-century Hindu temple. This temple may be the world's largest religious structure.

SMALL BUDDHAS
Statues of Buddha meditating under corbeled vaults line the corridors on the square terraces of Borobudur. The walls are carved with events from Buddha's life.

CORBELED DOMES

A simple corbeled dome is built by laying circles of stones flat on top of each other. One end of each stone juts out slightly over the room that is being domed. Pressure below and above one end of each stone holds it in place. A wide, heavy stone set on top locks all the layers below it in place.

At the top
A statue is hidden under the highest stupa.

ANANDA
This cluster of stupas in Pagan, Burma, partially hides Ananda, a white marble stupa rising in tiers above Pagan. This stupa shelters Buddhist relics.

The goal
At each compass point, pilgrims can look up a long flight of steps and glimpse their goal at the top.

WAT PRA KEO
The Royal Pantheon stands at the center of Wat Pra Keo in Bangkok, Thailand, the Buddhist area in the grounds of the royal palace. Ceremonies are held in the Royal Pantheon, which has eight gold statues of kings inside.

Discover more in Foundations of Religions

THE GREAT WALL OF CHINA
In the third century BC, the Chinese completed their first wall to keep out invaders from the north. This wall was rebuilt in the fourteenth century during the Ming dynasty. Five horses could walk side by side along the top. The wall still stretches for 1,500 miles (2,400 km) across northern China.

BEAMS AND BRACKETS

The roofs of Chinese temples are supported by a number of crossbeams. They rest on short posts set on the beam below, so that fewer posts clutter up the floor. Each column has brackets on top. A bracket is like a pair of arms that reach out from the sides of a post. Each "hand" of the bracket supports a beam. Sometimes a bracket "hand" holds another bracket to reach even further out from the post.

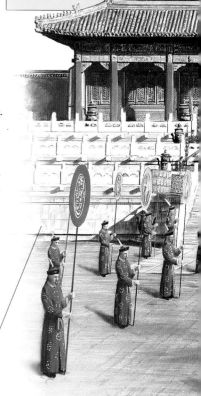

• EMPIRES OF THE EAST •

Center of the Universe

China is a unique country with a single civilization that has flourished for centuries in an area as large as Europe. The Chinese are known for their silk and porcelain and for their philosophies, Confucianism and Taoism. Philosophy and building have close ties in China. Both deal with how a person finds his or her place in the universe. Everyone is at the center of their own universe. A family's house marks the center of the family's universe. The palace of the Emperor stood at the center of China and of the universe as a whole. The Chinese were also influenced by other countries. Traders and travelers brought Buddhism from India along with Buddhist techniques for building with wood. Few ancient buildings survive today but we can still see what they looked like because important wooden buildings were later copied in stone. These buildings have elaborate wooden roofs covered with glazed tiles.

TEMPLE OF HEAVEN
At the beginning of each spring, the Emperor prayed for an abundant harvest at this round hall with its three tiled roofs.

The Earth
The hall, like all important buildings, stands on a platform that represents the Earth.

HALL OF SUPREME HARMONY
The Emperor arrives at the Hall of Supreme Harmony in the Forbidden City. The hall was originally built in the fifteenth century during the Ming dynasty as part of the Imperial Palace. The building as it stands now was rebuilt in 1696 during the Qing dynasty. The hall and the Emperor's seat face south because the Chinese believe that a south-facing seat shows honor and respect.

THE FORBIDDEN CITY
The Imperial Palace was called the Forbidden City because few people were allowed inside its powerful fortifications. This 600-year-old painting shows government officials gathering outside the gates of the city.

The heavens
The wide eaves are turned up at the ends and seem to make the roof float above the hall. The elaborate roof of a Chinese building represents the heavens.

In Harmony with Nature

HIMEJI CASTLE
Castles were built for the nobility in the sixteenth century. Himeji, in Hyogo, has a tall central tower, or keep, surrounded by smaller towers linked by corridors. Soldiers, called samurai, defended the castle with guns and arrows.

T he Japanese have learned to appreciate the beauty of natural things from a religion called Shinto—the way of gods. Shinto teaches that simple things in nature, such as tree or waterfall, may embody the forces of nature. The Japane have also learned from the Chinese. In the sixth century, Buddhism reached Japan from China by way of Korea. Chinese and Korean carpenters brought woodworking skills with them, which the Japanese soon adapted to their own taste. The Japar Buddhists also embraced the Shinto love of nature. Japanese wooden buildings are very delicate and have complicated deta Houses and temples are designed so they blend into nature, no stand apart from it. The people inside a building never feel cut off from the outdoors. A wall is often built so that it can be pushed to the side to open the room to a garden outside.

DID YOU KNOW?
When early Japanese governments moved to a new capital, they ordered that the most sacred temples be taken apart, moved and reassembled at the new location.

Brackets
These simple brackets make it possible to build this wall between the two roofs with just a few wooden posts set far apart.

Standing tall
A mast, which stands on a stone over the Buddhist relics, holds up this pagoda and its five wide roofs supported by brackets.

ORYUJI TEMPLE COMPLEX
ese Buddhist temples in Nara were uilt in about the year 700 and are the dest surviving wooden buildings in the orld. This pagoda marks the place where mbolic Buddhist relics are buried and onored. The Golden Hall on e left shelters a statue Buddha.

PHOENIX HALL
This villa in Uji opens onto surrounding gardens and pools. It became a temple of the Pure Land sect of Buddhists, who like to meditate in places that resemble the paradise their faith promises.

MADE TO MEASURE

For centuries, the Japanese have constructed buildings with standard parts made in just a few sizes. The distance between the pillars in a home or tea house fits the standard-size mats on the floor.
The frame for each panel of the wall is the same size as a mat. Paper covers each frame to form a panel of the wall. These panels slide to the side to make two rooms into one or they open a wall to the outside.

Lever arm
A complicated yet beautiful system of interlocking brackets and levers makes it possible for posts inside the building to support the weight of the wide eaves.

Heaven Meets Earth

Christians believe in Jesus, the son of God, and their religion is based on his life and teachings. Christians were persecuted for many years during the Roman Empire, but in AD 313 Emperor Constantine made the religion legal. He then left the city of Rome and moved east to Byzantium and established a new Christian capital named Constantinople, which is now Istanbul in Turkey. The Roman Empire later split into east and west. The western empire collapsed after it was invaded many times by nomadic tribes from central Asia, but the eastern part survived to become the Byzantine Empire. Christianity as it developed there is called Orthodox Christianity. Hagia Sophia was the magnificent Orthodox church in Constantinople and it inspired builders of Orthodox churches for centuries. The great dome at the center of the church represented the heavens. The floor below represented life on Earth.

DID YOU KNOW?

For many hundreds of years the dome of the Pantheon was the largest in the world. It measures 142 ft (43 m) across and is the same in height. Walls 16 ft (5 m) thick buttress the base of the dome.

THE PANTHEON
For many years, the dome of the Pantheon, in Rome, Italy, baffled modern engineers. They did not know how the ancient Romans managed to build such a large dome. Then they discovered that the dome was made of concrete that becomes lighter as it gets higher because each level is mixed with lighter stones such as volcanic pumice.

A new technique
Byzantine architects learned to construct round domes over square rooms. They used four pendentives, triangles cut from a circle, to provide a round base on which the dome rests. The pendentives shift the weight of the dome to the four supports below.

THE CHURCH TODAY
Four towers called minarets surround Hagia Sophia. They were added when the Islamic Ottoman Turks, founders of modern Turkey, conquered the Byzantine Empire and converted the church into a mosque.

Central dome
The large, lightweight dome is built of a single layer of brick and is 107 ft (33 m) wide. It has a row of arched windows cut into its base.

THE CONGREGATION
There were no seats in Hagia Sophia. Worshippers stood in the space beyond the columns—the men in the aisle below and the women in the gallery above—to listen to the singing of the Orthodox church service.

Half domes
A half dome at each end lengthens the nave to 250 ft (76 m) and buttresses the main dome by pressing against its base.

DECORATING WITH MOSAICS

A mosaic is a design or picture made up of small pieces of colored glass or stone that are mounted on a wall or ceiling. Mosaics seem to glow in the dimmest light. At one time, many colorful mosaics covered the ceilings of Hagia Sophia. Jesus (right) and other great leaders and heroes of Christianity were portrayed in mosaics against a gold background, which symbolized Heaven.

HAGIA SOPHIA
Byzantine architects began this church in Constantinople in 532, during the reign of Emperor Justinian. They finished it six years later and it soon became the model for future Orthodox churches. The clergy, as God's representatives, met the emperor, the worldly ruler, under the great domes, where the teachings of Jesus were read.

CHURCH OF THE NATIVITY
This church stands in an open-air museum of buildings near the city of Novgorod. Timber corbels support the gallery and demonstrate the remarkable skills of Russian carpenters. Although simpler in construction, it has much in common with St. Basil's.

Corbels

TRINITY ST. SERGIUS MONASTERY
Tsar Ivan the Terrible built the blue-domed cathedral for this monastery after the monks helped to fund his war against the Tartars. It was the most powerful of the Russian monasteries that were built inside fortifications, and it housed soldiers.

• EAST MEETS WEST •

The Russian Heritage

The first Russian people lived in the forests west of the Ural Mountains, where Europe meets Asia. Russian merchants traveled down the long rivers and across the Black Sea to trade furs with their powerful neighbor, the Byzantine Empire. They later adopted the religion of Byzantium and became Orthodox Christians. Mongol Tartars, nomads from Asia, conquered the area in the thirteenth century and ruled it for 200 years before the Russians succeeded in regaining their independence. In the sixteenth century, Tsar Ivan the Terrible attacked two Tartar states and took over their lands. He then set out to make Russia a great power. Russian carpenters were skilled builders of wooden houses and boats and learned from the Byzantines how to build with stone and brick. Both Russian and Byzantine churches have many domes, but Russian domes are mounted high above the roofs and shaped like onions to shed the heavy snow and rain that falls so far north.

THE KREMLIN
The city of Moscow grew out from this kremlin, or fortress. Palaces and cathedrals stand within its walls, as does a tall bell tower built by Tsar Ivan the Terrible.

DID YOU KNOW?
St. Basil's is named after Basil the Fool, a holy man who dared to criticize Tsar Ivan the Terrible. He was so popular that Ivan did not dare punish him.

A LOOK INSIDE
Frescoes of plants in colorful abstract patterns flow across the walls and ceiling of St. Basil's. These frescoes were rediscovered in 1954, hidden beneath layers of plaster.

198

Central tower
The tall towers in the center of early Russian churches were inspired by the high-roofed tents of the earliest Russians.

ST. BASIL'S CATHEDRAL
When Tsar Ivan the Terrible conquered the Tartars he celebrated by ordering his architects to build a cathedral that would be a "hymn of joy." Construction on St. Basil's in Moscow began in 1554. This colorful building was originally painted white.

Onion domes
Eight colorful domes, each with a unique shape, surround the tower. Each dome crowns a small chapel.

MAKING FRESCOES

A fresco painter spreads wet plaster on a wall or ceiling then paints it quickly so the paint sinks into the plaster before it dries. The only way to correct a mistake is to scrape off the layer of plaster and begin again. Sunlight slowly bleaches the color from frescoes, and moisture can cause the plaster to flake off. Shown here are frescoes painted on the outside of Voronet Monastery in Moldavia, Romania. They are unusual because they have survived the weather for more than three centuries.

Chapels
Access to the chapels is from a gallery around the cathedral, which is reached by two covered stairways.

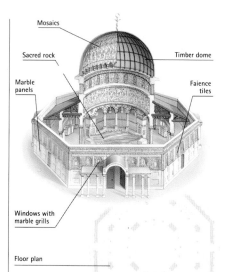

Mosaics

Sacred rock

Marble panels

Timber dome

Faience tiles

Windows with marble grills

Floor plan

DOME OF THE ROCK
This mosque in Jerusalem, Israel takes its name from the high dome built over a rock at a site that is sacred to Muslims. Pilgrims kneel to pray under the low roofs surrounding the rock. The mosque was completed in 691 and is the oldest surviving Islamic building. The decoration has been added in more recent centuries.

RESTING STOPS
Resting places for caravans were built along trade routes and in cities. Camels, donkeys and horses rested in the stables while merchants showed their goods.

Birth of Islam

In the seventh century, Mohammed, an Arab trader, founded a new religion called Islam, which means "surrender to the one God." Mohammed urged his followers to care for the poor and weak. He spent his life teaching in the cities of Medina and Mecca and converted most Arabs to his beliefs by the time of his death. Those who believe in Islam are called Muslims, and they stop whatever they are doing five times each day to pray. A leader calls them to prayer from the minaret, or tall tower, of the nearest mosque, an Islamic place of worship. Mosques are decorated with flowing Arabic script and geometric patterns. Pictures of animals or people never appear on a mosque because Mohammed was determined to stop the worship of false gods. The tall, arched doorways of mosques and their high domes are often pointed or in the shape of a horseshoe.

Garden paradise
The tomb opens onto a garden because Mohammed, who lived in a desert land, pictured paradise as a beautiful garden cooled by fountains.

PROTECTIVE TILES

Since ancient times, people in the Middle East have made tiles from baked clay. They glazed or covered the tiles with a mixture of liquid and glass and baked them again. These tiles were waterproof and were first used to protect sun-dried brick buildings from the rain. This picture is made up of faience tiles, which are tiles painted with pictures or other patterns before they are glazed.

THE MEMORIALS
A marble screen, carved to look like delicate lace, surrounds the memorials to Shah Jahan and his wife, who are buried below.

Call to prayer
Each minaret has a staircase that winds up to a balcony at the top of the tower. A crier calls Muslims to prayer from the balcony.

Double dome
An 80-ft (24-m) high dome sits inside the 200-ft (61-m) high pointed dome. The space between the two domes is empty.

[T]AJ MAHAL
[S]hah Jahan ruled an Islamic state [in] northern India. When his wife [M]umtaz Mahal died in 1630, he [b]uilt a magnificent tomb for [h]er—the Taj Mahal [in] Agra.

[P]assing into the tomb [th]e tall doorway set [de]ep in the wall is [de]corated with colored [m]arble that is cut and [pl]aced together like [pi]eces of a puzzle.

Tower of the Ladies

Hall of the Two Sisters

TALAKARI MADRASA
This high archway leads into an Islamic
university, or madrasa, built in the seventeenth
century in Samarkand, Uzbekistan. The dome
rises over the mosque. Glazed tiles decorated
and protected buildings, many of which
were built of sun-dried brick.

• EAST MEETS WEST •

Spread of Islam

From the eighth century, the Islamic faith spread out along
the trade routes. Islam reached China along with the camel
caravans that brought Chinese jade and silk west along the
Silk Road through Samarkand in the deserts of Central Asia. Islam
spread along the north coast of the Mediterranean Sea where Arab
traders exchanged Indian cotton and spices for glass and cloth to
sell in India, where the new faith also took root. Strong Islamic
states grew up along these trade routes. Rulers there built powerful
fortresses on hills overlooking their cities. The luxurious palaces
were designed to be cool during the heat of the long summers. They
had large courtyards filled with colorful flowers, pools of water and
fountains. The spray of the fountains kept the air fresh and cool.
Shady rooms opened onto the courtyards and were separated from
the outside by rows of columns.

COURT OF THE LIONS
Shaded walks surround this courtyard in the Alhambra. The fountain is surrounded by carved lions. The ruler held court in the Hall of Judgment at the end.

THE ALHAMBRA
The Alhambra, or red castle, was built in the fourteenth century on a high ridge above the city of Granada in Spain. Low buildings and garden courtyards form a palace at its center. Complicated geometrical patterns and religious sayings in graceful Arabic script are carved into the stucco on the walls.

CARVED DECORATION

Patterns carved into stucco decorated many surfaces. During this time, stucco was made from marble dust, wet lime and egg white. It was spread on a surface then allowed to dry before additional layers were added. The rows of small stalactites on the underside of the arch seen here were carved into seven layers of stucco.

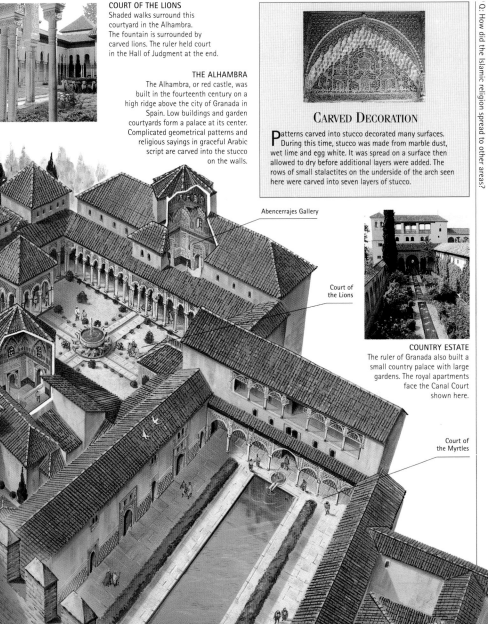

Abencerrajes Gallery

Court of the Lions

COUNTRY ESTATE
The ruler of Granada also built a small country palace with large gardens. The royal apartments face the Canal Court shown here.

Court of the Myrtles

Starting Over

L ife in the Roman Empire had become increasingly
insecure, even frightening by AD 300, as entire nations
of migrating nomads invaded the area. Christian
worship, which had been illegal, was permitted in the hope
that the Christians would convert the people of Rome and
unite them to face the invaders. The first churches were
rectangular buildings and could hold large crowds. The
churches looked much like the Roman emperor's own imperial
court, but the emperor's statue was noticeably absent. In its
place was a mosaic of Jesus. These early churches were built
with inexpensive trussed wooden roofs. Their stone columns
often came from abandoned buildings and did not always
match in height or style. The nomads gradually adopted
Christianity and the religion spread out across Europe.
People in the forest regions used wood to build their first
churches, which resembled their pre-Christian temples.

Legendary beasts
Dragons, which the Vikings
had always carved on their
boats and houses, appear
here on the gables of
St. Andrew's Church. They
are carved in wood and sit
alongside the Christian
symbol of the cross.

ST. ANDREW'S CHURCH
Built about 1150, this early
Christian church in Borgund,
Norway, is nearly 50 ft (15 m)
high. Norwegians built churches
the same way they built their
boats. Flat boards, or staves,
were attached to a wooden
frame to form the walls. Here
12 tall masts support the
highest of the three roofs.
A second roof covers a low
aisle and the lowest roof
shelters the porch.

BENEATH THE ROOF
Triangles are used in several
ways to strengthen the trussed
roof of St. Andrew's against the
wind and snow. Just under the
peak of the roof, two timbers
cross to form what is called a
scissor brace.

TRUSSED ROOFS

A wooden roof built with trusses will span a wide room without posts in between. Trusses consist of triangles that are rigid because the angles of a triangle cannot change unless the length of its sides changes first. A trussed roof may form one large triangle or many small ones. The roof at St. Botolph's church in Norfolk, England, stands on two short beams called hammer beams, carved with angels. The wall holds one end of a hammer beam. A brace attached to the other end completes the triangle.

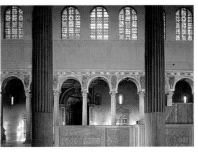

SANTA SABINA

This early church in Rome, Italy, was built shortly after the nomadic Visigoths captured the city in 410, destroying many buildings. This view across the church shows the Corinthian columns, which came from an abandoned building.

Jesus shown as a shepherd

COLORFUL MOSAICS

Early Christian churches were plain on the outside. But on the inside, the upper walls were decorated with mosaic portraits of great Christians and scenes from stories about Christianity. These mosaics are from early churches in Ravenna, Italy.

Three kings bring gifts to the baby Jesus

DID YOU KNOW?

Dragons are imaginary beasts that appear in the mythology of many different cultures and are found carved on buildings all over the world.

Discover more in Heaven Meets Earth

A Monastic Life

Religious communities lived in monasteries or abbeys and these were the chief centers of art and learning in Europe between the tenth and twelfth centuries. A single community often included several hundred men called monks, or women called nuns, who lived in a walled settlement. The monks and nuns divided each day between worship, study and work. Monasteries were often located in the frontier areas of Europe among various nomadic tribes. Monks built churches that looked like fortresses because they were seen as strongholds of God in an evil world. People came there seeking peace from the violence and wars around them. Living areas of a monastery opened off a cloister—a covered walkway built around a square garden. After the fall of the Roman Empire in the fifth century, many building techniques were forgotten. Stonemasons had to rediscover how to build arched stone vaults so the churches had fireproof roofs. These vaults were like those built by the Romans, so the style is called Romanesque.

Dormitory
In the winter, the monks sat by a fire in the warming room then went to bed in the unheated dormitory upstairs. A door in the dormitory led into the church because the monks worshipped in the middle of the night.

Refectory
Twice a day, monks sat down in the refectory to eat their simple meals.

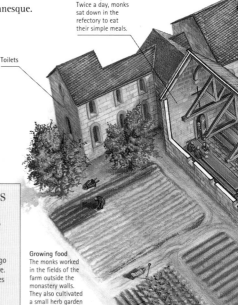

FEEDING THE COMMUNITY
On feast days, the monks roasted a wild boar over a fire in the center of the floor of this kitchen at Glastonbury Abbey in Somerset, England. They cooked other dishes for the large community over the four fireplaces in the corners of the room.

Toilets

MAKING PILGRIMAGES

People rarely traveled in these times, but they did make a trip, or pilgrimage, to pray at the burial place of a Christian saint. Some pilgrims walked hundreds of miles to reach their goal, such as Santiago de Compostela in Spain shown here. They slept in monastic guest houses and prayed at churches along the way. Pilgrims brought home new ideas from their travels, including new ways to build churches.

Growing food
The monks worked in the fields of the farm outside the monastery walls. They also cultivated a small herb garden where they grew the plants used to make medicines.

DID YOU KNOW?

People liked living near a monastery. It often provided the only hospital or school in an area and travelers stayed at guest houses located within the monastery.

Cellar
The monks made cheese and candles, cured hams and brewed ale to stock their cellar with all the things the community needed.

MARIA LAACH ABBEY
This twelfth-century Romanesque abbey west of Koblenz, Germany, has six towers decorated with dark stone. This scene reconstructs a typical monastery cloister next to the abbey church.

SEGOVIA ALCAZAR
Large towns often built
fortified castles to protect
them. This alcazar, or
castle, in Segovia, Spain,
guarded the town from the
top of an isolated rock.

CONWAY CASTLE
King Edward I of England built
Conway castle in Conway,
Wales. A workforce of 1,500
men completed most of the
castle between 1283 and 1287.
The king often arrived at the
castle's water gate by boat,
while townspeople and knights
entered across the drawbridge.

Motte

Keep

Bailey
This section is the
outer courtyard of
the castle.

NEUSCHWANSTEIN CASTLE
King Ludwig of Bavaria was
fascinated by castles. He built this
country palace, which looks like a
medieval castle, in the 1800s.

Palisade

Ditch

MOTTE AND BAILEY CASTLE
A simple castle was built by digging a ditch around a piece of
land then surrounding it with a wooden fence made of stakes
called a palisade. A hill, or motte, was built with dirt from the
ditch and might also have a palisade or ditch around it. The
knight of the castle lived on top of the motte in the keep.

A place to sleep
Royal bedchambers took up
two floors of the king's
tower. Treasure was hidden
in a cellar reached through
a trap door in the floor.

• THE RISE OF EUROPE •

Royal Fortresses

For hundreds of years after the collapse of the Roman Empire in the fifth
century, Europeans were often at war. Tribes fought tribes, and knights
fought among themselves until strong kings conquered them. The first
royal fortresses, or castles, were built of wood. The earliest stone castles were
single square towers called keeps, built on high ground and surrounded
by fences and ditches. As weapons changed, so did the design of
castles. People built stone walls 16 ft (5 m) thick to shield them from
battering rams, arrows, stones and a kind of burning tar. They filled
their ditches with water, turning them into moats, so the enemy could
not dig a tunnel under the wall and make it cave in. Archers
positioned themselves in round towers that bulged out from the walls.
This enabled them to fire their arrows on attackers from three sides.

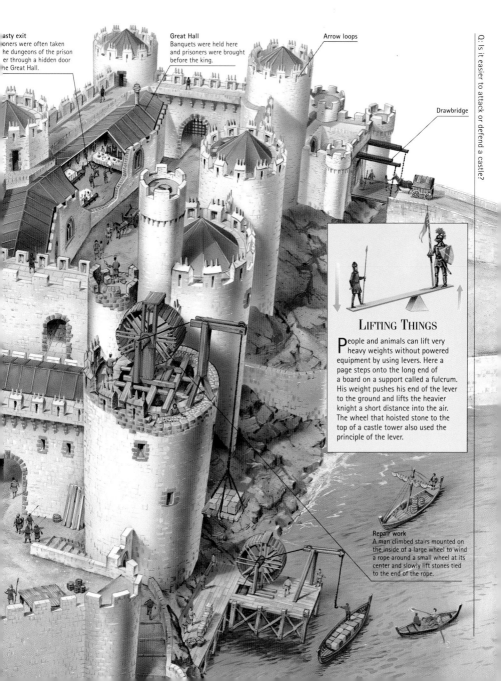

asty exit
oners were often taken
he dungeons of the prison
r through a hidden door
he Great Hall.

Great Hall
Banquets were held here
and prisoners were brought
before the king.

Arrow loops

Drawbridge

LIFTING THINGS

People and animals can lift very
heavy weights without powered
equipment by using levers. Here a
page steps onto the long end of
a board on a support called a fulcrum.
His weight pushes his end of the lever
to the ground and lifts the heavier
knight a short distance into the air.
The wheel that hoisted stone to the
top of a castle tower also used the
principle of the lever.

Repair work
A man climbed stairs mounted on
the inside of a large wheel to wind
a rope around a small wheel at its
center and slowly lift stones tied
to the end of the rope.

SALON DE MARS
Louis XIV received his court of royal attendants three evenings a week in a suite of six rooms including the Salon de Mars. Mars was the Roman god of war, and battle scenes decorate the ceiling of the room.

PALACE CHAPEL
The king sat in this private balcony in the palace's own church. Members of the court gathered below him before the altar.

Grand Palaces

By 1660, a century of destruction from invasions and civil wars had come to an end in Europe. In several countries, kings took all political and military power for themselves, and claimed the credit for peace. These rulers were called absolute monarchs as they claimed absolute, or unlimited, power over their people. Absolute monarchs were popular if they used their power to bring peace. Kings moved out of the cities that had rebelled against them, and built huge palaces in the countryside for their governments. These palaces had magnificent staterooms built for court ceremonies. They were decorated with tapestries, paintings and statues that praised the king's victories or recalled the glorious deeds of powerful Roman emperors and mythical Greek gods. The most skilled workers in each nation decorated their ruler's palace, which became a showplace of the country's finest products.

PALACE OF VERSAIL
The Palace of Versailles was the first unforti residence built in Europe since the fall of Roman Empire. In 1661, Louis XIV begar build the palace around a hunting lodge loved to visit as a boy. He demanded that most powerful men in the kingdom live ur his control there. Festivities and mili ceremonies took place in the courtya as shown in this paint

THE SUN KING

Louis XIV, France's absolute monarch, chose the sun as his symbol. He believed his government was as valuable to France as the sun is to the Earth, where life is only possible because of the sun's light and heat. Scientists had only recently recognized that the sun and not the Earth was the center of the solar system. At Versailles, all roads and garden paths radiate from the palace, much like the rays of the sun.

HALL OF MIRRORS

The long, narrow stateroom in the Palace of Versailles has tall windows on one long wall matched by mirrors on the opposite wall. The bright sunlight and reflections moving across the mirrors dazzle the eyes so that it is hard to see what is at the opposite end of the room. This room inspired a hall of mirrors in every palace in Europe for more than a century.

A GRAND ENTRY

Rulers of small countries tried to look more powerful by building impressive palaces much like the Palace of Versailles. A coach pulled by four horses could drive right into this palace at Würzburg in Germany, where its occupants would descend and sweep up the stairs to the staterooms above.

Discover more in Age of Happiness

THE AMALIENBURG
In the 1730s, the ruler of Bavaria built this hunting lodge in the gardens of the Nymphenburg Palace in Munich, Germany for his wife Maria Amalia. A stairway led from her bedroom to her shooting terrace on the roof.

NYMPHENBURG PALACE
Three cube shapes linked by bridges make up this palace in Munich, Germany, which is built in the Baroque style. The Amalienburg is hidden among the trees to the right.

• THE RISE OF EUROPE •

Age of Happiness

The eighteenth century was an optimistic and light-hearted age. New ideas in science had convinced people that famine and disease could be conquered. Happiness was the highest goal in life. Statues of saints dancing or angels swinging from vines sometimes decorated churches. People no longer feared nature and they enjoyed the ever-changing plant and animal life around them. Some retreated from the busy city life to houses in the countryside. They invited friends there to enjoy the surroundings, listen to music, discuss science or play games. These country retreats were built in a delicate new architectural style called Rococo. Pink, yellow and other pastel colors made every room look cheerful. Rococo architects wanted buildings to look light and weightless. One way they achieved this was by covering walls and ceilings with delicate vines made of stucco—a type of plaster they could mold by hand into shapes. Stucco birds and butterflies flew across ceilings.

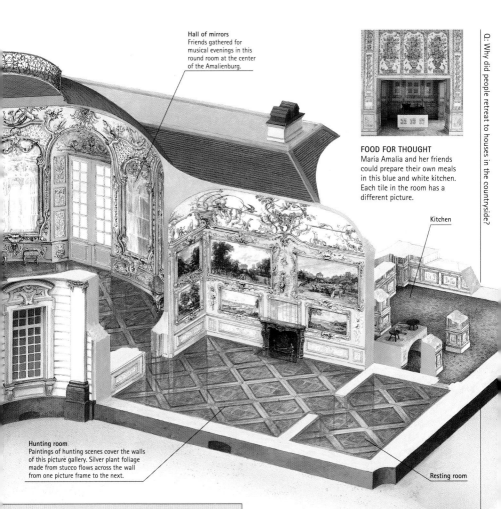

Hall of mirrors
Friends gathered for musical evenings in this round room at the center of the Amalienburg.

FOOD FOR THOUGHT
Maria Amalia and her friends could prepare their own meals in this blue and white kitchen. Each tile in the room has a different picture.

Kitchen

Hunting room
Paintings of hunting scenes cover the walls of this picture gallery. Silver plant foliage made from stucco flows across the wall from one picture frame to the next.

Resting room

DECORATING WITH MIRRORS

Glass was invented in Mesopotamia in ancient times. The Romans were the first to use glass for windows. In the seventeenth century, the French made plate glass by pouring liquid glass out onto a table and rolling it flat. Once the glass hardened, it was ground smooth and polished to make mirrors and large windowpanes. Mirrors often decorated the inside of Rococo buildings.

DE LUXE KENNELS
Hunting dogs slept in the kennels at the base of the walls in this room in the Amalienburg. Guns were stored in the cabinets above.

Discover more in Inspired by Nature

COALBROOKDALE BRIDGE
The world gained a new construction material when inexpensive iron was developed. In 1779, the English built Coalbrookdale bridge in Shropshire, which was the first iron bridge to be constructed.

DID YOU KNOW?
The first skyscraper was built in 1884 in the city of Chicago, Illinois. It was only ten-stories high.

EIFFEL TOWER
This iron and steel tower was built for the Paris Exposition of 1889. When radio was invented, the tower began its long career as an antenna. It carried the first transatlantic radio–telephone call.

Reach for the Sky

Skyscrapers are a product of the Industrial Revolution, which began in England in the eighteenth century. New inventions revolutionized the way people lived. Steam engines, and later electricity, made enough energy available to do many more times the work that people and animals had done before. A new method of smelting iron produced huge quantities at low prices. Other inventions gave builders steel, a material even stronger than iron. Cities grew and skyscrapers provided a solution to the problems of overcrowding because they take up little space on the ground. Skyscraper frames were first built with iron, then with steel. New engines powered elevators to hoist people to the top. The weight of a tall building can easily cause it to sink or lean, so the early skyscrapers were usually built on solid rock. This is why so many were built on Manhattan, a rocky island in New York City.

| St. Peter's Basilica, Italy 1612 453 ft (138 m) | Great Pyramid of Khufu, Egypt 2700 BC 479 ft (146 m) | Eiffel Tower, France 1889 984 ft (300 m) | Empire State Building, USA 1931 1,250 ft (381 m) | Sears Tower, USA 1974 1,453 ft (443 m) | CN Tower, Canada 1976 1,804 ft (550 m) |

THE TALLEST OF THEM ALL

For more than 4,500 years, the Great Pyramid of Khufu in Egypt was the tallest building in the world. Then the Eiffel Tower was built in France in 1889. In many parts of the world today, skyscrapers and towers continue to grow taller and taller.

Crown
The Art Deco style, a novelty of the 1930s, inspired the triangle-shaped windows. These are set within tiers of arches on the crown.

LIFE AT THE TOP
Native Americans were some of the earliest construction workers on skyscrapers. They worked at great heights while standing only on 8-in (20-cm) wide steel beams.

OING UP?
team engines powered e first elevators, which ere used only for eight. The first assenger elevators were stalled in 1857 after a ay was found to stop em from falling if a able broke. By 1889, ey were powered by ectric motors. The evator doors of the hrysler Building (above) e decorated in the Art eco style.

Core
A strong frame is built inside the building for the elevators. This frame also helps the building resist the pushing and twisting forces of the wind.

CHRYSLER BUILDING
Walter Chrysler built this 77-story skyscraper in New York City during the worldwide Depression of the 1930s. It provided much-needed employment for many construction workers. The building was the headquarters for his automobile empire and a monument to his success.

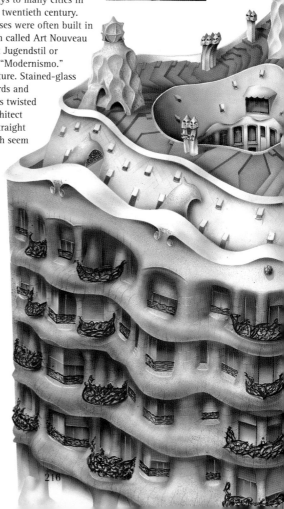

Inspired by Nature

New inventions had brought conveniences such as street lights and railways to many cities in Europe by the turn of the twentieth century. Railway stations, parks and houses were often built in an original new style the French called Art Nouveau or "New Art." Germans named it Jugendstil or "Style of Youth" and Spaniards simply said "Modernismo." Art Nouveau architects were inspired by nature. Stained-glass ceilings came alive with brightly colored birds and flowers, and the iron in balcony railings was twisted and tangled like vines. The Modernismo architect Antonio Gaudí copied nature by avoiding straight lines and right angles in his buildings, which seem to have been sculpted from lumps of clay. Such unique shapes were possible because the Industrial Revolution gave builders iron, steel and concrete to work with.

CASA MILÁ
In 1906, Gaudí began work on an apartment building, the Casa Milá in Barcelona, Spain. His friends said it looked like a wave that had turned to stone. Stables and parking for horse-drawn carriages were located in the basement.

BY THE SEA
Stairs curve up the lower wall of the two courtyards in Casa Milá. The apartment doors open onto the landings of the stairs. Gaudí designed the stairway to look as though it had been hollowed out by the pounding of waves from the sea.

COME INSIDE

In this Casa Milá apartment, the walls of the parlor curve around corners and up into the ceiling. The ceiling was inspired by the ripples left by the tide in the sand.

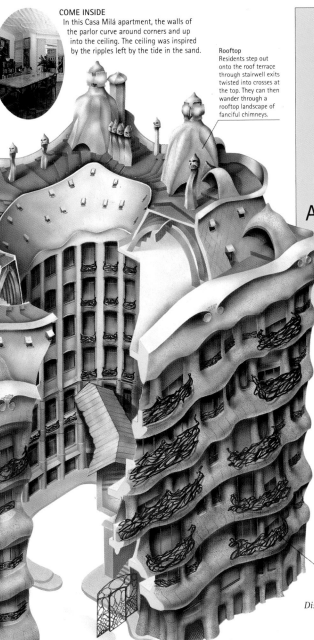

Rooftop
Residents step out onto the roof terrace through stairwell exits twisted into crosses at the top. They can then wander through a rooftop landscape of fanciful chimneys.

SCULPTING A BUILDING

Art Nouveau architects rebelled against the practice of copying historic buildings such as the Parthenon in Greece. They studied nature and copied the shapes they saw there. The Belgian Victor Horta, a founder of Art Nouveau, imitated plant life in glass and metal. At Horta's Tassel House in Belgium, shown here, light from a colorful glass dome shines down on the twisted metal vines in the stairwell.

NOTRE DAME DU HAUT

The shape of the surrounding hills inspired the design of Notre Dame du Haut in Ronchamp, France. The Swiss architect Le Corbusier built this church in the 1950s.

Balconies
Casa Milá was inspired by the seashore. The iron guards on the balconies look like bundles of seaweed.

Discover more in Age of Happiness

FALLINGWATER

The architect of the Guggenheim, Frank Lloyd Wright, also designed houses that nestle into their natural surroundings. Fallingwater in Bear Run, Pennsylvania, copies the shelves of rock over which the waterfall beneath it flows.

GUGGENHEIM MUSEUM

The Solomon R. Guggenheim Foundation built its museum of modern art between 1956 and 1959 in New York City. The museum is constructed from a spiral of reinforced concrete in the shape of a hollow funnel. Its unusual shape contrasts sharply with the straight lines of the skyscrapers built around it.

New addition
A tower was added to t[ ] original museum in 199[ ] It provides additional o[ ] and exhibition space.

• THE INDUSTRIAL WORLD •

Adventurous Shapes

After the Second World War, many nations around the world made great economic recoveries. People felt confident and full of adventure. Architects of the 1950s and 1960s designed buildings with unusual shapes to reflect this new confidence. Some buildings have simple geometric shapes, while others look like huge abstract sculptures. Many of these structures could not have been built without the invention of a new material called reinforced concrete. A roof of reinforced concrete will bridge a wide room without any other supports in between. Steel and reinforced concrete are so flexible that walls and ceilings can be built into any shape. Sometimes the walls of a room or ceiling were built so they curved away from the people inside the building. This was done to give people a feeling of exhilaration–the spirit of the times in which these buildings were designed and constructed.

ROCK AND ROLL HALL OF FAME

This museum in Cleveland, Ohio, was designed by I. M. Pei and opened in 1995. It has rectangular bridges and circles and triangles of glass, granite and white-painted steel.

At the top
Museum visitors ride elevators to the top of the Guggenheim where they step out into the huge hollow shell of the museum. Sunlight from the glass dome floods down to every level of the museum.

ON EXHIBIT
Visitors to the museum wind their way down the long spiral ramp and stop to look at the modern paintings, hung on walls that lean outwards.

A Better Building Material

Reinforced concrete is made by pouring concrete into molds around steel rods or wire mesh. This kind of concrete is no longer brittle, so roof supports can be set farther apart. Concrete reinforced with wire mesh is used to build thin, lightweight ceilings and walls and can be easily molded into any shape desired. The bowl shape of the assembly room in the Palace of the National Congress in Brasilia, Brazil (below), is possible because of reinforced concrete.

Walking on air
The reinforced concrete ramps have no supports below them at all. Only the walls hold them in place.

Discover more in A New Design

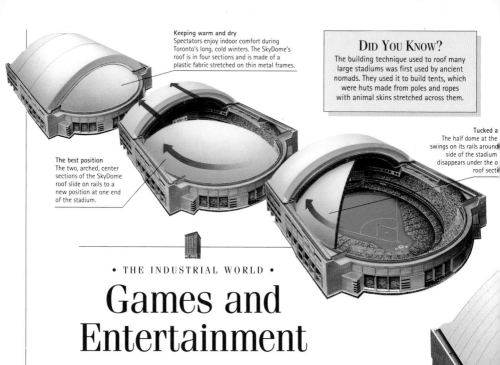

Keeping warm and dry
Spectators enjoy indoor comfort during Toronto's long, cold winters. The SkyDome's roof is in four sections and is made of a plastic fabric stretched on thin metal frames.

The best position
The two, arched, center sections of the SkyDome roof slide on rails to a new position at one end of the stadium.

Tucked a
The half dome at the swings on its rails around side of the stadium disappears under the o roof secti

• THE INDUSTRIAL WORLD •

Games and Entertainment

Since ancient times, people have gathered in large public stadiums or arenas for entertainment. Crowds still flock to these large buildings to watch sporting competitions, concerts and other special events. Many stadiums are open to the sky and spectators are at the mercy of the weather. They face the heat in summer and the freezing cold in winter. Events can be canceled if it rains or snows. But people in many countries no longer have to consult the weather report before an event, because some stadiums are now covered by roofs. The development of new synthetic materials, particularly plastics, has made these roofs possible. Tough, lightweight plastics are stretched tight on thin frames, much like umbrellas. These roofs, which come in many shapes and sizes, can cover even the largest stadiums. They shelter the crowds and the players or performers, and no-one's view is blocked by roof supports standing on the playing field.

SPORTS COMPLEX
Many large stadiums are built originally for special events. This stadium in Seoul, South Korea, was built for the 1988 Summer Olympic Games. It seats 100,000 spectators.

OLYMPIC STADIUM
Spectators in the Olympic Stadium in Munich, Germany, sit under a clear canopy of panels made from glass and plastic. The panels hang from a square mesh made of steel cable and are attache to cables that are stretched between 56 reinforced concrete poles and the ground.

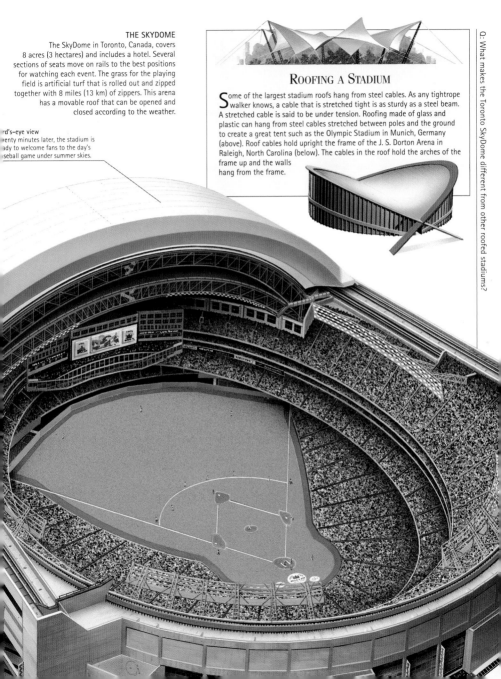

THE SKYDOME

The SkyDome in Toronto, Canada, covers 8 acres (3 hectares) and includes a hotel. Several sections of seats move on rails to the best positions for watching each event. The grass for the playing field is artificial turf that is rolled out and zipped together with 8 miles (13 km) of zippers. This arena has a movable roof that can be opened and closed according to the weather.

rd's-eye view
enty minutes later, the stadium is
ady to welcome fans to the day's
seball game under summer skies.

ROOFING A STADIUM

Some of the largest stadium roofs hang from steel cables. As any tightrope walker knows, a cable that is stretched tight is as sturdy as a steel beam. A stretched cable is said to be under tension. Roofing made of glass and plastic can hang from steel cables stretched between poles and the ground to create a great tent such as the Olympic Stadium in Munich, Germany (above). Roof cables hold the frame of the J. S. Dorton Arena in Raleigh, North Carolina (below). The cables in the roof hold the arches of the frame up and the walls hang from the frame.

Q: What makes the Toronto SkyDome different from other roofed stadiums?

SYDNEY OPERA HOUSE

In 1957, a Danish architect Jørn Utzon won a contest to design the Opera House in Sydney, Australia. But it took him six years and the help of engineers and early computers to come up with a way to actually build it. Here the building is shown during different stages of construction.

Giant cranes
Three cranes arrived from France. Each required 30 trucks to transport it to the building site where they were assembled.

Roof ribs
Computers showed that the roof originally planned might have collapsed, so the design changed. The prestressed concrete roof was made by casting concrete pieces that were placed on the building before steel cables were threaded through them and pulled tight.

• THE INDUSTRIAL WORLD •

A New Design

The construction of an innovative building is difficult and often requires new techniques and special building materials. Many unexpected problems arise no matter how careful the advance planning may be. The architects and engineers building the Opera House in Sydney, Australia, faced major obstacles. The design was so innovative that it took several years for engineers to work out a way to actually build it. Specialists in sound, called acoustical engineers, advised on how the chosen building materials would affect the quality of sound. Metal, plastic and glass from around the world was used in the building. Manufacturers designed essential new equipment and construction workers learned new skills to build the Opera House. There were many unexpected costs and delays in construction. Sixteen years later, the architect's imaginative design became a unique masterpiece, which today is recognized throughout the world.

PRESTRESSED CONCRETE

The Opera House roofs were designed to be made of prestressed concrete much like the Trans World Airline building at Kennedy Airport in New York City, seen here. The steel in prestressed concrete is stretched tight so that it squeezes the concrete around it. The Trans World Airline building has a thin, lightweight roof, which was made by pouring concrete over a tightly stretched wire mesh.

Concrete sections
Each rib was assembled from concrete sections cast at the building site in reusable molds.

222

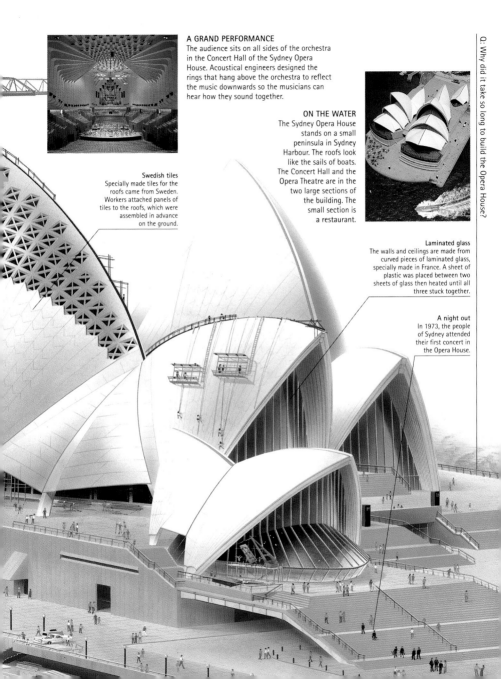

A GRAND PERFORMANCE
The audience sits on all sides of the orchestra in the Concert Hall of the Sydney Opera House. Acoustical engineers designed the rings that hang above the orchestra to reflect the music downwards so the musicians can hear how they sound together.

ON THE WATER
The Sydney Opera House stands on a small peninsula in Sydney Harbour. The roofs look like the sails of boats. The Concert Hall and the Opera Theatre are in the two large sections of the building. The small section is a restaurant.

Swedish tiles
Specially made tiles for the roofs came from Sweden. Workers attached panels of tiles to the roofs, which were assembled in advance on the ground.

Laminated glass
The walls and ceilings are made from curved pieces of laminated glass, specially made in France. A sheet of plastic was placed between two sheets of glass then heated until all three stuck together.

A night out
In 1973, the people of Sydney attended their first concert in the Opera House.

A Challenging Future

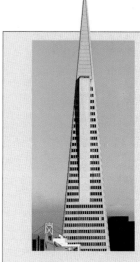

Large international corporations are building impressive and functional buildings. Many of these giant buildings are in California and Japan and other countries that border the Pacific Ocean. These areas are threatened by earthquakes that can shake buildings to pieces and destructive winds that can twist them apart. To protect the people inside, builders in these areas built low-rise buildings out of lightweight materials. As space became scarce, corporations needed high-rise buildings, so architects and engineers designed skyscrapers to withstand earthquakes. These now dot the skylines of many cities in the world. But buildings are not only threatened by natural disasters. Modern lifestyles also affect their future. Pollution from the fuel used to power cars and heat homes, for example, corrodes and weakens concrete. Each new generation of architects and engineers will face many new obstacles and technological opportunities. They will create different building materials, methods and architectural styles to meet these challenges.

HOW TO BEAT AN EARTHQUAKE

The TransAmerica Building in San Francisco is wide at the bottom and narrow at the top so that it will not topple over if the earth shakes. A high-rise building can also be built on top of a thick cushion made of rubber and steel, which absorbs earthquake shocks. Some buildings have steel tubes that push and pull the walls to keep them in their normal place even while the building shakes.

DID YOU KNOW?

Many perfectly sound buildings have been torn down because no-one could find a use for them. Today's buildings are more flexible. Even walls can be moved so a building can change as often as a business does.

STACKING TRIANGLES
Destructive winds cannot twist Hong Kong's Bank of China out of shape because its frame is a series of rigid triangular braces. The bank blends into the sky, which is seen reflected in its mirror-covered walls.

CORPORATE MONUMENT
Lloyd's of London conducts its insurance business from this modern corporate building in London. Stairs, elevators and other services occupy the many towers, which surround a large atrium. People can escape by the stair towers on the outside if there is a fire in the building.

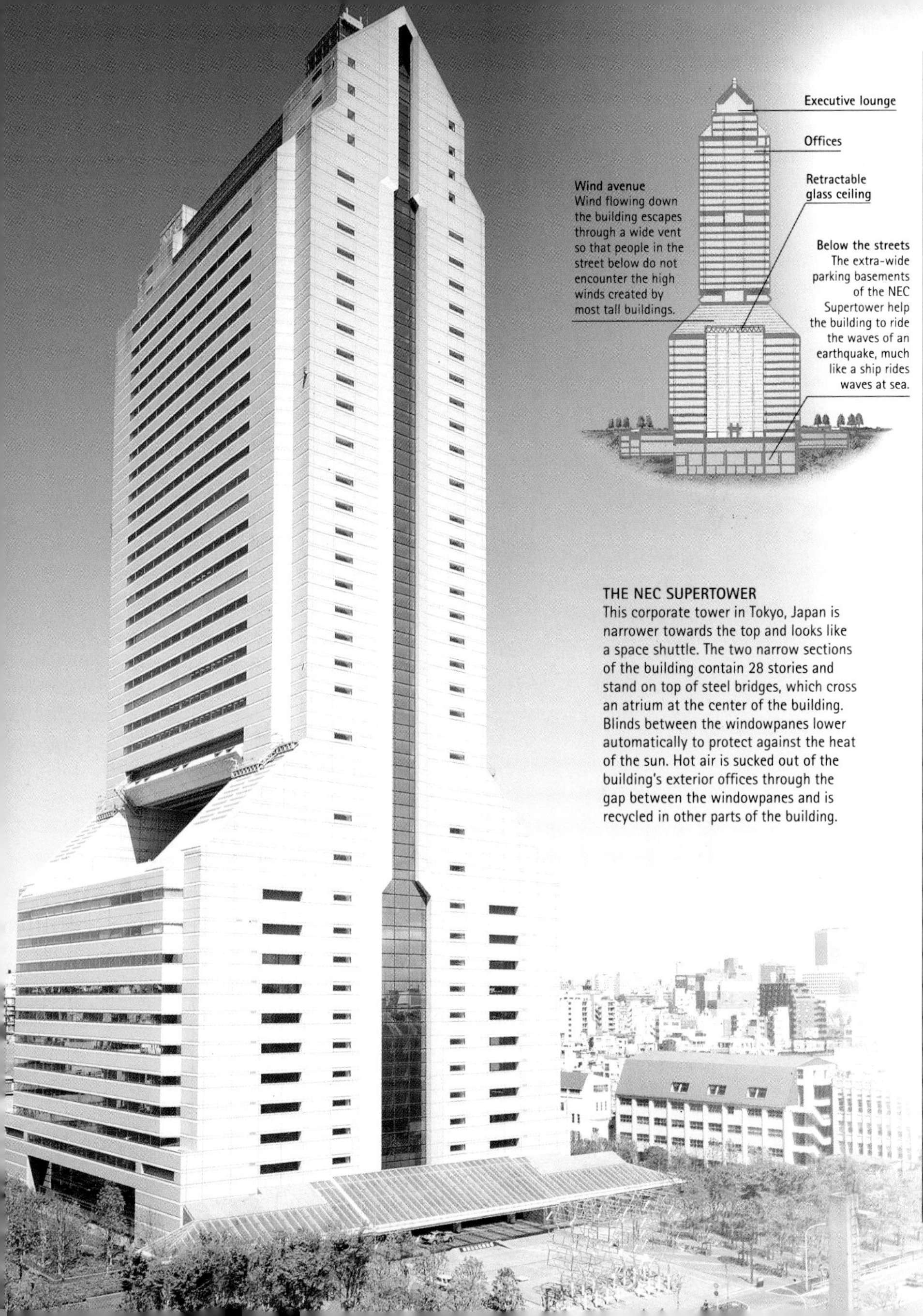

Executive lounge

Offices

Wind avenue
Wind flowing down
the building escapes
through a wide vent
so that people in the
street below do not
encounter the high
winds created by
most tall buildings.

**Retractable
glass ceiling**

Below the streets
The extra-wide
parking basements
of the NEC
Supertower help
the building to ride
the waves of an
earthquake, much
like a ship rides
waves at sea.

THE NEC SUPERTOWER

This corporate tower in Tokyo, Japan is
narrower towards the top and looks like
a space shuttle. The two narrow sections
of the building contain 28 stories and
stand on top of steel bridges, which cross
an atrium at the center of the building.
Blinds between the windowpanes lower
automatically to protect against the heat
of the sun. Hot air is sucked out of the
building's exterior offices through the
gap between the windowpanes and is
recycled in other parts of the building.

— A Global View —

A great building can reflect many different ideas and styles and tells us about the beliefs and values of the people who designed and built it. You may not like the way a particular building looks or even understand why it is thought to be a great building. But once you learn about the lives and thoughts of the people who built it and about the time and place in which it was built, you may find something about the building that you do like. Throughout history, people have traveled from country to country, across waterways and mountains, carrying ideas about building with them. The pictures on this map show the locations of the major buildings that are featured in this book. But these are just some of the great buildings in the world. There are many others to discover.

St. Andrew's, Borgund, Norway

St. Basil's, Moscow, Russia

Notre Dame, Paris, France

Palace of Versailles, Paris, France

Maria Laach Abbey, Koblenz, Germany

Amalienburg, Munich, Germany

Conway Castle, Wales

EUROPE

Toronto SkyDome, Canada

Casa Milà, Barcelona, Spain

Ishtar Gate Babylon, Ir

Chrysler Building, New York City, USA

NORTH AMERICA

Parthenon, Athens, Greece

Guggenheim Museum, New York City, USA

Baths of Caracalla, Rome, Italy

Hagia Sophia, Istanbul, Turkey

Alhambra, Granada, Spain

St. Peter's, Rome, Italy

Pyramid of the Magician, Uxmal, Mexico

AFRICA

ATLANTIC OCEAN

SOUTH AMERICA

In Ancient Times

Ancient buildings around the world looked different because they were shaped by the building materials available. Each material inspired a different construction method. People in the forest areas of Europe built in wood. Wood was scarce but stone was plentiful along the Mediterranean coast. The Egyptians there built great stone pyramids and temples, which still stand today. The ancient Greeks built their temples of white marble and the ancient Romans developed concrete to construct their huge buildings. The people who lived near the deserts between the Mediterranean Sea and the Indian Ocean learned to mold earth into bricks and melt sand to make glass. China and Japan built their early temples with timber. The brackets that supported their temple roofs were first used on ancient temples in India. Early people in parts of America used brick and stone as well as a type of concrete to construct their temples.

ASIA

Hall of Supreme Harmony, Beijing, China

NEC Supertower, Tokyo, Japan

Horyuji Temple, Nara, Japan

j Mahal, gra, India

Kandariya Mahadeo, Khajuraho, India

PACIFIC OCEAN

Thatched hut, Trobriand Islands, Papua New Guinea

Borobudur, Java, Indonesia

DIAN CEAN

Sydney Opera House, Australia

AUSTRALIA

Sports
and Games

How did the Olympic Games begin?

Which president of the United States
was an accomplished wrestler?

Who was the first female golfer in history?

Contents

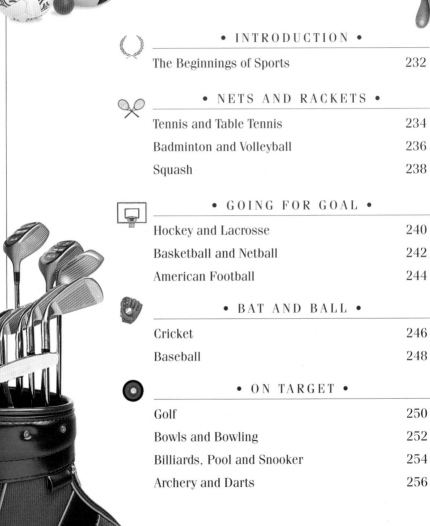

• INTRODUCTION •

The Beginnings of Sports

People have always enjoyed competing against each other. In prehistoric times, a hunter who could run fast and throw a stone or spear accurately was a valued member of the tribe. The ancient Greeks included athletics in many of their religious festivals. In 776 BC, they held a festival of sport to honor Zeus, the greatest of the Greek gods. Athletes from all over the country gathered in a stadium in the valley of Olympia to test their speed, strength and skill in the first Olympic Games. These games were held every four years, and for a long time only males were allowed to watch and take part in the races. The games continued for several centuries after the Romans conquered Greece, until the Roman Emperor Theodosius I ended them in AD 394. Almost 1,500 years later, the ruins of the Olympic stadium were discovered by archaeologists. Frenchman Baron Pierre de Coubertin suggested holding a modern, international Olympic Games. The first games of the new era of the Olympics were held in Athens in 1896.

HAIL THE CHAMPION
Today's victorious Olympic champions receive gold medals. Sporting heroes of ancient Greece were crowned with wreaths made from laurel leaves, as shown on this vase.

SPORTING SPECTACULAR

In 680 BC, four-horse chariot races were added to the program of the 25th Olympic Games. As many as 40 chariots crashed, jostled and maneuvered their way around the course marked out in the hippodrome.

THE OLYMPIC TORCH

The lighting of the flame is the high point of the opening ceremony at the modern Olympic Games. Since 1936, this custom has served as a reminder of the beginnings of this festival. A lighted torch is carried by relay runners from Olympia in Greece, site of the original Olympics, to the city where the modern games are to be held. This torch is used to light the Olympic flame that burns above the stadium throughout the festival. The flame is seen as a symbol of nations and athletes competing peacefully in the spirit of sport.

DISCUS HERO

Among the sports to have survived from the earliest Olympics is discus throwing. This ancient Greek vase shows a discus thrower placing or withdrawing the peg that is used to mark the distance the discus has been thrown.

COURT ACTION
Each major tennis tournament is divided into men's or women's programs with championships decided in singles and doubles play. Often, there is also a mixed doubles program in which men and women play together. Here, on a grass court, a women's doubles match is in progress.

• NETS AND RACKETS •

Tennis and Table Tennis

The knights who returned to France from the Crusades in the Middle East 800 years ago brought with them a ball game we now know as tennis. Europeans played this game indoors, often inside a monastery or palace. It was known as royal, real or court tennis. At first, the ball was struck with the bare hand. Later, gloves and then a bat or paddle were used. The introduction of a mesh of strings in the sixteenth century enabled royal tennis players to hit the ball with more power. A game called sphairistike, or lawn tennis, was introduced into England by Major Walter Wingfield in 1874. But the real birth of modern tennis was at Wimbledon, England, in 1877 when the All England championship, or Wimbledon tournament, was held for the first time. Today, tournaments usually take place on clay or synthetic playing surfaces, although a few are still held on grass. Tennis can be played as singles—with one player at either end of the court, or as doubles—with two players teaming up to play another two.

TENNIS EQUIPMEN
Tennis players use a stringed racket and felt-covered ball while table-tenni players use a solid, woode paddle and a lightweigh celluloid bal

SPEED AND SKILL
With a net and miniature court area, table tennis offers the excitement of tennis on a smaller scale.

Tennis

Table tenni

DID YOU KNOW?

Until the 1960s, most tennis players were amateurs and could not earn money from the game. Today, top players can win millions of dollars in prize money by playing on the world professional circuit.

PLAYING PING PONG

Table tennis, or ping pong, was developed in the late nineteenth century as a miniature indoor version of tennis. The trademark Ping-Pong was introduced by a supplier of early table-tennis equipment to describe the noise the ball made: "ping" when it hit the bat and "pong" when it bounced off the table. The first balls were made of rubber or cork covered with cloth. The introduction of the celluloid ball helped to make table tennis an exciting test of reflexes and skill. Today, it is played all around the world. It is especially popular in China where children learn the game on concrete tables built in schoolyards.

TENNIS COURT
A tennis court is 77.96 ft (23.77 m) long and 35.98 ft (10.97 m) wide. The net is 2.98 ft (0.91 m) high at the center.

Doubles side line
Singles side line
Service court
Center service line
Service court
Net
Base line

A GAME FOR KINGS
Royal tennis was popular from the twelfth century on. In the sixteenth century, King Henry VIII of England enjoyed the game so much that he built a court at his Hampton Court Palace.

235

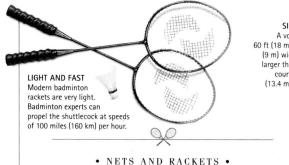

LIGHT AND FAST
Modern badminton rackets are very light. Badminton experts can propel the shuttlecock at speeds of 100 miles (160 km) per hour.

SIMILAR GAMES
A volleyball court is 60 ft (18 m) long and 30 ft (9 m) wide. This is much larger than a badminton court, which is 44 ft (13.4 m) long and 20 ft (6.1m) wide.

Back line and service line

Attack line

Center line and net

Attack line

Side line

Boundary and lon singles service line

Doubles long service line

Short service line

Net

Singles side line

Center line

Badminton court

Volleyball court

• NETS AND RACKETS •

Badminton and Volleyball

In the game of badminton the players use stringed rackets to hit a shuttlecock high over a net. A player scores a point by grounding the shuttlecock on their opponent's side of the net. Games similar to badminton were played in ancient China and India. The sport was not given its name, however, until the 1870s when the Duke of Beaufort, a famous English sportsman, invited guests to play the game at his country estate—Badminton. The game of volleyball also has a high net, but its players do not use rackets. Instead, they can use any part of the body above the waist to play the ball. Most shots, however, are played with one or both hands, and players must not hold or throw the ball. There are two forms of the game. Beach volleyball is played outdoors with two players on each side. The standard version of volleyball is held on indoor courts with six players on each side.

TEAM PLAY
The beach volleyball player on the left has hit the ball in a shot known as a "spike." One defensive player on the opposing side is leaping to block the shot, while the other positions herself to retrieve the ball if the block fails.

SLAMMING THE SHUTTLE
Badminton is a fast game. Players have to be very agile and have quick reflexes. It can be played by two players (singles) or as a doubles contest with four players, as shown here.

WORLD STAGE
Volleyball is very popular around the world. National teams compete every four years the Olympic Games and every two years for the World Cup. Japan, China, Cuba and the United States are all leading volleyball nations.

THE SHUTTLECOCK

Instead of a bat or a racket, ancient Chinese of the Han Dynasty played a version of badminton using their feet. The earliest shuttlecock was made from a piece of cork in which feathers were embedded. It was so unsteady in the air that it was nicknamed "the wobbler." Designers tried to come up with a shuttle that would perform in a predictable manner. They invented the barrel-shaped shuttle in 1900 and then the straight-feather shuttle nine years later. The modern "birdie" (as it is called in the United States) is made of 16 goose feathers set in a domed cork. It has a small lead weight in the base to make it move freely through the air. Mass-produced shuttles are often made of plastic.

• NETS AND RACKETS •

Squash

The sport of squash can be traced back to the 1700s to a game called racquets. It was first played by prisoners serving a sentence in London's Fleet Prison for not paying their debts. Later, the English upper classes began to play the sport. When students at Harrow, an English Public School, played the game with a soft rubber ball that could be squashed, the game became even more popular. Later, indoor courts with four walls were built. In squash, the ball can come in contact with any of the four walls, providing it hits the front wall with every shot played. The ball can also bounce off the floor, but not the roof. The game is played socially and as a competitive sport, with world championships held each year. A game very similar to squash, called racquetball, was invented in 1950 by American Joe Sobek in Greenwich, Connecticut. Like squash, racquetball is played on a four-walled court, but it is considered easier to play because its larger ball bounces higher than the smaller squash ball.

WATCH AND PLAY
Major championship matches are played on courts with walls made of thick, transparent plastic. Spectators and television cameras have a clear view of the game, but the players have a clouded view of the outside so that they are not distracted by people moving around.

THE COURT

A squash court is a fully enclosed area, 21 ft (6.40 m) wide and 32 ft (9.75 m) long. The floor and all four walls are marked with lines 2 in (5 cm) wide that indicate the playing area.

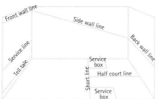

Front wall line
Side wall line
Back wall line
Service line
Service box
Tell tale
Short line
Half court line
Service box

A SPEEDING BALL

Pelota, or jai alai, is perhaps the world's fastest ball game. It is similar to squash but played on a much larger court. Instead of a racket, the players use a 3 ft (90 cm) curved basket that is attached to a glove strapped tightly at the wrist. This enables them to catch and fling the ball at extremely high speeds. Contestants sometimes wear helmets in case they are struck by the ball. The game is popular in Mexico and other Spanish-speaking countries. Although pelota means "little ball," the modern version is played with a ball the size of a baseball.

EQUIPMENT

A small ball and a lightweight racket with a long handle are used in squash. Racquetball players use a heavier racket with a shorter handle, and a larger ball.

Squash

Racquetball

HIGH ENERGY

Squash is a high-energy sport, and players need to be fit and agile. It is usually played by two people who score points by keeping the ball in play while hitting it against a wall. Players can use backhand and forehand strokes similar to tennis.

Discover more in Tennis and Table Tennis

• GOING FOR GOAL •

Hockey and Lacrosse

The game of hockey developed from the many basic games that used a stick and a ball. The ancient Greeks, Egyptians, Native Americans and Aztecs all played a similar game, as did the Irish who called it hurling. These early games were very rough, and often resembled a brawl more than a sport. The rules to the modern game of field hockey were officially introduced in England during the 1880s. It is played by two teams of men or women with 11 players on each side. Like soccer, the object of the game is to score by hitting the ball into the opposing team's goal. Ice hockey is a combination of field hockey and ice-skating and has just six players on each side. It is a very physical game and offenses, such as tripping and roughing, are penalized. The sport was devised in Canada. At first, a rubber ball was used. This was later replaced by a hard, flat rubber disk known as a puck, which can skim across the ice at great speed.

Field-hockey stick

Standard ice-hockey stick

Goal-tender's ice-hockey stick

Lacrosse stick

Side line

Goal
Goal line
Penalty spot
Shooting circle
Center line

Goal
Goal line
Face-off circle
Blue line
Red line

PLAYING AREAS
The field-hockey field (left) is 300 (91.4 m) long and 180 ft (54.8 m) wide. The ice-hockey rink (above) can be 184–200 ft (56–61 m) long and 85–98 ft (26–30 m) wide.

240

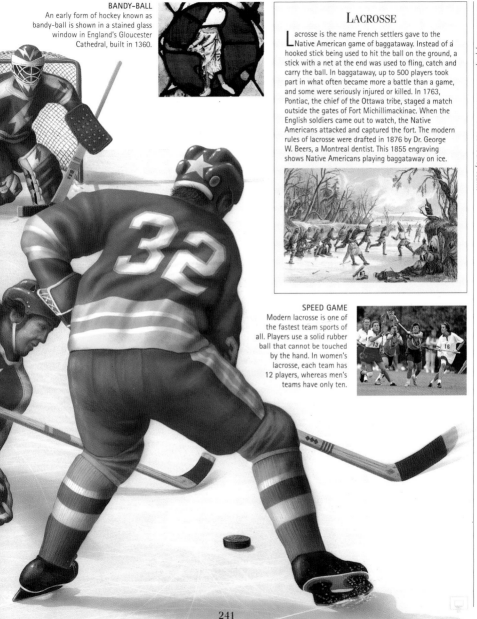

BANDY-BALL
An early form of hockey known as bandy-ball is shown in a stained glass window in England's Gloucester Cathedral, built in 1360.

LACROSSE

Lacrosse is the name French settlers gave to the Native American game of baggataway. Instead of a hooked stick being used to hit the ball on the ground, a stick with a net at the end was used to fling, catch and carry the ball. In baggataway, up to 500 players took part in what often became more a battle than a game, and some were seriously injured or killed. In 1763, Pontiac, the chief of the Ottawa tribe, staged a match outside the gates of Fort Michillimackinac. When the English soldiers came out to watch, the Native Americans attacked and captured the fort. The modern rules of lacrosse were drafted in 1876 by Dr. George W. Beers, a Montreal dentist. This 1855 engraving shows Native Americans playing baggataway on ice.

SPEED GAME
Modern lacrosse is one of the fastest team sports of all. Players use a solid rubber ball that cannot be touched by the hand. In women's lacrosse, each team has 12 players, whereas men's teams have only ten.

241

HEIGHT IS MIGHT
Being tall is a great advantage for a basketball player. It is not unusual to find players in the National Basketball Association who are more than 7 ft (2.1 m) tall.

Basketball and Netball

Canadian Dr. James Naismith wanted to find a sport that his students could play indoors during icy winters in Massachusetts. In 1891, he invented basketball. This fast-action game with five players on each side is now one of the world's most popular team sports. It is particularly well supported in the United States where the best basketball players play professionally in the National Basketball Association. The aim of basketball, and netball (a similar game), is to score goals by shooting the ball through a hoop or basket positioned about 10 ft (3 m) above the ground. Basketball players can advance the ball up the court by running and bouncing it (dribbling), or by passing it to a team-mate. Netballers can only move the ball by passing it. The top netballing nations compete for the World Championship. Basketballing nations take part in the World Championship as well as the Olympic Games.

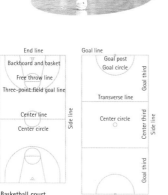

Basketball court — End line, Backboard and basket, Free throw line, Three-point field goal line, Center line, Center circle, Side line

Netball court — Goal line, Goal post, Goal circle, Goal third, Transverse line, Center circle, Center third, Side line, Goal third

COURT SIZE
A netball court is 100 ft (30.5 m) long and 49.8 ft (15.2 m) wide. A basketball court is 91.8 ft (28 m) long and 49.2 ft (15 m) wide.

PASSING ON
Netball, which is played mainly by women, demands great fitness and athleticism. Australia, New Zealand and Jamaica are among the leading nations in world netball.

KORFBALL

Korfball is a game like basketball and netball. Men and women in teams of six play on a court (sometimes indoors) that is as big as a soccer field.

DUNK TIME!

Spectacular plays like the dunk, in which the scorer's hand is above the basket to push the ball through the hoop, have made basketball one of the world's most exciting team sports.

POK-TA-POK

In ancient times, the Mayan and Toltec peoples of Central America played pok-ta-pok, where they tried to get a solid rubber ball through stone rings fixed nearly 25 ft (8 m) high on a wall. Players were not allowed to use their hands. Instead, they had to strike the ball with their knees, elbows or hips. It was not regarded as a sport, but as a serious ritual. The losing captains were occasionally beheaded. One of the giant stone courts on which these games were played is still standing in Chichen Itza, Mexico.

BALL SIZE

A basketball is larger and heavier than a netball. The circumference of a basketball is 29.2–30.4 in (75–78 cm), while a netball is 27.7 in (71 cm).

243

American Football

The most popular style of football played in the United States is American football. Based on a combination of the rules of rugby and soccer, American football developed in schools and universities during the 1860s. Walter Camp drafted the rules of the American football code in 1876, and the new sport soon became the nation's favorite winter game. It is a team sport with 11 players on each side. To score points, players must carry the ball or receive it over their opponents' goal line for a touchdown, or kick the ball through the upright goal posts for either a field goal or a conversion after a touchdown. University or college football attracts big crowds and is shown on television. The main professional organization is the National Football League (NFL) which began as the American Football Association in 1920. Today, the member clubs are multimillion-dollar business enterprises. They pay their players high salaries in an attempt to win the NFL championship.

THE SPECIALIST
Although there are only 11 players on each side at a time, as many as 40 may be used in a single game. They include specialist players such as the punt kicker who is called into play to kick the ball downfield.

THE GRIDIRON
The playing field is sometimes called the gridiron because it resembles the wire rack used for grilling food. It is 360 ft (110 m) long and 160 ft (49 m) wide. The field is divided into "yard lines," each about 15 ft (4.6 m) apart.

Side line

Goal posts
End zone
Goal line
Yard lines
Hashmarks
End line

FACE TO FACE
The teams face each other along an imaginary line called the line of scrimmage. After play begins, the defenders try to tackle the ball-carrier while his team-mates attempt to block them so the ball-carrier can run or pass.

POWER AND SPEED

American football is a very physical, body-contact game that involves blocking, tackling, running and passing the ball. Players at the top level need to be very fast and strong.

THE SUPER BOWL

The annual Super Bowl game to decide the NFL championship was introduced in 1967. It is one of the biggest events in American sports and is watched by millions of television viewers around the world, including people who may not even understand the rules of the game. They enjoy watching the entertainment program before the kick-off and during breaks in the game. This tradition started at college games with cheerleading squads and marching bands, but has since become a major form of sports entertainment.

BOWLING STYLES

There are two main methods of bowling—fast and slow. A slow bowler often spins the ball so that it bounces off the pitch in an unexpected direction. A fast bowler delivers the ball at a very high speed.

Slow bowling
The slow bowler has a short follow-through.

Fast bowling
The fast bowler has a much longer follow-through.

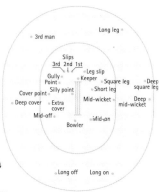

3rd man

Long leg

Slips
3rd 2nd 1st

Leg slip

Gully
Point

Keeper

Square leg

Deep square leg

Cover point

Silly point

Short leg

Deep cover

Extra cover

Mid-wicket

Deep mid-wicket

Mid-off

Mid-on

Bowler

Long off   Long on

IN THE FIELD

Cricket is played on an oval field that has a pitch 66 ft (20 m) long in the middle. The members of the fielding team are usually placed at any of the marked fielding positions to catch or stop the ball.

BAT AND BALL

The standard cricket bat is made of wood with a rubber grip. The ball is hard with an outer coating of red leather stitched into a seam.

Cricket

For centuries, the traditional game of cricket has been a part of village life throughout England. For the last 100 years, it has been played as an international sport by countries that were once colonized by the British. Although King Edward I of England mentioned cricket as early as 1300, nobody really knows how the sport began. The cricket bat of 350 years ago looked more like a hockey stick, and the ball was bowled along the ground. Instead of three stumps, there were two. As the ball often rolled between the two stumps when bowlers tried to hit them, a third stump was added. In the 1800s, the modern style of overarm bowling became popular. Cricket is played between two teams of 11 players. The batting team has two players on the field trying to score runs, while the fielding team (all 11 players) tries to get them out.

IN THE STREET

Children around the cricketing world have long played a street version of the sport with their own rules for batting, bowling and fielding.

246

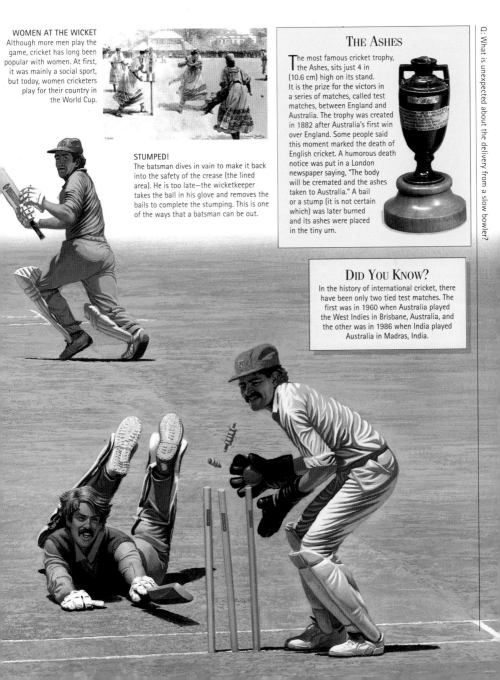

WOMEN AT THE WICKET
Although more men play the game, cricket has long been popular with women. At first, it was mainly a social sport, but today, women cricketers play for their country in the World Cup.

STUMPED!
The batsman dives in vain to make it back into the safety of the crease (the lined area). He is too late—the wicketkeeper takes the ball in his glove and removes the bails to complete the stumping. This is one of the ways that a batsman can be out.

THE ASHES
The most famous cricket trophy, the Ashes, sits just 4 in (10.6 cm) high on its stand. It is the prize for the victors in a series of matches, called test matches, between England and Australia. The trophy was created in 1882 after Australia's first win over England. Some people said this moment marked the death of English cricket. A humorous death notice was put in a London newspaper saying, "The body will be cremated and the ashes taken to Australia." A bail or a stump (it is not certain which) was later burned and its ashes were placed in the tiny urn.

DID YOU KNOW?
In the history of international cricket, there have been only two tied test matches. The first was in 1960 when Australia played the West Indies in Brisbane, Australia, and the other was in 1986 when India played Australia in Madras, India.

Baseball

Baseball is the national sport of the United States. It is also played in most parts of the world, and is especially popular in the Caribbean, Latin America and Japan. The game is believed to be based on rounders, a sport that flourished in England during the nineteenth century. Baseball is played on a diamond-shaped field with two teams of nine players. The aim of the game is to score more runs than the other team. A run is scored when a batter hits the ball thrown by the pitcher and completes a full circuit of the four bases. Batters can do this by stopping at any of the bases on the way, or by hitting the ball so far that they can complete the circuit without stopping. This is called a home run or a homer. One team bats while the other team fields. Both teams receive nine turns to bat, each of which is known as an inning.

Baseball

BATS AND BALLS
The standard baseball bat is usually heavier and wider than the softball bat. The T-ball bat is ideal for small children. Notice how much smaller the hard baseball is when compared to the softball.

Softball

T-ball

SITTING TARGET
T-ball is a modified version of baseball. Instead of the ball being thrown by the pitcher, it is placed on a stand. This makes it easy for the batter to hit.

SLIDE RULE
The batter races against the ball as he slides into the base. If the opposing player gets the ball and steps on the base or tags the runner before he reaches the base, the runner is out.

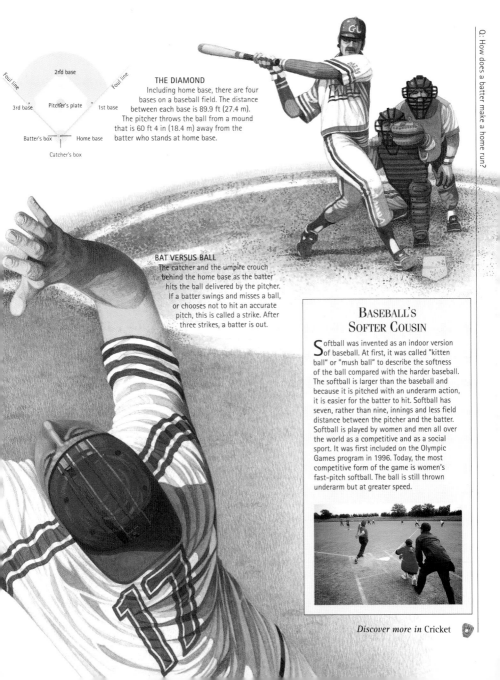

THE DIAMOND

Including home base, there are four bases on a baseball field. The distance between each base is 89.9 ft (27.4 m). The pitcher throws the ball from a mound that is 60 ft 4 in (18.4 m) away from the batter who stands at home base.

2rd base

Foul line
Foul line

3rd base   Pitcher's plate   1st base

Batter's box   Home base

Catcher's box

BAT VERSUS BALL

The catcher and the umpire crouch behind the home base as the batter hits the ball delivered by the pitcher. If a batter swings and misses a ball, or chooses not to hit an accurate pitch, this is called a strike. After three strikes, a batter is out.

BASEBALL'S SOFTER COUSIN

Softball was invented as an indoor version of baseball. At first, it was called "kitten ball" or "mush ball" to describe the softness of the ball compared with the harder baseball. The softball is larger than the baseball and because it is pitched with an underarm action, it is easier for the batter to hit. Softball has seven, rather than nine, innings and less field distance between the pitcher and the batter. Softball is played by women and men all over the world as a competitive and as a social sport. It was first included on the Olympic Games program in 1996. Today, the most competitive form of the game is women's fast-pitch softball. The ball is still thrown underarm but at greater speed.

Discover more in Cricket

Golf

A DRIVING SHOT
The player "tees off" for the first shot by pushing the tee into the ground, placing the ball on it and then striking the ball, usually with a wooden-headed club known as a driver.

N o-one really knows how golf began. Some have suggested that the first golfer was a shepherd in Scotland who hit stones with his hooked stick (called a crook). When he accidentally hit a stone into a rabbit hole, he began to take turns with a friend to see who was better at hitting stones into the hole. Many other cultures, such as Chinese, Celtic and Dutch, played games that were similar to modern golf. The aim of golf is to hit the ball from a starting point, known as the tee, into the hole on the green in the least number of shots. A round of golf is played over nine or 18 holes and the player who finishes with the fewest shots is declared the winner. The first golf balls were made of wood and the first clubs were tree branches. Later, club makers added separate heads made of wood, iron and steel. The rubber-core ball was introduced in the twentieth century.

THE BIRTH OF G
St. Andrews in Scotland is the birthplace of mod golf. The club built on this course, the Royal Ancient Club, laid down the first rules of the gal The British Open, one of the world's g tournaments, has often been held on this cou

250

TRAPS FOR THE UNWARY
Golf courses include hazards such as sand traps, or bunkers. If a ball lands in a bunker, it is difficult for the golfer to play a good shot out of the sand.

CADDIES

The word caddie comes from the French word "cadet" meaning a "little chief." In ancient Scotland it was used to describe pageboys or messengers. In time, boys who carried and cleaned the sticks of wealthy golfers were called caddies. Today, motorized carts are common on most golf courses, but caddies still carry the bags of the top players in tournament golf. They have a great knowledge of the game and their advice is highly valued by professional golfers.

BASIC EQUIPMENT
Golf bags are used to carry clubs, balls and tees. Clubs have different names and each is used for a different type of shot.

Putter
A putter is used for tapping, or putting, the ball into the hole.

Iron
An iron is used for medium-range shots.

Wood
A wood or driver is used for long shots from the tee.

Discover more in Billiards, Pool and Snooker

• ON TARGET •

Bowls and Bowling

The Romans introduced the game of bowls to the northern European countries we now know as France, Germany and Britain. The Italians still play this game, called bocce, with only minor differences. The aim of most forms of bowls is to toss or roll the bowl or bowls near to the target bowl, which is called the jack. The winning player or team is the one that rolls the greatest number of bowls nearest the jack. The game can be played between single players or teams of two to four players. Lawn bowling is played on flat greens, but in crown green bowling the lawn slopes gently up to the center, or crown, of the green. Indoor, or carpet bowls, is very similar to lawn bowling but it is played indoors on a carpet or synthetic green. The most famous bowler in history was the English sailor Sir Francis Drake. Legend says that in 1588, he refused to fight the warships of the invading Spanish Armada until he had finished his game of bowls.

PLAYING BOWLS
By the nineteenth century, variations on the game of bowls were played throughout Europe. The Italians enjoyed bocce, the English played lawn bowling and the French liked to relax with a game of boules, or pétanque, such as the one being played here.

CURLERS AND SOOPERS

Players in the game of curling
slide a stone on the ice from one target area
to another that is 112 ft (34 m) away.
Sweepers called soopers sweep the ice
in front of the stone with brooms to
remove any objects in its path and
to help it travel faster.

MARBLES

The game of marbles is similar to bowls. Its
players are rewarded for hitting the jack or
knocking an opponent's marble out of the way.

DIFFERENT BALLS

The tenpin bowling ball can weigh up
to 16 lb (7.2 kg). A lawn ball
weighs up to 3 lb 8 oz (1.59 kg),
and a boules ball weighs
only 2 lb 3 oz (1 kg).

Tenpin ball

Lawn ball

Boules ball

BIASED BOWLS

The bias on a bowling ball makes it
swerve to one side. Here, lawn bowlers roll the
ball on a curved path to get closer to the jack.

PINS, STRIKES AND ALLEYS

In tenpin bowling, a large ball is rolled down a
wooden alley in an attempt to knock down a set of
pins set up in a triangular formation. The original
tenpin bowling alley was a church aisle and the game
was played by European churchmen. It was introduced
to the United States by the Dutch in the seventeenth
century. It became very popular in the 1950s when
the automatic pinsetter was invented. This replaced
the fallen pins in the correct positions ready for the
next bowler. Knocking down all the pins with one
bowl is known as a strike.

Chalk
Players rub chalk on
the end of the cue to
prevent it from sliding
on the cue ball.

Curved rest
Players reach the cue
ball safely by placing a
curved rest over a ball
that is in the way.

Crosspiece rest
Players can rest the
cue on the crosspiece
if the cue ball is out
of their reach.

The cue
A standard cue
is a tapered st
that must no
less than 3 f
(95 cm) lon

The bridge
Players spread their hand
and use their thumb and
index finger to guide them
when they play a shot.

IN THE STREETS
Croquet grew from an old game
called pall mall, where people
used a mallet to drive a ball
through iron rings placed at
various distances along an
alley or street. One
famous London street
where the game was
popular was later
renamed Pall Mall.

Cue ball

SETTING UP
At the start of a game of pool,
the players place 15 balls inside a
frame in a triangular formation. The
remove the frame and play the firs
shot of the match by striking this
formation with the white cue ball.

• ON TARGET •

Billiards, Pool and Snooker

EYEING THE BA
Snooker was invented in India in 1875 by Brit
army officers who wanted a more exciting versi
of billiards. Today, professional snooker players ea
huge prize money in televised tournamen
including the World Championsh

B illiards is thought to have started when bad weather drove
avid bowlers and croquet players indoors. In the beginning,
players used a wooden pole to move the ball across the
floor. Later, the pole was replaced by a tapered stick called a cue
and the game was played on an elevated platform or table covered
in green cloth to represent lawn. From billiards, the games of
snooker and then pool developed. By the 1940s, these two games
became more popular than billiards. Billiards is played with three
balls, one red and two white. Snooker is played with 15 red balls,
six balls in other colors and one white cue ball. Pool, also called
pocket billiards, has a white cue ball and 15 numbered, colored
balls. Today, all three games are played on a rectangular table that
has pockets in the corners and sides. Points are scored by
knocking the balls into the pockets with the cue ball.

PLAYING THE SHOTS
A billiards player can score points by sinking one of the three balls, as the player here is attempting, or by playing a "cannon." The cannon occurs when the cue ball hits the other two balls in succession. When a ball is sunk in a pocket, it is returned to the table for further play.

THE RISE AND FALL OF CROQUET

By the beginning of the eighteenth century, the sport of pall mall had nearly disappeared. By the 1850s it was back under the new name of croquet. At first the game was popular in France and Ireland. It soon spread to England where rules were drawn up in the 1860s and the All England Croquet Club was formed in 1868. Croquet, a gentle game of skill rather than vigor, was played by people of all ages. Since the beginning of the twentieth century, however, the number of croquet players and greens has fallen dramatically.

PLAYING POOL
In the game of pool, the balls numbered from one to eight are plain colored. The balls numbered from nine to fifteen are striped. The number eight ball (colored black) must be sunk last.

Archery and Darts

Before the invention of gunpowder, bows and arrows were the main weapons used in Europe in the Middle Ages. Legends were created about the skill of famous archers, such as Robin Hood. When armies replaced their bows and arrows with firearms, archery was practiced mainly as a sport. There are two types of archery—target and field. In target archery, the archers shoot at a fixed target. Field archery takes place in forests or on open ground, with targets of different sizes set at various distances. In both, the aim is to score points by firing arrows as closely as possible to the center of the target. The innermost circle of the target is known as the bull's-eye. Archery contests can take place at indoor and outdoor ranges, and since 1931, world championships have been held every two years. Archery is also featured on the Olympic Games program.

IN FLIGHT
Arrows have a vane—a tail of feathers or plastic—to help them fly better. They are usually tipped with points so that they will stick into the target.

BOWS
Most bows are made out of carbon or wood. The powerful compound bow is used for field archery, while the recurve bow can be used for both field and target competition.

Recurve bow

Compound bow

A TEST OF ACCURACY
Modern archers practice the same basic skills as their ancestors of hundreds of years ago. Here, an archer is taking part in target archery, the most common form of competition archery.

...hough bows and arrows were no longer
...apons of war in nineteenth-century
...rope, archery was enjoyed by both
...n and women as a competitive
...d social sport.

DID YOU KNOW?

Anne Boleyn, the second
wife of Henry VIII of
England, gave him a set of
jewel-encrusted darts as
a birthday present.

THREE DARTS AND A BOARD

Darts can be traced back to fifteenth-century England
where the game was introduced as an indoor activity.
It is played with three weighted metal darts each about
5 in (14 cm) long. The darts are thrown at a board 1 ft 6 in
(45.7 cm) in diameter. The player throwing
the dart stands 8 ft 10 in (2.7 m)
from the board. There are many
ways to play darts, and one of
the most popular methods
of scoring is to start with
a certain number and to
keep subtracting the points
scored. The winner is the
first player to reach zero.

SMALLER SHOTS
Like archery, the popular indoor game of darts
is also a test of accuracy. Instead of using a
bow to fire the darts at the target, players
simply throw them at the dartboard.

WINNING HOLDS

In an amateur wrestling match, points are scored for pinning an opponent to the mat, keeping him in a hold, or escaping from a hold. A freestyle wrestling match is won by a number of techniques, including pinfalls, knockouts or submissions.

Outside ankle pick
The weight of his opponent and the upward pressure on his ankle pins the fallen wrestler to the mat.

Leg and arm lock
By using his body weight and a grip on both leg and arm, a wrestler attempts to pin his opponent.

Fireman's carry
One wrestler has exerted his superior strength and skill by lifting his opponent off the mat.

THE REFEREE

In a tournament match, the referee wears a red band on one arm and a blue band on the other to signify the colors of each contestant.

• BODY SKILLS •

Wrestling

Wrestling is the oldest known competitive sport. Greeks and Romans included wrestling in festivals such as the ancient Olympic Games. Native Americans, Chinese, Mongolians and Japanese all had their own forms of wrestling. They all involved two unarmed opponents who tried to secure a fall by means of a body grip, strength or skill. Many different forms of wrestling are practiced today. In Greco-Roman wrestling holds are limited to above the waist, and legs and feet cannot be used to trip an opponent. In freestyle wrestling, there are fewer restrictions on the number of holds used. Both freestyle and Greco-Roman wrestling are Olympic sports with contestants taking part in a series of elimination rounds. Wrestlers compete against opponents of similar weight. The weight limit for a light flyweight (the lightest division) is 106 lb (48 kg). The limit for the giants of the super heavyweight division is 286 lb (130 kg). Professional wrestling uses the freestyle form but is considered more of an entertainment than a sport.

AN ARTIST'S VIEW

Wrestling was very popular in ancient societies such as Greece, Rome and Egypt. Many works of art that have survived from these times show figures wrestling.

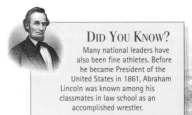

DID YOU KNOW?

Many national leaders have also been fine athletes. Before he became President of the United States in 1861, Abraham Lincoln was known among his classmates in law school as an accomplished wrestler.

258

SUMO SURVIVES

In modern Japan, the ancient art of sumo wrestling, which was once part of the samurai warrior's training, is more popular than ever. Each sumo wrestler attempts to score a fall by moving his opponent, who may weigh more than 442 lb (200 kg), from the wrestling ring.

WRESTLING DRESS

In yagli, a Turkish form of wrestling, competitors smear themselves with grease to make it difficult for opponents to grip them. Russian jacket wrestlers use each other's clothing to apply a grip, while Japanese sumo wrestlers are permitted to take hold of each other's belt. In the Middle East, where a belt was once a sign of a person's importance, the entire contest involved the wrestlers gripping each other's belt in their efforts to force a fall. The Greek bronze statuette (above), which is from the fourth century BC, shows naked (and probably greased) wrestlers.

Blue corner

Passivity zone

Center

Protection zone

Red corner

IN THE RING

Wrestling contests are held on a mat at least 1.6 in (4 cm) thick within a circle 29 ft 6 in (9 m) in diameter. The red circle, called the passivity zone, warns the wrestlers they are near the edge of the mat. One corner is marked in red and one in blue to signify the color of the singlet worn by each wrestler.

KENDO
Kendo was first practiced by samurai warriors in Japan. They used real swords, but modern contestants fight with bamboo swords known as "shiani."

JUDO
In a judo bout, contestants are judged on their ability to throw an opponent to the floor. This technique is known in Japanese as "nagewaza."

• BODY SKILLS •

Martial Arts and Fencing

Many centuries ago, people in Asia combined physical exercise with the art of self-defense. This form of training is known as martial arts, and it includes ju-jitsu, kendo and tae kwon do. Judo, which in Japanese means "the gentle way" because it uses as little force as possible, developed from ju-jitsu and was accepted as an Olympic sport in 1964. Contestants use throws, locks and holds to force their opponent to submit. In self-defense karate, fighters kick and strike each other with their hands, knees and elbows. In competition karate, actual blows are not allowed. The word "karate" means "empty hand" in Japanese because the sport does not use any weapons. Fencing, or the sport of sword fighting, is another ancient military skill. It has been part of the Olympics since the first modern games in 1896. Swords were eventually replaced by specially made fencing items such as the foil, épée and light fencing saber.

FENCING EQUIPMENT
All fencing weapons today have a flexible steel blade. With the foil and the épée, a hit is scored with the point. With the saber, the point and cutting edge can be used. A protective mask and a glove must be worn.

Protective mask

Saber

Glove

Épée

Foil

TAE KWON DO
Blocking, punching and parrying are used in tae kwon do, but this popular form of hard-contact martial arts is best known for its range of kicking techniques.

A FLYING KICK

Kicks, blows and punches are all scoring techniques in the competitive combat sport of karate. In competition, fighters pull back their blows before they make contact with their opponents to prevent injuring them. Because karate developed from a form of self-defense designed to stop an attacker, many karate techniques are too dangerous to be used in competition. Contestants score points by using correct karate methods on the scoring parts of an opponent's body. A fighter who uses excessive physical force is disqualified.

ON GUARD!
For French musketeers of the seventeenth century, dueling with swords was not just a popular sport. Sometimes they fenced to settle a disagreement, or defend their honor. Swordsmen were often injured or died in these contests.

PARALYMPICS
Every four years, the
best athletes with a
disability from around
the world compete at the
Paralympics. In the track
program, athletes in
wheelchairs take part in
sprint, middle-distance
and long-distance events.

THROWING THE DISC
Competitors in the discus
event try to throw a wooden
disc edged in metal the
longest distance. They
must make their throw
from inside a circle
8 ft 2 in (2.5 m) across.

• BODY SKILLS •

Track and Field

oot races held in Greece thousands of years ago were probably the first organized athletics. Today, running, walking, jumping and throwing are all part of the world of track and field. Both categories of events—track and field—demand a wide range of skills from contestants. At a championship meet, the track program is made up of sprint, middle-distance, distance, relay, hurdle and walking races. The field events include jumping competitions, such as the high and long jumps; and throwing events, such as the javelin, discus, hammer and shot put. The decathlon for men brings together many of the track and field skills in one competition made up of ten events. The heptathlon is a seven-event contest for women. The two most important track and field meets are the World Championships and the Olympic Games, both of which take place every four years.

GREAT LEAP
The high jump is a field event in which the competitor uses strength and agility to jump over the crossbar without dislodging it.

HURDLES
Hurdle events combine the two skills of running and jumping and are part of the track program at a track and field meet. If competitors knock over a hurdle, it slows them down but they are not penalized.

POINT FIRST
The javelin is a throwing event based on the skill of throwing a spear in hunting or warfare. The javelin must land point first and the winner is whoever can throw it the longest distance.

THE OLYMPICS

Track and field events formed the basic program for the ancient Olympics and are just as important to the modern Olympics. Held in the main Olympic stadium, the track and field program brings together the world's best athletes. After a series of heats and elimination contests, the finest performers are left to compete for the gold medals in the Olympic finals. Because athletes value an Olympic gold medal above all other sporting trophies, they train hard to be at the very peak of their performance for the games. Often, world records are broken during the finals as the track and field contestants strive to go higher, faster or farther. These events are watched by thousands of spectators in the Olympic stadium, and the pictures are beamed live around the world to billions of television viewers.

THE HORIZONTAL BAR

The horizontal bar is 9 ft (2.75 m) high and allows competitors to perform such exercises as swings and single-handed handstands. Great strength and control is required to perform on the bar. It is used only in men's competitions.

Gymnastics

The Egyptians, Greeks, Chinese, Minoans, Etruscans and Romans all practiced acrobatics as a form of entertainment and a way to celebrate the fitness and flexibility of the human body. These acrobatic displays were the forerunners of competitive gymnastics. One of the pioneers of modern gymnastics was Friedrich Ludwig Jahn. In 1811, he created the first open-air gymnasium and introduced such devices as the parallel bars and rings. Gymnastics became very popular in Europe, and it was one of the first sports included on the program for the first modern Olympic Games in 1896. Gymnastics takes in many kinds of movement and dance. There are three separate forms of competitive gymnastics: artistic gymnastics, rhythmic gymnastics and sport acrobatics. Each discipline requires different types of equipment and styles of performance.

THE FLOOR

The floor routine is common to both men's and women's gymnastics. In women's competition, the gymnast must show dance steps as well as acrobatic and gymnastic elements in her routine.

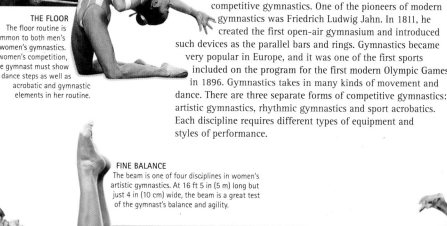

FINE BALANCE

The beam is one of four disciplines in women's artistic gymnastics. At 16 ft 5 in (5 m) long but just 4 in (10 cm) wide, the beam is a great test of the gymnast's balance and agility.

THE RINGS

The rings are one of six disciplines in men's artistic gymnastics. They hang 9 ft (2.75 m) above the floor and competitors perform handstands, swings and somersaults. Gymnasts need great strength and control to perform on the rings.

Janssen/fritsen

TRAMPOLINING

Trampolining is an exhilarating form of exercise that developed from circus acrobatics. The trampoline is a webbed mat hung by springs from the sides of a frame set about 3 ft (1 m) above the floor. By jumping on the webbed mat, the trampoline gymnast gains great bounce from the springs. There are trampoline competitions for individuals, teams of five and synchronized pairs (pictured). Judges award marks for the difficulty of the exercises and the way they are performed. Marks are deducted for loss of height, breaks in the routine and loss of rhythm.

DID YOU KNOW?

The vaulting horse was developed from acrobatic routines practiced on real horses. When it was first used in Germany, the vaulting horse had a horse's head as well as a tail. Over the years, however, the horse lost its head and tail.

ACROBATIC SKILL

In this time-lapse photograph, the gymnast makes his run up to the springboard, performs a vault over the vaulting horse and completes a perfect landing on the other side.

• BODY SKILLS •

Cycling

The sport of cycling is almost as old as the bicycle itself. Races were held on street courses until 1868 when the first recorded track race took place in Paris with competitors racing over 3,936 ft (1,200 m). Cycle tracks sprang up all over Europe, and by 1889 the top cyclists were being paid to race in front of large crowds. Great Britain was the most powerful cycling nation until 1900 when other European countries formed the Union Cycliste Internationale. There are two main types of competition racing: track racing and road racing. In track racing, cyclists race over sprint or longer distances on a banked circuit, or velodrome, which has curved, sloping sides. Road racing is held on public roads and may take many days and sometimes weeks to complete. During this time, the riders cover hundreds or even thousands of miles. Other types of bike racing are BMX (bicycle motocross), mountain-bike events and cyclo-cross, held over open country.

THE TOUR MAP
The route for the Tour de France can change each year. This was the route for 1995. However, the race always crosses the Pyrenees, where riders climb to heights of more than 8,200 ft (2,500 m), and it always finishes in Paris.

THE GREAT RA
The most famous road race of all is the Tour de Fra It is held over three weeks and covers more t 1,860 miles (3,000 km). People turn out to cheer riders who come from all parts of the world. leader at the end of each section has the ho of wearing a yellow jersey the next

DIFFERENT BIKES
The mountain bike has a strong frame for rough conditions and is fitted with brakes and gears. The road bike has brakes as well as a large selection of gears for hill climbing. The track bike has neither brakes nor gears.

Mountain bike   Road bike   Track bike

IN PURSUIT
A velodrome (right) is
used for cycling events,
such as the team pursuit (top).
This race is held over 13,120 ft
(4,000 m) between two teams
that have four riders each. The
winning team is decided by the
finishing times of the first
three riders of each team.

EARLY DAYS OF CYCLING

The invention of the
chain-driven bicycle in the
1870s revolutionized this new
means of transportation and
enabled cyclists to travel at
greater speeds. Cycling was soon
recognized as a healthy form of
exercise and a way to see the
countryside. The Pickwick Cycle
Club of London was the world's
first club for people who wished
to enjoy this new sport. Later,
bicycle touring clubs were
introduced to Europe and the rest
of the world to promote the new
pastime and to protect the rights
of cyclists on the roads. Here,
children in the Netherlands enjoy
cycling on an extra large bicycle.

SWIMMING CAP AND GOGGLES
A close-fitting cap helps streamline the swimmer. Goggles protect the eyes and make it easier to see underwater.

Backstroke

Breast stroke

FLYING START
At the beginning of major swimming r except backstroke, competitors dive fr their starting block into the water.

• IN THE SWIM •

Swimming and Diving

S wimming is one of the most popular sports in the world. The earliest swimming competition on record was held in Japan in 36 BC. Today, swimmers compete at all levels—from races held in school pools to long-distance swimmers battling the seas of the English Channel. The fastest and strongest swimmers in the world race against each other at the World Championships and the Olympic Games. In a pool that is 164 ft (50 m) long and divided into eight lanes, they usually swim either freestyle, butterfly, backstroke or breast stroke, though some specialize in more than one stroke. Races are set over distances of 164 ft (50 m) to 4,920 ft (1,500 m). Diving is also an Olympic sport and is separated into springboard and platform events. Divers perform a set number of dives, which are rated for their degree of difficulty. Judges award points on how well the dives are performed.

WATER POLO
Water polo originated in England and was included on the program of the second modern Olympics held in 1900. It is a combination of soccer and volleyball. Water-polo players need to be excellent swimmers.

Freestyle

Butterfly

SWIMMING STROKES

Top-class swimmers usually specialize in one of the four swimming strokes. But some can swim all strokes very well. They compete in an event called the medley, which is made up of equal distances of each stroke.

WATER BALLET

Synchronized swimmers perform acrobatic and dancelike routines to music. They move with grace and rhythm, and judges award them marks for the quality of their movements in the water. Synchronized swimmers compete in solo, duet or team performances in a pool that is at least 10 ft (3 m) deep. It has underwater speakers so that performers can hear the music while they twist and turn under water. Australian Annette Kellermann and American Esther Williams are famous synchronized swimmers.

HIGH DIVER

In his dive from the 33-ft (10-m) platform, an international competitor displays the classic pike position—bent at the hips, feet together and toes pointed. He will enter the water with his arms extended in front of his head and his body in a straight line.

Discover more in Gymnastics

BODY-SURFING
Not all surfers need a board to catch a wave. A body-surfer swims onto the breaking wave and uses his body and hands to glide towards the shore.

SAILBOARDING
In wave sailing, sailboarders ride their boards up the face of a breaking wave and use it as a ramp to make spectacular aerial maneuvers high above the surface of the water.

Catamaran
The lightweight catamaran has twin hulls. It is a popular form of racing and leisure craft.

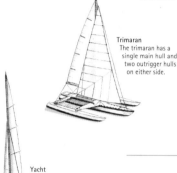

Trimaran
The trimaran has a single main hull and two outrigger hulls on either side.

SAILING CRAFT
There are many types of sailing craft, such as catamarans, trimarans and yachts.

Yacht
The keel, which is part of the yacht's single hull, gives the boat extra stability.

• IN THE SWIM •

Surfing and Sailing

WIND POWER
Under the power of the wind, a fleet of racing skiffs sails around a marker buoy that forms part of the course. The skiffs are trying to complete the course in the shortest time.

The English explorer Captain James Cook saw people riding waves in their canoes when he was in the Polynesian islands in 1771. Eight years later, other visitors to the islands noted how the natives also lay on long, narrow surfboards as they rode the waves to the shore. Those boards were made of wood. Today's modern surfboards are crafted from foam and fiberglass and can be seen on beaches in such surfing locations as Australia, Hawaii, California, South Africa and some of the islands of Indonesia. A sailboard is like a surfboard with a mast and sail attached. Sailboarding can be performed in the surf or on still water. It is just one of more than 3,000 types of competitive sailing craft in the world. Whether on a tiny one-person dinghy or on a mighty maxi yacht with a crew of 24, sailing requires skill and experience.

THE TUBE

[A] surfer positions his board in the [ho]llow part of the breaking wave [kn]own as the tube or the barrel. [Th]e tube ride is one of the highest [sco]ring maneuvers in competitive [su]rfing. By riding as close as [po]ssible to the breaking section, [th]e surfer can gain maximum [spe]ed from the ride.

WATER-SKIING

A contestant in a slalom water-skiing competition throws a wall of spray as she banks through a tight turn. As a competitive sport, water-skiing involves jumping, slalom and trick riding. In jumping, skiers are judged on the length of the jump they make after skiing over a steep ramp with the boat traveling at a set speed. The slalom skiers must zigzag past six buoys in the fastest time. In the trick event, judges award points for tricks performed. Individual awards are made for each event, but the winner of the championship is the best skier over the three events.

SAFETY EQUIPMENT
White-water and slalom racing can be hazardous. Helmets protect paddlers from contact with rocks and other obstacles. Flotation devices prevent the craft from sinking.

DOWN RIVER
A two-person canoe heads down river. Its design is similar to a craft used by Native Americans thousands of years ago.

ESKIMO ROLL
Perhaps the most important safety technique a paddler must learn is the Eskimo roll. If a kayak capsizes, this maneuver allows the paddler to return to an upright position in seconds.

• IN THE SWIM •

Rowing and Canoeing

EIGHT PLUS ONE
The crews in the event known as the eights pull away from the start at a rowing regatta. The cox or coxswain, often the smallest crew member, sits in the stern to help keep the rowers on course and in perfect time.

People from the Stone Age hollowed out logs to make canoes, the Inuit (Eskimos) of North America stretched animal hides over frames to construct kayaks, and Polynesians of the Pacific Islands made outrigger canoes. These paddle-driven craft were among the most ancient forms of water transportation and were the forerunners of today's sports models such as racing kayaks and Canadian class canoes. Olympic racing canoes are made of plywood but other models use fiberglass or aluminium. Kayak and canoe racing can take place on still water or white water, or over rapids. Rowing, which is held on still water, is also an Olympic sport. Sculling is a type of rowing, but the rower uses two oars (called sculls) instead of one. Rowing tournaments are called regattas and are held for crews of one, two, four or eight people. The most important rowing regattas are held at the World Championships and the Olympic Games.

DIFFERENT CRAFT
Both the canoe and the kayak are propelled by paddles and can be used in still-water and white-water competitions. The two craft, however, have some important differences, as shown.

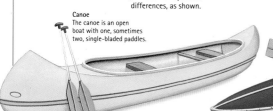

Canoe
The canoe is an open boat with one, sometimes two, single-bladed paddles.

Kayak
The kayak is an enclosed craft with double-bladed padd

KAYAK RACING

Kayak racing over slalom and white-water courses tests the competitor's ability to control the craft in difficult conditions. In slalom racing, the kayak must pass through several gates set up over the course. A white-water racing course must be close to 2 miles (3 km) long.

DRAGON-BOAT RACES

Hundreds of paddles churn through the water to the sound of beating drums as the fearsome-looking craft in this dragon-boat race surge across Hong Kong Harbor. The most spectacular form of paddling craft, the dragon boats have 20 rowers, a crew member in the stern to steer and a drummer to help the paddlers keep time. Dragon-boat racing was originally part of an ancient Chinese festival. It has now been adopted by other countries and crews regularly take part in international competitions.

Q: What are the two main differences between a canoe and a kayak?

Helmet
Ski poles

DOWNHILL DASH
Downhill skiing is a high-speed race against the clock. Competitors wear tight-fitting aerodynamic suits, and helmets to protect them in case they fall.

Ski suit

Ski gloves

Ski boots

Skis

TWISTING TURNS
Pairs of flags mark the course for the slalom skier. In a slalom race, skiers must complete the course twice. Their two recorded times are added together.

Skiing and Tobogganing

The people of the snow-covered lands of Scandinavia in northern Europe were the first people to ski. Thousands of years ago, they skied for practical reasons—to move around during winter—rather than for sport. Today, the sport of skiing falls into two main categories: nordic and alpine. Nordic include cross-country, ski-jumping and biathlon, which consists of two events—skiing, and shooting at targets. Alpine takes in the faster events, such as slalom and downhill racing. As winter sports gained popularity, people soon adapted the sled and toboggan for racing. These had already been used for years to carry goods and passengers across snow and ice. The first bobsled races were held in Switzerland during the 1880s. Bobsled teams consist of two or four people racing a heavy sled down a high-speed, ice-walled track. The luge toboggan is built for one or two riders. Unlike the bobsled, it does not have brakes or a steering device.

BOBBING ABOUT
The bobsled is made of steel with a lightweight body constructed of fiberglass or a composite material. The driver steers the bob with a rope attached to the front axle.

READY FOR TAKE-OFF
Distance, style and technique are all important as a ski jumper heads for the take-off point. This sport requires courage, grace and strength.

DANGEROUS DESCENT
Dressed in a helmet and aerodynamic suit, a luge rider hurtles feet first down a specially made course that is more than 3,280 ft (1,000 m) long. By moving his body, the competitor can apply pressure to the runners and steer the luge through the sloped curves of the course.

ACROBATIC SKIING

Freestyle skiing is the most modern form of this sport. It includes three events: ballet, moguls (a high-speed run over a bumpy course) and the spectacular aerials. Skiers in the aerials launch themselves off a steep ramp to perform twists, and single, double or even triple somersaults before attempting to land safely on a slope of soft snow. Skiers are awarded points for the quality of their take-offs, their style or form in the air, the height they reach and the accuracy of their landings. A snowboard (far left) allows its rider to perform maneuvers similar to those of surfboard riding.

Ice-skating

FIT FOR ROYALTY
This Royal Albert skate was made in the nineteenth century for Prince Albert of England. It featured an extended blade in the shape of a swan's head.

L ong before the days of the Vikings, the Norse people of northern Europe made ice skates from the shank (shin) bones of animals. They used staffs, or poles, to push themselves along the ice. Later, wood and eventually iron runners secured with straps replaced bone. Following the invention of refrigeration, the first artificially frozen indoor ice rink was opened in London in 1876. Today, there are two types of competitive skating: speed skating and figure skating. Both are Winter Olympic sports. In the speed events, competitors are judged not on their technique, but on their speed around a set course. Speed skaters compete over distances ranging from 1,640 ft (500 m) to 16,400 ft (5,000 m). A figure-skating championship meet is divided into men's, women's, ice-dancing and pairs competitions. A judging panel awards points for the way the skaters perform their routines.

SOCIAL SKATI
Skating has been a popular winter sport fo long time, as shown by this scene fro fifteenth-century Dutch wood

ICE DANCING
Ice dancing is the most spectacular form of skating. It combines great artistic and athletic skill. The skaters perform a compulsory dance, an original dance and a free dance to accompanying music.

RACING
ith fingertips touching the
e to assist their balance, three
eed skaters take a tight corner
uring a race at the 1994
Vinter Olympics.

SKATING MANEUVERS

The single skater is judged on an original and a free-skating program. In the original program, each competitor performs eight compulsory maneuvers. Judges award two marks, one for the compulsory movements and one for the overall presentation. In the free-skating program, the skater chooses a number of movements such as jumps, spins and dance steps to perform with music. Judges award another two marks, one for technical ability and one for artistic expression. The skater placed first by a majority of judges is the winner.

FREE SKATING
The free-skating part of the pairs competition is considered one of the most challenging events. It allows the skaters to introduce their own moves and music, and to demonstrate their ability in all the main areas of skating, such as jumping and spinning.

Discover more in Gymnastics

277

Horse Riding

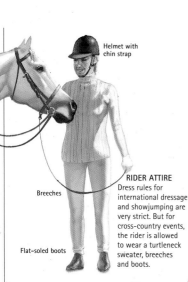

Helmet with chin strap

Breeches

Flat-soled boots

RIDER ATTIRE
Dress rules for international dressage and showjumping are very strict. But for cross-country events, the rider is allowed to wear a turtleneck sweater, breeches and boots.

SADDLES
Competitors in dressage events are required to use a kind of English hunting saddle. Show-jumpers, however, are allowed to use the saddle of their choice. The stock saddle shown here is a popular saddle for recreation and general use.

C ompetition on horseback, or equestrian sport as it is known, dates almost from the time humans first learned to ride. The earliest recorded showjumping event, however, was held in Dublin, Ireland, in 1864. Today, in the Olympic Games, there are three types of equestrian event—dressage, cross-country and showjumping. The Olympic Games program also includes a three-day event, which is a contest of all three disciplines. Dressage is a competition in which the horse and rider perform a complicated set of maneuvers within a limited area. Showjumping is a contest in which the horse and rider must jump over a series of obstacles. The winner is the horse and rider with either the lowest number of faults or the fastest time. The cross-country event tests the speed, endurance and jumping ability of the horse over a long, open course.

OVER THE JUMPS
A show-jumper clears an obstacle known as brush and rails. A typical international Grand Prix course would be made up of 15 jumps, including a water jump, triple vertical parallels, oil drums and poles, and a stone wall.

CROSS-COUNTRY
The aim of the cross-country event is to jump over the many obstacles on the course with as few mistakes as possible, and to keep within the time limit.

278

HARMONY OF HORSE AND RIDER
In international dressage, each
competitor performs a routine including
a number of paces (walk, canter and trot);
halts; changes of direction; and figures,
such as circles and figure eights.
Their performance is judged by
a jury of five officials.

THREE-DAY OLYMPIC EVENT

The three-day event is often the most exciting
equestrian competition on the Olympic program.
Held over three consecutive days, it includes dressage,
cross-country and showjumping. It is run both as
a team and an individual event. The team from each
competing nation is made up of four members and the
best three results are counted in the team's final
score. At the same time, each member of the team
is competing as an individual. Each rider must ride the
same horse throughout the competition.

Q: What type of saddle is used in dressage events?

Sporting Personalities

Michael Jordan

After leading the University of North Carolina to the national college championship and the American team to the 1984 Olympic gold medal, basketball player Michael Jordan began his professional career with the Chicago Bulls. He carried the Bulls to three consecutive National Basketball Association (NBA) championships, and became the first basketball player to be named Most Valuable Player (MVP) in three straight NBA finals.

Carl Lewis

Born in 1961, American athlete Carl Lewis gained international fame with three victories at the 1983 World Championships. The following year he was awarded four gold medals in the 1984 Olympic Games. He won the long jump, the 100-m sprint, the 200-m sprint and the 4 x 100-m relay. Lewis competed at two more Olympic Games and won four more gold medals, including one for the 100-m sprint in 1988 when the winner, Ben Johnson, was disqualified for using drugs.

Carl Lewis

Pam Burridge

Surfing champion Pam Burridge from Sydney, Australia, won her first major title in 1979. During the next three years she was almost unbeatable, but later slipped from the top rankings because of strong competition. In 1990, however, she rediscovered her competitive drive to win the world crown.

Pam Burridge

Babe Ruth

George Herman (Babe) Ruth was born in 1895 and died in 1948. Until 1974 when his career record of 714 runs was bettered, he was the best home-run hitter of American baseball. Ruth had started as a pitcher with the Boston Red Sox in 1914, but soon discovered his true genius as a batter. He moved to the New York Yankees in 1920 and to the Boston Braves in 1934. In 1936, he was elected to the Baseball Hall of Fame.

Babe Ruth

Mark Spitz

Born in California, Mark Spitz was 22 years old when he emerged from the Munich Olympic pool in 1972 as the greatest swimmer in the history of the Olympic Games. He had won seven gold medals. From two Olympics, in 1968 and 1972, he earned a total of 18 medals—nine gold, five silver and four bronze.

Heather McKay

This Australian squash player won the world's leading championship, the British Open, a record 16 times. When world championships were contested in 1976 and 1979, McKay won both, the second after coming out of retirement at 40 years of age. From 1962 until her final tournament in 1979, she was undefeated.

Jack Nicklaus

American Jack Nicklaus is often acclaimed as the world's greatest golfer. Nicklaus, nicknamed "the Golden Bear," won the US Masters championship six times. He has also won five US Professional Golf Association championships, four US Open championships and three British Open championships.

Jack Nicklaus

Donald Bradman

cket legend Don Bradman of Australia
ne greatest batsman in the history of the
ne. In an international career that started
1928 and ended in 1948, Bradman had a
ting average of 99.94 runs. His average
ild have been more than 100 if he had
been dismissed for 0 in his final
ings. No other player has come
se to that record.

Joe Montana

A champion quarterback, Joe Montana
won four Super Bowl championships
with the San Francisco 49ers. In
two of those finals he was selected
as the Most Valuable Player (MVP).
He was named the National
Football League's (NFL) MVP on
four occasions. He is the
highest rated quarterback
in NFL history.

Joe Montana

Sonja Henie

By the late 1920s, figure skating had been an
international sport for 40 years, but it was not until
Norway's Sonja Henie dazzled audiences that it became
widely popular. Henie won gold medals at the 1928, 1932
and 1936 Winter Olympics, and her skill and artistry
inspired generations of young skaters to take up the
sport. After retiring as a skater, she became a Hollywood
movie star.

Franz Klammer

In 1975, Austrian snow skier Franz Klammer won
eight of nine World Cup downhill races. At the 1976
Winter Olympics in Innsbruck, Austria, he almost
crashed several times at high speed in a desperate
attempt to beat the fastest time set by defending

Martina Navratilova

Franz Klammer

champion Bernhard Russi of Switzerland.
Klammer's courage and control gave him
the gold medal by one-third of a second
over Russi.

Martina Navratilova

Martina Navratilova was born in 1956
in Czechoslovakia and later became an
American citizen. She was one of the
world's outstanding tennis players of the
1970s and 1980s, and is one of the best
of all time. She introduced a new level of
power and fitness to women's tennis to win
nine Wimbledon singles finals. Navratilova
was named the 1980s athlete of the decade
and became the highest paid player in the
history of women's tennis.

Michael Chang

In 1987, at the age of 15 years and
6 months, Michael Chang became the youngest player to win
a match in the main draw at the United States Open tennis
championships. In 1988, he won his first Grand Slam
tournament with a victory in the French Open. He is a sporting
hero throughout Asia where he has won many tournaments.

Fanny Blankers–Koen

Fanny Blankers-Koen was 18 years old when she ran at the
1936 Olympics. Although she was the holder of six world
records at the time of the 1948
Olympics, the Dutch champion was
thought to be too old to compete.
She went on to win four of the
nine women's track and field
events in the program.

Roger Bannister

On May 6, 1954, England's
Roger Bannister became the first
person to run the mile in less than
four minutes. He used another runner
to set the pace for him through the first
half of the race. Bannister set a new
world record of 3 minutes 59.4 seconds.

Pelé

Legendary soccer player Edson
Arantes do Nascimento, known to
the fans as Pelé, led Brazil to three
World Cup victories in 1958, 1962
and 1970. Still regarded as the
greatest soccer player of all time,
Pelé scored more than 1,000
goals and starred with the Santos
club in Brazil before joining the
New York Cosmos to help promote
soccer in the United States.

Pelé

History

Ancient Egypt

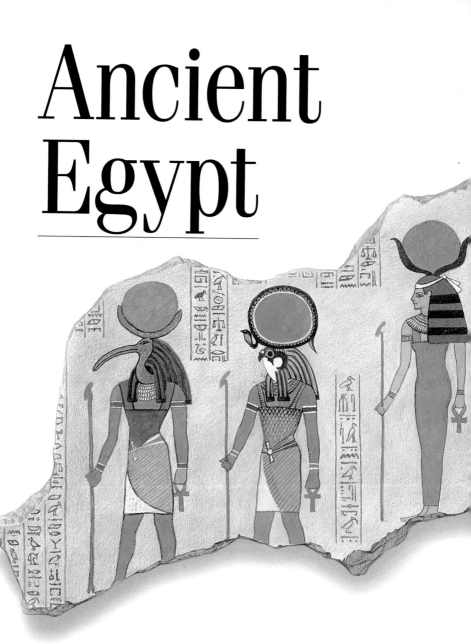

Why did ancient Egyptians wear cones of perfumed grease on their heads?

What Egyptian god had the head of a jackal?

Why did the Egyptians mummify cats?

Contents

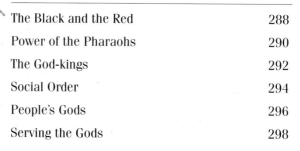

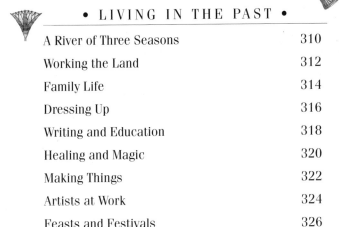

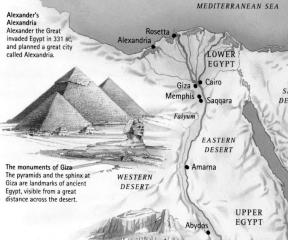

MEDITERRANEAN SEA

Alexander's Alexandria
Alexander the Great invaded Egypt in 331 BC, and planned a great city called Alexandria.

Rosetta
Alexandria

LOWER EGYPT

Giza • Cairo
Memphis • Saqqara

SINA DESE

Faiyum

THE CLIFFS OF THEBES
Limestone cliffs line the western boundary of the valley at Thebes. Pharaohs built temples on the edge of the floodplain and tombs in the hills beyond.

EASTERN DESERT

• Amarna

The monuments of Giza
The pyramids and the sphinx at Giza are landmarks of ancient Egypt, visible from a great distance across the desert.

WESTERN DESERT

LIBYA

UPPER EGYPT

Abydos •

Valley of the Kings • • Karnak
• Luxor
Thebes

Queen Hatshepsut's temple
Queen Hatshepsut, who ruled as pharaoh, built a terraced temple at Deir el-Bahri on the west bank of the Nile. She filled the gardens with sweet-smelling plants.

Edfu •

The Black and the Red

Temples at Karnak was important r center. Stor with elabor tops suppor heavy roofs huge templi

• Abu Simbel

Abu Simbel
Ramesses II ordered two huge temples to be built in the desert at Abu Simbel in Nubia. They were carved out of the sandstone cliffs.

NUBIA

NUBIAN D

People began to live beside the River Nile many thousands of years ago. The river cut through the desert and provided them with water. The valley of Upper Egypt in the south formed a long narrow strip; the delta of Lower Egypt in the north spread out across the river mouth. Every year, floods washed thick mud over the banks and left good soil behind. Early Egyptians called this the "Black Land" and used it for growing crops. Beyond it was the "Red Land," an immense stony waste where it hardly ever rained and nothing useful grew. Where the Black Land ended, the Red Land began. A person could stand with one foot on fertile ground and the other on dry sand. Wolves and jackals hunted along the edges of the desert, but human enemies were seldom able to cross it and attack ancient Egypt.

CUSH

MARSH HUNT
The hunter felled birds with his throwing stick after his trained cat had startled them from the papyrus reeds.

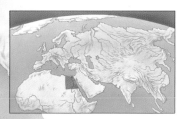

LAND OF THE LOTUS
People in modern times have likened ancient Egypt to a lotus plant, with its valley as the stem and its delta as the flower.

THE CIVILIZATION OF ANCIENT EGYPT

Hippopotamus hunt

New stone-age pottery

LAND OF TWO KINGDOMS
Begins about 3000 BC.
Ditches were dug to irrigate the land and villages became more established. In 3100 BC, Narmer united Upper and Lower Egypt.

Narmer's palette shows his victory

Tuthmosis IV

OLD STONE AGE
Before 12,000 BC.
The earliest Egyptians hunted lions, goats and wild cattle on land, and hippopotamuses and crocodiles in the river marshes.

NEW STONE AGE
Begins about 4500 BC.
During this period, people discovered fire for cooking. They learned to herd animals and to grow grain.

RULE OF THE PHARAOHS
2920 BC to 332 BC.
Egypt was strong for much of this time. Monuments were built and trade with foreign countries developed.

Discover more in A River of Three Seasons

Power of the Pharaohs

The civilization we call ancient Egypt started about 5,000 years ago, when the rule of the pharaohs began. They made Egypt a rich and powerful nation, admired throughout the ancient world. They also ordered the building of great temples for their gods and elaborate tombs for themselves. Some pharaohs, such as Pepy II, came to the throne when they were very young and stayed in power for many years. Sons inherited their father's throne. Pharaohs' wives were also important, but few women ever ruled the country. Teams of workers crafted beautiful objects for the pharaohs and their families. They used materials such as semi-precious stones and gold from the desert mines. The royal couple often displayed their riches in public. Processions, receptions for foreign visitors and visits to the temples were opportunities to show the power of the pharaohs.

A ROYAL JOURNEY
The magnificent royal barge gliding down the river reminded people of the wealth and importance of their god-king and his "Great Royal Wife."

The Dynasties of the Pharaohs

ARCHAIC PERIOD
2920 BC to 2575 BC.
Upper and Lower Egypt were united. Building programs included impressive monuments in Saqqara and Abydos.

Stone vase

OLD KINGDOM
2575 BC to 2134 BC.
This period was also known as the Age of Pyramids. Crafts and architecture developed. Picture symbols, called hieroglyphs, were used to write the texts inside the pyramids.

Female brewer

FIRST INTERMEDIATE PERIOD
2134 BC to 2040 BC.
At the end of the sixth dynasty, a series of weak pharaohs ruled. Local officials called nomarchs struggled for more power. Low Nile floods caused widespread famine.

MIDDLE KINGDOM
2040 BC to 1640 BC.
Strong pharaohs uni the country again a trade revived. The twelfth-dynasty pharaohs organized canals and reservoi for better irrigation

King Mentuhotpe II

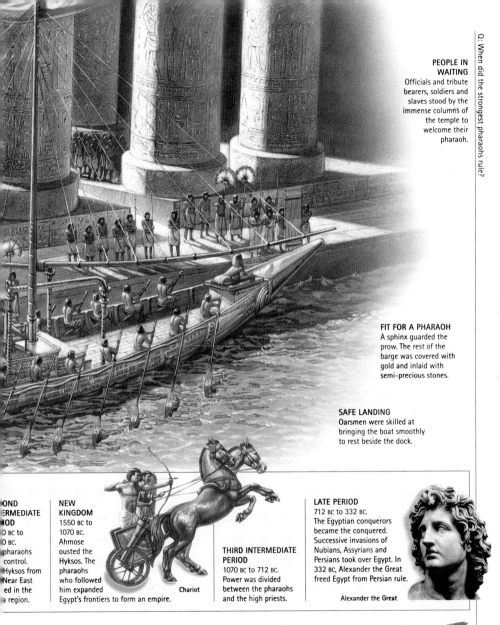

PEOPLE IN WAITING
Officials and tribute bearers, soldiers and slaves stood by the immense columns of the temple to welcome their pharaoh.

FIT FOR A PHARAOH
A sphinx guarded the prow. The rest of the barge was covered with gold and inlaid with semi-precious stones.

SAFE LANDING
Oarsmen were skilled at bringing the boat smoothly to rest beside the dock.

OND ERMEDIATE IOD
0 BC to 0 BC.
pharaohs control.
Hyksos from Near East ed in the a region.

NEW KINGDOM
1550 BC to 1070 BC.
Ahmose ousted the Hyksos. The pharaohs who followed him expanded Egypt's frontiers to form an empire.

Chariot

THIRD INTERMEDIATE PERIOD
1070 BC to 712 BC.
Power was divided between the pharaohs and the high priests.

LATE PERIOD
712 BC to 332 BC.
The Egyptian conquerors became the conquered. Successive invasions of Nubians, Assyrians and Persians took over Egypt. In 332 BC, Alexander the Great freed Egypt from Persian rule.

Alexander the Great

Discover more in Defending the Kingdom

The God-kings

An ancient Egyptian creation myth tells how the god Osiris was sent by Re, the sun-god, to rule the country. The Egyptians believed that all pharaohs were god-kings. The god-kings took part in many ceremonies. They had to dress, eat and even wash in a special way, and every day they went to the temple to offer food to their ancestors. People expected pharaohs to be physically strong, expert at hunting and able to lead the army to victory in battle. Their subjects thought the god-kings controlled the flowing and flooding of the Nile and the growth of crops, as well as the country's success in foreign trade. Everyone knelt and kissed the ground when they approached the royal person. The pharaohs continued to be worshipped even after they had died and joined the god Osiris in the kingdom of the dead.

DID YOU KNOW?

When Queen Hatshepsut's husband died, she took over government and ruled for her stepson Tuthmosis III, who was only five. She held power for about 20 years. Statues show her wearing the false beard of kingship.

Haremhab

MARK OF A PHARAOH
Oval shapes containing hieroglyphs were called cartouches. Two of them make up a pharaoh's name. Cartouches have helped Egyptologists decipher the ancient Egyptian language.

THE GREAT ROYAL WIFE

Pharaohs' wives were also regarded as gods, and shared their husbands' wealth. This painted limestone bust of Queen Nefertiti shows her wearing a crown and necklace rich with jewels. Nefertiti was the wife of Akhenaten. She helped him establish a new city at Amarna on the east bank of the River Nile in Middle Egypt. Women rarely ruled the country unless it was for a short time at the end of the dynasty when there were no men to take over. Hatshepsut was the only strong woman ruler.

COURT VISITORS
Foreigners, such as this group from the Middle East, often appeared at the pharaoh's cour They came to offer gifts or to discuss trade agreements.

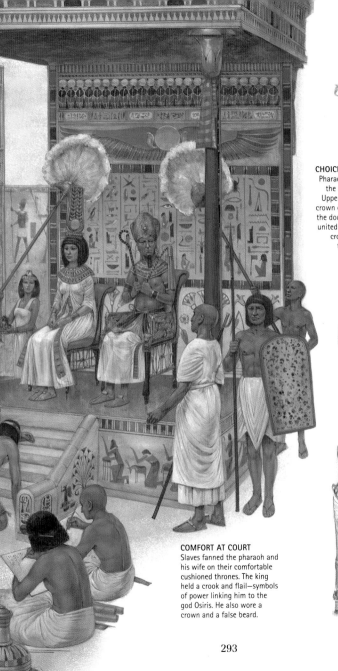

White crown

Red crown

CHOICE OF CROWNS
Pharaohs might wear the white crown of Upper Egypt, the red crown of Lower Egypt, the double crown of a united Egypt, the atef crown of Osiris or the blue crown.

Double crown

Atef crown

Blue crown

COMFORT AT COURT
Slaves fanned the pharaoh and his wife on their comfortable cushioned thrones. The king held a crook and flail—symbols of power linking him to the god Osiris. He also wore a crown and a false beard.

ROYAL SEAT
Tutankhamun's wooden throne, covered in gold leaf, pictured the young king with his wife Ankhsenamon.

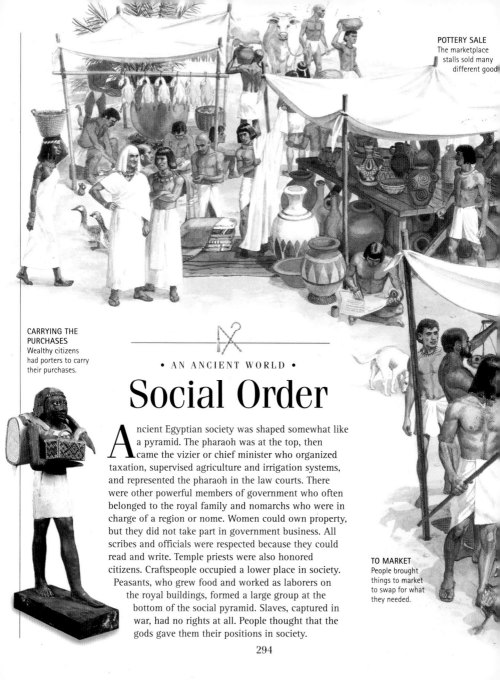

CARRYING THE
PURCHASES
Wealthy citizens
had porters to carry
their purchases.

• AN ANCIENT WORLD •

Social Order

Ancient Egyptian society was shaped somewhat like a pyramid. The pharaoh was at the top, then came the vizier or chief minister who organized taxation, supervised agriculture and irrigation systems, and represented the pharaoh in the law courts. There were other powerful members of government who often belonged to the royal family and nomarchs who were in charge of a region or nome. Women could own property, but they did not take part in government business. All scribes and officials were respected because they could read and write. Temple priests were also honored citizens. Craftspeople occupied a lower place in society. Peasants, who grew food and worked as laborers on the royal buildings, formed a large group at the bottom of the social pyramid. Slaves, captured in war, had no rights at all. People thought that the gods gave them their positions in society.

TO MARKET
People brought
things to market
to swap for what
they needed.

294

COUNTING THE CATTLE

An ancient Egyptian's wealth was measured in cattle rather than in coins. Meketre, who was once mayor of Thebes, had a healthy herd of animals. This model from his tomb shows them being counted in front of their master and other officials. This was a yearly event of great importance. Government scribes came to record the numbers carefully on papyrus. Later, they would calculate how many beasts Meketre must give the pharaoh as a tax payment.

WOMEN'S WORK
Some tasks, such as grinding grain, baking and harvesting, spinning and weaving flax, were done more often by women than by men.

FRESH FOOD
In Egypt's hot climate, people had to shop daily for fresh provisions. Few foods were dried for storage.

WEIGHING THE GOODS
Ancient Egyptians did not use money. Stallholders weighed produce in copper weights, called debens, and priced goods accordingly. A goat worth 1 deben could be exchanged for vegetables weighing 1 deben.

People's Gods

HOME HELP
Bes had the ears, mane and tail of a lion. He brought happiness to the home and protected it from evil.

THOTH'S OTHER DISGUISE
Thoth, god of writing and knowledge, was sometimes shown as a baboon. At other times he was represented by a man with an ibis's head.

LIONESS-HEADED
The goddess Sakhmet the Powerful was shown with a woman's body and the head of a lioness.

PAPYRUS PASTURE
The cow in the marsh is the goddess Hathor in disguise. She was protector of fertility and childbirth.

Religion was a very important part of the lives of ancient Egyptians. They worshipped hundreds of gods. Some, such as the sun-god Re or Amun-Re, were honored by everyone throughout the land in a festival that lasted for a month in the flood season when farmers did no work in the fields. In addition, each of the 42 regions (nomes) adopted a different god to look after its affairs. At home, people turned to lesser gods for help with everyday problems. Many gods were depicted as animals—for example, Bastet the cat, goddess of love and joy—or as human figures with the heads of animals and birds, such as ibis-headed Thoth, god of knowledge. The gods had families too. Osiris and Isis were husband and wife with a son named Horus.

Hathor

Re

Thoth

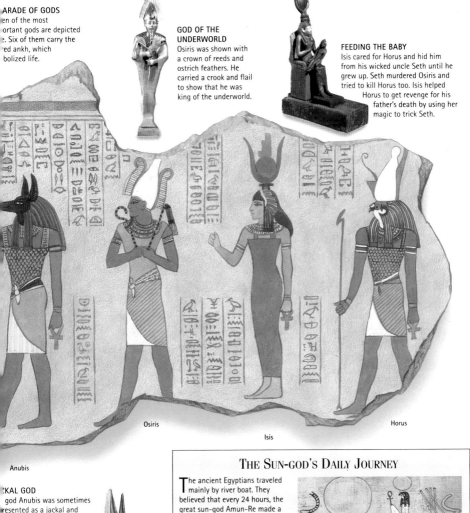

PARADE OF GODS
Ten of the most
important gods are depicted
here. Six of them carry the
sacred ankh, which
symbolized life.

GOD OF THE
UNDERWORLD
Osiris was shown with
a crown of reeds and
ostrich feathers. He
carried a crook and flail
to show that he was
king of the underworld.

FEEDING THE BABY
Isis cared for Horus and hid him
from his wicked uncle Seth until he
grew up. Seth murdered Osiris and
tried to kill Horus too. Isis helped
Horus to get revenge for his
father's death by using her
magic to trick Seth.

Osiris

Isis

Horus

Anubis

JACKAL GOD
The god Anubis was sometimes
represented as a jackal and
sometimes as a man with a
jackal's head. He looked after
the place of mummification
and supervised the priests'
embalming work.

THE SUN-GOD'S DAILY JOURNEY

The ancient Egyptians traveled
mainly by river boat. They
believed that every 24 hours, the
great sun-god Amun-Re made a
voyage across the sky as though
he were on the waters of the
Nile. At night, he sailed through
the underworld of the spirits and
emerged from this dark place at
sunrise each day.

Discover more in Journey to Osiris

Serving the Gods

FLAG BEARER
Horus, the falcon-god,
sits on top of a pole
that once held
a flag in
a temple
procession.

Ancient Egyptians believed that the spirits of the gods dwelt within the temples. Many people were employed to look after these enormous buildings, which were the focus of every community. An inner sanctuary in the heart of each temple protected the statue of the god. Only the pharaoh and the high priest were allowed to enter this sacred place. The people could leave written prayers outside the temples, but they never saw the statues of the gods. Even in processions, portable shrines hid the figures from public view. Women played some part in temple ritual, but the high priests were men. They washed, dressed and applied make-up to the statues as though they were alive. The priests lived by strict rules of cleanliness. They bathed four times a day, shaved their heads and bodies and wore fine, white linen gowns.

PROVISIONS FOR A GOD
The high priest or the pharaoh carried food and drink to the sanctuary three times a day. Before each meal, they washed the statue and clothed it in fresh linen.

MUSIC FOR THE GODS
A rattle, called a sistrum, was used in temple rituals. It was the symbol of the goddess Hathor.

RITUAL OFFERING
This relief, a shallow carving in stone, shows Ramesses II offering incense to Horus

ENTERTAINING THE GODS
In the temple courtyards, women sang, danced and put on acrobatic displays for the gods.

298

ON FESTIVAL DAYS
The priests placed the god's statue in a shrine and carried it in procession around the outside of the temple.

SEALING THE SANCTUARY
When the high priest left the shrine, he sealed the doors with a mud seal.

TEMPLE FOR THE SUN-GOD

Treasure from successful military expeditions helped pay for the temple of Amun-Re at Karnak. Central columns, higher than a nine-storey building, are crowned with carved papyrus heads. The walls are inscribed with records of Sethos I's battle victories. The grounds once included gardens, orchards and living quarters for temple workers.

FIRST SERVANT OF THE GOD
The high priests represented the pharaoh. They also supervised other priests who attended to the daily routines of the temple.

THE WINGS OF ISIS
One corner of Ramesses III's stone sarcophagus is carved with the protective figure of Isis with outstretched wings.

TUTANKHAMUN'S MASK
This life-size mask, engraved with magic spells, protected the head of Tutankhamun's mummy. The mask is made from gold inlaid with semi-precious stones.

• THE WORLD BEYOND •

Preparing for the Afterlife

The ancient Egyptians enjoyed life and wanted all their earthly pleasures in the afterlife. They believed that every person had vital spiritual parts. The *Ka* was the life force, created at birth and released by death; the *Ba* was like the soul. In order to live forever, the *Ka* and *Ba* had to be united with the body after death, so it was important to preserve the corpse. Poor people were buried in the desert where the sand dried their bodies. Food, tools and jewelry were laid beside them for use in the kingdom of Osiris. The wealthy could afford to have their bodies mummified and placed with their possessions in special tombs. The coffins were enclosed in large stone boxes, which were later called sarcophagi, to protect them from tomb robbers or attacks from hungry wild animals.

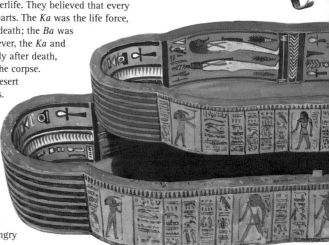

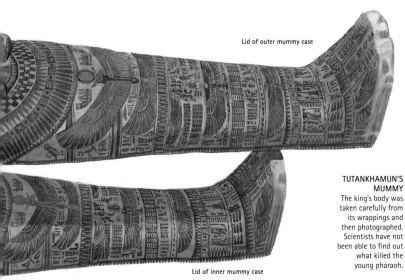

Lid of outer mummy case

Lid of inner mummy case

TUTANKHAMUN'S MUMMY
The king's body was taken carefully from its wrappings and then photographed. Scientists have not been able to find out what killed the young pharaoh.

Wrapped mummy with mask

Bottom of inner mummy case

BA BIRD
This picture of the hovering *Ba* bird is from the scribe Ani's *Book of the Dead*. It shows the spell that will return Ani's important *Ba* to his mummified body.

MUMMY CASES
Coffins made of wood or cartonnage, a kind of papier-mâché, were painted with pictures of gods, spells and many hieroglyphs praising the owner. The inner coffin fit inside one or two outer cases.

of outer mummy case

MUMMIES AND MODERN SCIENCE

Modern technology allows scientists to examine mummies without opening the coffins or damaging the bodies. At St Thomas's Hospital in London, England, the body of Tjentmutengebtiu, a 20-year-old woman, was analyzed with the help of an advanced X-ray process called CAT scanning.

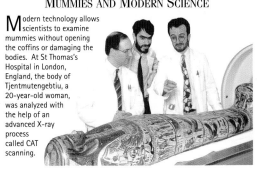

Discover more in Artists at Work

Mummies in the Making

EYES OF GOLD
In the last years of ancient Egypt, lifelike representations of eyes made from gold leaf were placed over the eye-sockets of corpses.

Mummification is the process of slowly drying a dead body to stop it from rotting. In ancient Egypt, the process took about 70 days. Embalming priests removed the liver, lungs, stomach and intestines and stored them in four special little coffins called canopic jars. Later, these were placed in the tomb beside the mummy. The priests also removed the brain, but left the heart to be weighed by the god Anubis. They washed the corpse in palm wine and covered it with a natural salt called natron to absorb the moisture. After 40 days, embalmers rubbed the skin with oils, packed the body with spices, linen, sawdust and sand to reshape it, and wrapped it in layer upon layer of linen bandages that had been soaked in resin. They placed magic spells and good luck charms between the strips. Finally, they sealed the mummy in its case.

AIR PURIFIER
The priests burned incense to sweeten the air while they prepared the mummy, working as quickly as possible.

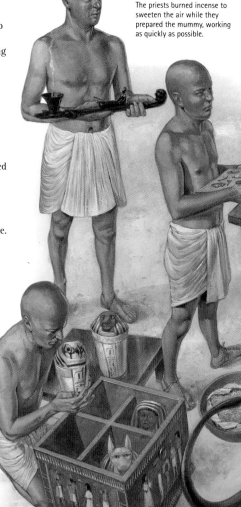

EMBALMING WORKSHOPS
Teams of embalming priests mummified bodies in workshops where all the special tools and equipment were kept.

A BAD JOB
Not all embalmers were good at the job. This queen's puffed and cracked face was the result of overstuffing.

CANOPIC JARS
The Sons of Horus protected different organs: Imset for the liver, Ha'py for the lungs, Duamutef for the stomach, Qebehsenuf for the intestines.

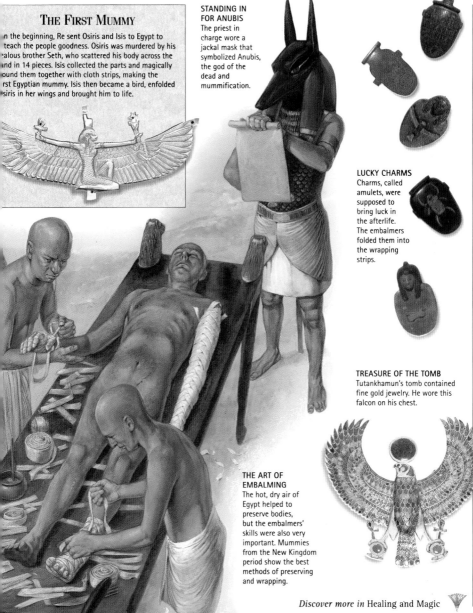

THE FIRST MUMMY

In the beginning, Re sent Osiris and Isis to Egypt to teach the people goodness. Osiris was murdered by his jealous brother Seth, who scattered his body across the land in 14 pieces. Isis collected the parts and magically bound them together with cloth strips, making the first Egyptian mummy. Isis then became a bird, enfolded Osiris in her wings and brought him to life.

STANDING IN FOR ANUBIS
The priest in charge wore a jackal mask that symbolized Anubis, the god of the dead and mummification.

LUCKY CHARMS
Charms, called amulets, were supposed to bring luck in the afterlife. The embalmers folded them into the wrapping strips.

TREASURE OF THE TOMB
Tutankhamun's tomb contained fine gold jewelry. He wore this falcon on his chest.

THE ART OF EMBALMING
The hot, dry air of Egypt helped to preserve bodies, but the embalmers' skills were also very important. Mummies from the New Kingdom period show the best methods of preserving and wrapping.

Discover more in Healing and Magic

303

Journey to Osiris

BOOK OF THE DEAD
Sheets of papyrus covered with magic spells for the afterlife were placed in people's tombs.

The "Opening of the Mouth" was one of the most important funeral ceremonies in the final preparation of the mummy. The deceased's family recited spells while priests sprinkled water and used special instruments to touch the mummy on the lips. Without this ritual, they believed, the dead person would not be able to eat, drink or move around in the afterlife. Before anyone could qualify for eternal life, the stern judge Anubis weighed their heart against the Feather of Truth to see how well they had behaved on Earth. Anubis threw the unworthy hearts to the monster Ammit, "Devourer of the Dead," and the owners went no further. Those who passed the test continued on the dangerous and difficult journey to the kingdom of Osiris. Magic symbols painted on the mummy's case were designed to protect the traveler along the way.

WORKERS FOR ETERNITY
Nobody wanted to work in the afterlife. Model servants called shabtis were placed in the tombs to obey the god Osiris's commands.

Hunefer   Anubis   Anubis   Ammit, Devourer of the Dead

ESSENTIALS FOR THE AFTERLIFE

We know much about the lives of ancient Egyptians from the contents of their tombs. Tutankhamun's tomb contained his childhood toys, 16 baskets of fruit, 40 jars of wine and boxes of roast duck, bread and cake. Musicians were buried with their instruments and women with their beauty kits. Eye make-up was included for everyone. Model boats provided transportation to the kingdom of Osiris.

STRANGE BUT TRUE

Thousands of mummified cats were found in many tombs. In the nineteenth century, about 300,000 cat mummies were shipped to England where they were ground up into garden fertilizer.

MUMMIFIED MENAGERIE
People thought that animals were messengers of the gods, and many were mummified, including calves, crocodiles and cats.

WEIGHING THE HEART
This is the mummy's most challenging moment. Ammit waits beneath the scales, hungry for sinful hearts. The scribe god, Thoth, takes notes. If all goes well, Osiris will welcome the newcomer, Hunefer.

Hunefer   Horus   Osiris

Set in Stone

Many ancient Egyptian stone monuments still stand in the desert today. Although moisture, wind, sandstorms and tourists have damaged them, the pyramids, tombs, temples and colossal statues tell us much about the ideas, beliefs and technology of the people who built them. These incredibly complicated projects required expert skills and a huge workforce. Astronomers studied the stars to determine the best sites, mathematicians and architects calculated the measurements, stonemasons shaped the blocks, and overseers organized the teams of several thousand laborers. Craftsmen worked soft stones with bronze and copper chisels. They pounded harder rocks with balls of dolerite, then rubbed the surface smooth with quartz sand. The Great Pyramids at Giza were faced originally with gleaming white Tura limestone and may have been capped with gold.

SHIPPING THE STONE
Carpenters built cargo vessels at the Nile shipyards to carry stone blocks from the quarries to the building sites.

THE PYRAMIDS OF GIZA
Khufu's Great Pyramid, the biggest of three massive pyramids at Giza, is the largest stone building in the world. It is 479 ft (146 m) high and contains nearly two-and-a-half million blocks of limestone.

ALSE DOOR

mbs had false doors
corated with prayers
d the owners' names.
ey were sacred places
r the living to leave
ferings for the dead.

THE PYRAMIDS AND THE STARS

We know from hieroglyphs on the pyramid walls that the ancient Egyptians likened their gods to the stars. Some scientists think that the arrangement of the three Great Pyramids on Earth matches Orion's belt in the sky. The buildings are placed in a line with the smaller one slightly to the left, just as the three stars in the constellation are aligned.

Orion

THE GIZA SPHINX

This huge sphinx, cut from rock, guarded the pyramids at Giza. The statue had a human head (representing intelligence) on a lion's body (a sign of strength). Together they symbolized royal power.

DID YOU KNOW?

Building measurements on the pyramids are very precise. The stone slabs on the outside of the Great Pyramid fit so snugly side by side that a hair cannot be pushed into the joints between them.

FAMILY AFFECTION

A painted limestone statue of the priest Meresankh and his two daughters was found in his tomb.

COLORED COLUMNS
The temple of the goddess Is stands on the island of Phila When this lithograph wa made in 1846, some color sti remained on t columns in the hall.

• THE WORLD BEYOND •

Great Temples

ON A GRAND SCALE
Massive granite statues of Ramesses II stood inside and outside his temple at Abu Simbel. A single foot was taller than an adult. Shallow reliefs, carved on the north and south walls, record Ramesses II's battle victories.

Many pharaohs ordered temples to be constructed for themselves as well as for the gods. Some of the temples were attached to pharaohs' tombs, erected in separate places or added to other buildings such as the one at Karnak. Temple complexes included huge statues, soaring columns, school rooms, storehouses and workshops, and spacious gardens. By the time Ramesses II came to power in 1290 BC, many magnificent monuments had already been built throughout ancient Egypt. He added several others during his reign of more than 60 years. The most impressive one was at Abu Simbel in the Nubian desert. The laborers chipped away the side of a hill to make the south front and then hollowed out a vast space behind it for the interior. Hatshepsut, Amenophis III, Sethos I and Ramesses III were also great temple builders.

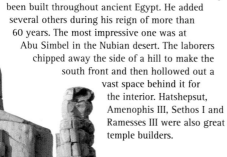

THE COLOSSI OF MEMNON
Two colossal stone statues are all that remain of Amenophis III's monument on the Nile's west bank.

DID YOU KNOW?
Twice a year, the shadowy interior of Ramesses II's temple is pierced by the rays of the rising sun, which illuminate the four statues in the temple's sanctuary.

RESCUING ABU SIMBEL
When the Aswan Dam was built across the River Nile in the 1960s, it created Lake Nasser. Many of the Nubian temples were moved to prevent them from being flooded.

A River of Three Seasons

TRAVELING SOUTH
The hieroglyph "to travel south" was a boat in full sail catching the northerly wind to help propel the craft upstream.

TRAVELING NORTH
River transport and walking were the main means of traveling. The hieroglyph "to travel north" was a boat with the sail down.

When the Greek traveler Herodotus saw ancient Egypt he called it "the gift of the Nile," and nobody has ever described it better. The river was a highway for transport and trade. It provided fish and larger game in the form of hippopotamuses and crocodiles. It sustained marshes where papyrus reeds and lotus plants grew and where waterfowl could be caught for food. It supplied water for drinking and washing. Every year, almost without fail, floodwaters from the lakes and mountain springs of eastern Africa, which fed the Nile's tributaries, washed down fertile silt. The river divided the farmers' calendar into three seasons. The flood time, the "time of inundation" when all work stopped, lasted from July to October. The "time of emergence," allotted to plowing and sowing, ran from November to February. Finally, the "time of harvest" occupied March to June.

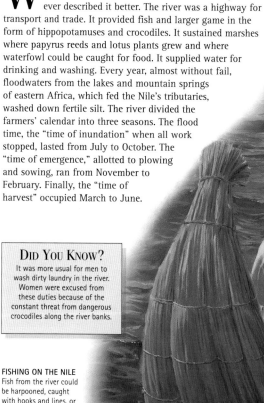

DID YOU KNOW?

It was more usual for men to wash dirty laundry in the river. Women were excused from these duties because of the constant threat from dangerous crocodiles along the river banks.

FISHING ON THE NILE
Fish from the river could be harpooned, caught with hooks and lines, or swept up in nets made from papyrus twine. They were part of the diet of ancient Egyptians.

PAPYRUS FOR MANY PURPOSES

Many years ago, papyrus grew plentifully along the banks of the Nile and in the delta swamps. The ancient Egyptians used it to make excellent paper. Some river boats were crafted from bundles of papyrus stalks; others were rigged with strong ropes made from its twisted fibers. This reed could also be woven into boxes, baskets, mats, sieves and sandals.

HUNTING GEESE
This scene from a tomb shows the sport of wealthy Egyptians—hunting with throwsticks.

SPEAR FISHING
This statuette shows Tutankhamun about to hurl his spear, probably into a hippopotamus. In his left hand he holds a cord for binding his prey.

Working the Land

The ancient Egyptians depended on the yearly cycle of flooding, sowing, and harvesting. Low floods and insufficient soil for the crops meant famine. During the growing season, a network of canals and ditches carried water to the fields. Farmers cultivated barley, emmer wheat, vegetables and fruit. Flax was another important crop. Birds and insects often invaded the fields, and sometimes violent wind storms flattened the ripening grain. Reapers, who were always men, harvested beneath the hot sun. They listened to flute music and prayed to Isis as they worked. Women never handled tools with blades. They winnowed grain, tossing it into the air so that the wind blew away the light stalks and the heavier seeds fell to the ground. Women also helped to make wine and beer and pressed oil from nuts and plants. Besides crops, farmers raised cattle, sheep, goats, ducks and geese for food. Tax assessors came every year to gauge the amount of produce that was owed to the government.

WHAT DOES THE GARDEN GROW?
The trees with the small fruit are date palms. The ancient Egyptians used dates to sweeten food.

HARVESTING
At harvest time, every healthy villager worked in the fields. Men cut the crop with sickles. Women and children bound the stalks into sheaves and separated the grain from the chaff.

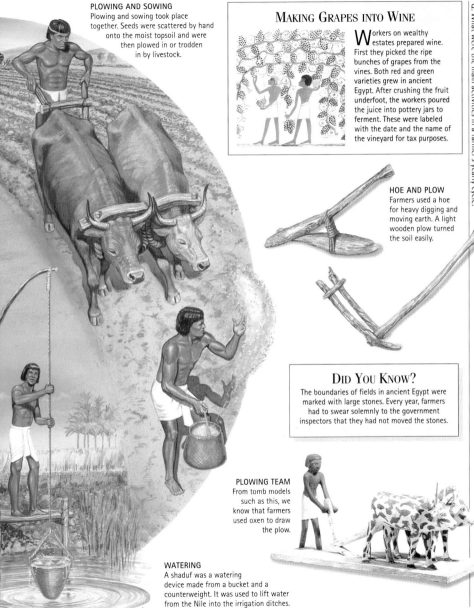

PLOWING AND SOWING
Plowing and sowing took place together. Seeds were scattered by hand onto the moist topsoil and were then plowed in or trodden in by livestock.

MAKING GRAPES INTO WINE

Workers on wealthy estates prepared wine. First they picked the ripe bunches of grapes from the vines. Both red and green varieties grew in ancient Egypt. After crushing the fruit underfoot, the workers poured the juice into pottery jars to ferment. These were labeled with the date and the name of the vineyard for tax purposes.

HOE AND PLOW
Farmers used a hoe for heavy digging and moving earth. A light wooden plow turned the soil easily.

DID YOU KNOW?

The boundaries of fields in ancient Egypt were marked with large stones. Every year, farmers had to swear solemnly to the government inspectors that they had not moved the stones.

PLOWING TEAM
From tomb models such as this, we know that farmers used oxen to draw the plow.

WATERING
A shaduf was a watering device made from a bucket and a counterweight. It was used to lift water from the Nile into the irrigation ditches.

313

IN SERVICE
Servants worked on wealthy
estates. They did housework
and tended the gardens,
crops and livestock.

AT HOM
The home was
place in which th
family could rela
Bigger houses ha
spacious living quarte
with painted walls ar
high windows, withou
glass, to help keep th
air cool. All homes ha
places for statues c
the household god

FAMILY CELEBRATION
This tomb painting is like a
family photograph. It shows
Inhirkha with his wife, son
and grandchildren.

• LIVING IN THE PAST •

Family Life

Most people in ancient Egypt lived in villages of
sun-baked brick houses crammed close together.
These had square rooms with small windows
and flat roofs that were often used for cooking. The rich,
who were able to employ servants, lived in grander
homes with gardens. Twice a day, women fetched water
and filled the huge clay vessels that stood in the
courtyard or by the doorway of every house. People
had very little furniture, especially in the poorer homes.
Stools, beds and small tables were the most common
pieces; chairs were a sign of importance. Ancient
Egyptians usually married within their own social
group. Girls became brides when they were about 12;
boys married at about 14.

TIME FOR BED
Beds were made from wood
and woven reeds, and had
wooden head-rests for
pillows. People did not
use bed sheets.

STRANGE BUT TRUE
Animals were part of the
household. When a pet cat died,
the whole family shaved off their
eyebrows as a sign of mourning.

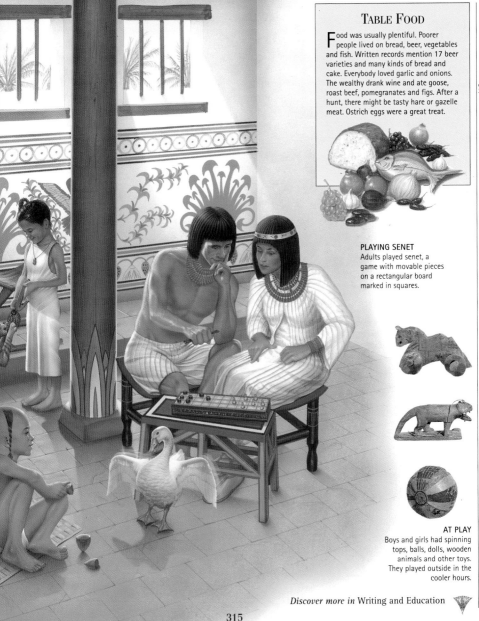

TABLE FOOD

Food was usually plentiful. Poorer people lived on bread, beer, vegetables and fish. Written records mention 17 beer varieties and many kinds of bread and cake. Everybody loved garlic and onions. The wealthy drank wine and ate goose, roast beef, pomegranates and figs. After a hunt, there might be tasty hare or gazelle meat. Ostrich eggs were a great treat.

PLAYING SENET
Adults played senet, a game with movable pieces on a rectangular board marked in squares.

AT PLAY
Boys and girls had spinning tops, balls, dolls, wooden animals and other toys. They played outside in the cooler hours.

Discover more in Writing and Education

315

Dressing Up

People in ancient Egypt dressed in light linen clothing made from flax. Weavers used young plants to produce fine, almost see-through fabric for the wealthy, but most people wore garments of coarser texture. The cloth was nearly always white. Pleats, held in place with stiffening starch, were the main form of decoration, but sometimes a pattern of loose threads was woven into the cloth. Slaves or servants, who came from foreign lands, had dresses of patterned fabric. Men dressed in loincloths, kilts and tunic-style shirts. Women had simple, ankle-length sheath dresses with a shawl or cloak for cooler weather. Children usually wore nothing at all. Men and women, rich and poor, owned jewelry and used make-up, especially eye paint. Everybody loved perfume and rubbed scented oils into their skin to protect it against the harsh desert winds.

ROYAL SANDALS
The upper soles of Tutankhamun's sandals showed Egypt's enemies. He crushed them as he walked.

EYE COLOR
Favorite eye shadows were green powdered malachite and black crushed lead ore.

REFLECTIONS
Polished bronze or copper mirrors were valued possessions. Their roundness and brightness suggested the sun's life-giving power. Poorer people checked their reflections in water.

THE MEANING OF SPOTS
Some priests wore leopard skins over their shoulders to show their importance. Here, Princess–priestess Nefertiabt is wearing a black-spotted panther skin.

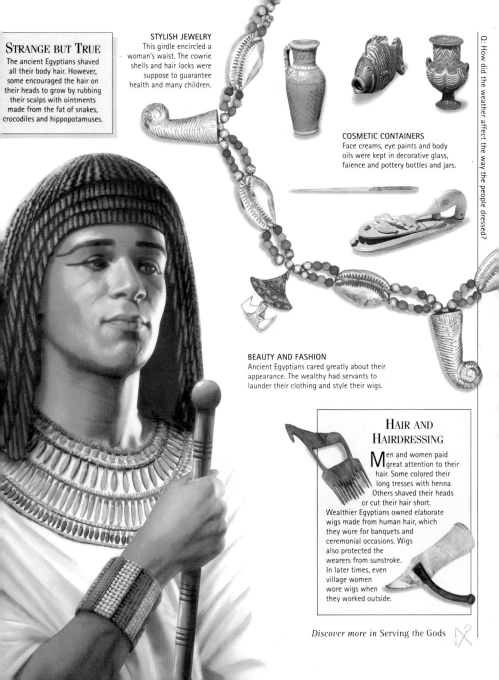

STRANGE BUT TRUE

The ancient Egyptians shaved all their body hair. However, some encouraged the hair on their heads to grow by rubbing their scalps with ointments made from the fat of snakes, crocodiles and hippopotamuses.

STYLISH JEWELRY
This girdle encircled a woman's waist. The cowrie shells and hair locks were suppose to guarantee health and many children.

COSMETIC CONTAINERS
Face creams, eye paints and body oils were kept in decorative glass, faience and pottery bottles and jars.

BEAUTY AND FASHION
Ancient Egyptians cared greatly about their appearance. The wealthy had servants to launder their clothing and style their wigs.

HAIR AND HAIRDRESSING

Men and women paid great attention to their hair. Some colored their long tresses with henna. Others shaved their heads or cut their hair short. Wealthier Egyptians owned elaborate wigs made from human hair, which they wore for banquets and ceremonial occasions. Wigs also protected the wearers from sunstroke. In later times, even village women wore wigs when they worked outside.

Discover more in Serving the Gods

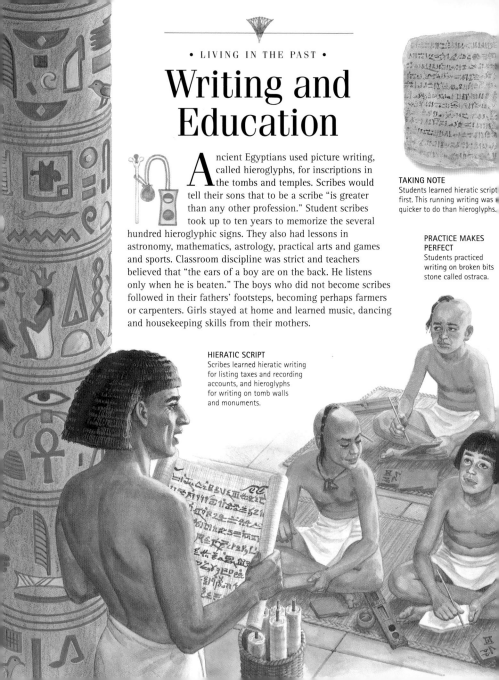

Writing and Education

Ancient Egyptians used picture writing, called hieroglyphs, for inscriptions in the tombs and temples. Scribes would tell their sons that to be a scribe "is greater than any other profession." Student scribes took up to ten years to memorize the several hundred hieroglyphic signs. They also had lessons in astronomy, mathematics, astrology, practical arts and games and sports. Classroom discipline was strict and teachers believed that "the ears of a boy are on the back. He listens only when he is beaten." The boys who did not become scribes followed in their fathers' footsteps, becoming perhaps farmers or carpenters. Girls stayed at home and learned music, dancing and housekeeping skills from their mothers.

TAKING NOTE
Students learned hieratic script first. This running writing was quicker to do than hieroglyphs.

PRACTICE MAKES PERFECT
Students practiced writing on broken bits stone called ostraca.

HIERATIC SCRIPT
Scribes learned hieratic writing for listing taxes and recording accounts, and hieroglyphs for writing on tomb walls and monuments.

DID YOU KNOW?

Egypt officially converted to Christianity when the Roman Empire took over in AD 324. Egyptian writing was banned because the Romans considered it to be pagan. People forgot how to write hieroglyphs and nobody learned how to read them. As a result, hieroglyphs became a lost language.

EVERLASTING WRITING
Carved in hard granite, this cross-legged scribe will write forever on the papyrus spread between his knees.

PAPER FROM REEDS
Paper was made from thinly sliced papyrus stems. One layer was placed on another and the plant's juices glued them together.

Owl

Water

Bread

Man

Arm

Reed

Mouth

Flax

Basket

READING THE STONE

Inscriptions on the Rosetta Stone were the key to reading the pyramid texts and other ancient Egyptian writing. The top band is in hieroglyphs. The middle band is in demotic script, a later form of hieratic writing. The bottom band is in Greek. The stone was discovered in 1799. By 1822, Jean-François Champollion, a French scholar, had deciphered some of the letters.

WRITING KIT
A scribe's tools consisted of a palette of colored paints and brushes made from reeds.

SCRIBE SCHOOL
The boys learned to read and write in groups by copying and reciting texts with wise messages that taught them how to behave properly.

Discover more in Social Order

319

TAWERET
Taweret protected pregnant women. She was depicted as part hippopotamus and part woman, with the legs of a lion.

CROOKED BACK
Some statues and paintings show bone deformities. This hunchback may have had tuberculosis of the spine.

SWEET DREAMS
This hippopotamus ivory wand was used to protect a sleeper from attacks by poisonous night creatures.

EYE OF HORUS
As the two gods struggled for power, Seth tore out Horus's eye. It was magically restored and became a symbol of protective watchfulness.

• LIVING IN THE PAST •

Healing and Magic

D octors in ancient Egypt set broken bones with wooden splints bound with plant fibers, dressed wounds with oil and honey, and performed surgery with knives, forceps and metal or wooden probes. They had cures for many diseases, some of which they thought were caused by worms, such as the "hefet" worm in the stomach or the "fenet" worm that gnawed teeth. Physicians knew that the heart "spoke" through the pulse, but they also thought it controlled everything that happened in the body and all thoughts and feelings. They did not realize that the brain was important. Plant remedies were popular. Garlic was prescribed for snakebite, to gargle for sore throats and to soothe bruises. Doctors used a vulture's quill to apply eyedrops containing celery juice. When practical medicine failed, physicians turned to magic. People wore amulets to ward off accident and sickness. They also thought some of the gods had healing powers.

A CRIPPLING DISEASE
Priest Remi's shorter, thinner leg was probably the result of polio. He would have had this illness as a child.

GOOD LUCK CHARMS

People in ancient Egypt believed that amulets protected them from harm. They wore them as personal jewelry and were buried with them for use in the afterlife. The fish amulet on the girl's braid guarded her against water accidents, such as drowning or being taken by crocodiles. The Bes amulet around her neck saved her from household dangers.

MOTHER AND CHILD
This wooden amulet shows a mother with her baby. It was supposed to ensure a safe childbirth.

WHAT MUMMIES CAN TELL US

Seqenenre II's preserved head shows the severe wounds from which he died. Through autopsies and X-ray examinations of mummies, experts have found out much about the health problems of the ancient Egyptians. They suffered from many of the sicknesses we do, but had no immunization against infectious diseases such as smallpox and polio. Some mummies have badly decayed teeth. Grit and sand in their bread may have worn away their teeth's outer surface, or perhaps ancient Egyptians ate too many cakes sweetened with honey and dates.

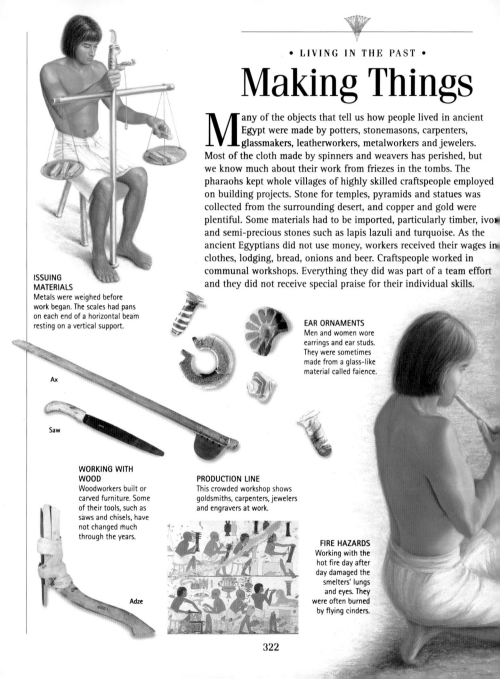

Making Things

Many of the objects that tell us how people lived in ancient Egypt were made by potters, stonemasons, carpenters, glassmakers, leatherworkers, metalworkers and jewelers. Most of the cloth made by spinners and weavers has perished, but we know much about their work from friezes in the tombs. The pharaohs kept whole villages of highly skilled craftspeople employed on building projects. Stone for temples, pyramids and statues was collected from the surrounding desert, and copper and gold were plentiful. Some materials had to be imported, particularly timber, ivory and semi-precious stones such as lapis lazuli and turquoise. As the ancient Egyptians did not use money, workers received their wages in clothes, lodging, bread, onions and beer. Craftspeople worked in communal workshops. Everything they did was part of a team effort and they did not receive special praise for their individual skills.

ISSUING MATERIALS
Metals were weighed before work began. The scales had pans on each end of a horizontal beam resting on a vertical support.

Ax

Saw

EAR ORNAMENTS
Men and women wore earrings and ear studs. They were sometimes made from a glass-like material called faience.

WORKING WITH WOOD
Woodworkers built or carved furniture. Some of their tools, such as saws and chisels, have not changed much through the years.

Adze

PRODUCTION LINE
This crowded workshop shows goldsmiths, carpenters, jewelers and engravers at work.

FIRE HAZARDS
Working with the hot fire day after day damaged the smelters' lungs and eyes. They were often burned by flying cinders.

322

STRANGE BUT TRUE

The first recorded strike in history took place near Thebes, where builders had waited two months for their wages. They refused to work and chanted "we are hungry" until they were paid.

FAIENCE BOWL

This bowl is patterned with fish swimming between lotus buds. The lotus, which opens at sunrise and closes at sunset, symbolizes rebirth.

NING BRIGHT

lters had to heat the metal ore in a ainer to burn off the impurities before they could use the molten metal. They made the fire burn fiercely by blowing on it through hollow reeds tipped with clay nozzles.

WOMEN AT WORK

Women did most of the weaving in ancient Egypt. This wooden model from chancellor Meketre's tomb shows the activity in the textile workshop on his estate. Some workers are walking around spinning linen thread from flax fibers. This will be woven into cloth. The weavers squatting down are operating the two horizontal looms on the floor.

BURIED TREASURE

This solid gold vase was found near the temple of ancient Bubastis. It is hammered in a corn-cob pattern.

BLOWN GLASS

Glassmakers did not learn the blowing technique until the Roman period. Before that, molten glass was molded around a core.

Discover more in The God-kings

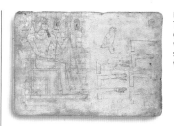

PERFECTING THE ART
This board shows a practice drawing of Tuthmosis III within a grid. The artist seemed to have trouble with the arm hieroglyph.

HEADS RIGHT, EYES FRONT
Artists drew people's eyes and shoulders as if they were seeing them from the front. All other parts of the body were drawn sideways. The left leg was always shown in front of the right.

PAINTING SEQUENCE
First, a stonemason smoothed the wall and covered it with a layer of th plaster. This surface was ma with a square grid made by string dipped in red paint.

• LIVING IN THE PAST •

Artists at Work

Ancient Egyptian paintings told stories about people's lives and what they expected to happen to them after they died and met their gods. Artists painted detailed scenes on houses, temple pillars and the vast walls of tombs, where well-organized teams worked by lamplight in difficult and stuffy conditions. They followed carefully prepared plans of what to paint and strict rules about the way to show figures and objects. The outline scribes always drew important people larger than anyone else who appeared in the picture. Painters used pigments made from crushed rocks and minerals—green from powdered malachite, red from iron oxide—which they mixed with egg white and gum arabic. The colors in many of the tombs and temples are still as fresh and brilliant as when they were first brushed on the walls more than 5,000 years ago.

PAPYRUS PAINTING
The spells and texts for a person's *Book of the Dead* were painted in bright colors on sheets of papyrus.

GRACEFUL WOMAN
Paint was applied to the plaster-coated cloth that covered the wood on this coffin lid.

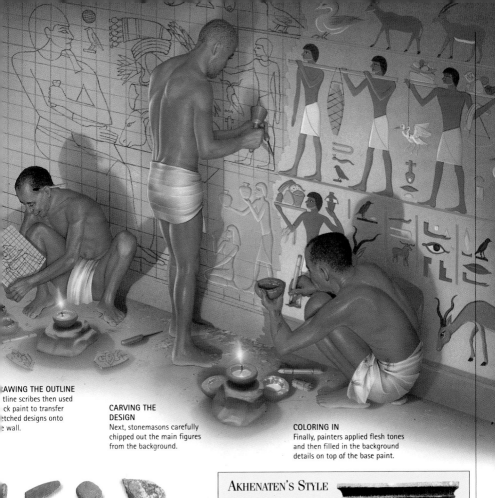

AWING THE OUTLINE
tline scribes then used
ck paint to transfer
tched designs onto
e wall.

CARVING THE DESIGN
Next, stonemasons carefully
chipped out the main figures
from the background.

COLORING IN
Finally, painters applied flesh tones
and then filled in the background
details on top of the base paint.

PAINTING EQUIPMENT
Artists' tools included palettes for
diluting paint and wooden brushes
with split ends for painting.

AKHENATEN'S STYLE

Pharaoh Akhenaten
introduced some changes
to the style of art during his
17-year rule. In his city at
Amarna, dedicated to the
sun-god he called Aten, the
artists and sculptors used a
more lifelike way of portraying
people. This scene, carved in
relief, shows Akhenaten giving
an earring to his daughter.
Queen Nefertiti holds two
younger daughters.

KEEPING THE BEAT
These broken pieces of clappers, made of ivory, were once part of percussion instruments musicians used to beat out dance rhythms.

GRACEFUL GYMNASTICS
Dancers performed somersaults, back bends and high-kicks. Weighted discs on the ends of their pigtails swung with the rhythmic movement.

• LIVING IN THE PAST •

Feasts and Festivals

One hieroglyphic inscription says "be joyful and make merry." Wealthy people loved to invite friends to their homes to share great feasts. The food was plentiful and the wine flowed freely. The hosts hired storytellers, dancers and other entertainers. Ancient Egyptian musicians played many instruments including flutes, clarinets, oboes, lutes, harps, tambourines, cymbals and drums. Poorer people enjoyed themselves on holidays for royal occasions, such as the crowning of a pharaoh, and at yearly harvest and religious festivals. Huge crowds gathered for a "coming forth" when the statue of a god was carried in procession outside the temple. Music and acrobatic displays were part of these parades. People made bouquets, garlands and collars from fresh flowers for private banquets and public festivals.

PERFUMED AIR
Scented cones on the hair melted slowly during a banquet. The perfumed grease released a pleasant fragrance as it ran down over wigs and clothes.

TABLE MANNERS
Banquet guests sat around low tables. Th ate with their fingers and afterwards servants brought wat to wash their hands.

LCOME GUESTS
vants handed out
cented hair cones
the door. Children
often joined in
the festivities.

FESTIVAL OF OPET

The Opet Festival, a yearly holiday, took place during the Nile flood. By the time Ramesses III reigned in 1194 BC, it lasted for 27 days. The statue of the great sun-god Amun-Re, attended by the pharaoh and priests, was carried in procession from Karnak temple south to Luxor temple, as shown above. After special ceremonies, the statue was returned to Karnak.

DID YOU KNOW?

Men and women never danced together in ancient Egypt. Dance routines included graceful acrobatics and gymnastics.

THE CAT GODDESS
Bastet, the cat goddess, was the daughter of Re. Her yearly festival was celebrated throughout the land.

AMUSING THE GUESTS
Storytelling and poetry recitals often began the feast. The entertainment became noisier and more energetic as the feasting progressed.

Discover more in Serving the Gods

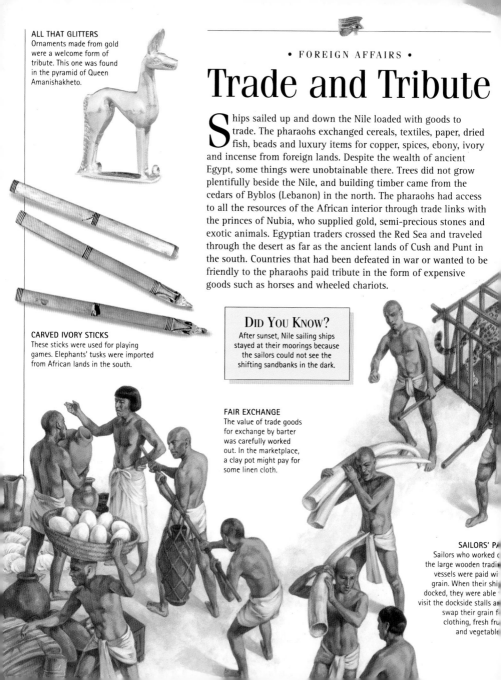

ALL THAT GLITTERS
Ornaments made from gold were a welcome form of tribute. This one was found in the pyramid of Queen Amanishakheto.

Trade and Tribute

Ships sailed up and down the Nile loaded with goods to trade. The pharaohs exchanged cereals, textiles, paper, dried fish, beads and luxury items for copper, spices, ebony, ivory and incense from foreign lands. Despite the wealth of ancient Egypt, some things were unobtainable there. Trees did not grow plentifully beside the Nile, and building timber came from the cedars of Byblos (Lebanon) in the north. The pharaohs had access to all the resources of the African interior through trade links with the princes of Nubia, who supplied gold, semi-precious stones and exotic animals. Egyptian traders crossed the Red Sea and traveled through the desert as far as the ancient lands of Cush and Punt in the south. Countries that had been defeated in war or wanted to be friendly to the pharaohs paid tribute in the form of expensive goods such as horses and wheeled chariots.

CARVED IVORY STICKS
These sticks were used for playing games. Elephants' tusks were imported from African lands in the south.

DID YOU KNOW?
After sunset, Nile sailing ships stayed at their moorings because the sailors could not see the shifting sandbanks in the dark.

FAIR EXCHANGE
The value of trade goods for exchange by barter was carefully worked out. In the marketplace, a clay pot might pay for some linen cloth.

SAILORS' PA
Sailors who worked o the large wooden tradi vessels were paid wi grain. When their shi docked, they were able visit the dockside stalls a swap their grain f clothing, fresh fru and vegetable

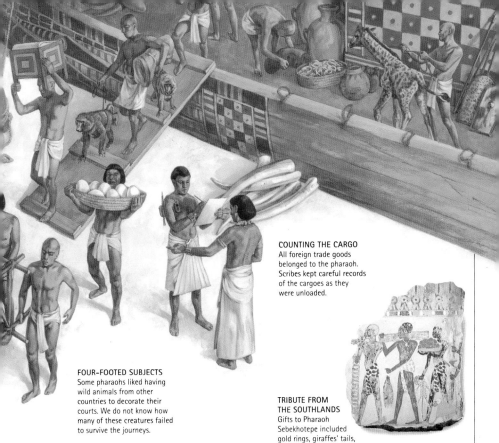

COUNTING THE CARGO
All foreign trade goods belonged to the pharaoh. Scribes kept careful records of the cargoes as they were unloaded.

FOUR-FOOTED SUBJECTS
Some pharaohs liked having wild animals from other countries to decorate their courts. We do not know how many of these creatures failed to survive the journeys.

TRIBUTE FROM THE SOUTHLANDS
Gifts to Pharaoh Sebekhotepe included gold rings, giraffes' tails, ebony logs, jasper, a leopard skin and live baboons on leashes.

SEA VOYAGES

The methods of travel in ancient times meant that trading expeditions could take several years. The ancient Egyptians carried their boats in pieces across the desert and assembled them on the shores of the Red Sea. This scene from Queen Hatshepsut's temple shows incense trees being loaded onto ships visiting the land of Punt. The roots were placed in baskets for the long voyage back to Egypt and then were planted in the temple gardens.

GOLDEN BRACELETS
Queen Ahhotep's bracelets were set with imported semi-precious stones—turquoise, carnelian and lapis lazuli.

Discover more in Working the Land

Defending the Kingdom

CLOSE COMBAT
Daggers and short swords were deadly weapons for hand-to-hand fighting at close range. The blades were riveted to the handles.

F or centuries, ancient Egyptians needed no permanent army. Egyptians seldom had to defend themselves against enemies, other than Libyan tribes who attacked occasionally from the Western Desert. After the Middle Kingdom, however, the Hyksos from the Near East seized Lower Egypt. They had curved swords, strong bows, body armor and horse-drawn chariots. The Egyptians copied these weapons and began training efficient soldiers. Their new army drove out the hated Hyksos and pushed them back through Palestine and Syria. Prisoners of war were forced to join the army or work as slaves. Egyptians built mudbrick forts, with massive towers surrounded by ditches, to defend their borders. Later, Ramesses III formed a navy of wooden galleys powered by oars and sails and trapped the slower sailing ships of pirates invading from the Mediterranean Sea.

TUTANKHAMUN'S CHARIOT
This scene shows a victorious Tutankhamun alone in his chariot. In real life he would have had a driver with him.

PRISONERS OF WAR
Enemies of the pharaoh were always shown with their hands bound or handcuffed.

DID YOU KNOW?
Lions represented courage in ancient Egypt. One poem said that Ramesses II fought "like a fierce lion in a valley of goats."

CEREMONIAL AX
This ceremonial ax belonged to the pharaoh Ahmose. The blade shows scenes celebrating his success in driving the Hyksos out of Egypt.

THE RULES OF
A king in the an
world led his arr
fight the enem
an open battle
The army waite
a signal to b
There were
surprise attacks
no battles after

OFF TO WAR

These wooden soldiers, marching in step and carrying shields and lances, represent a troop from one of the nomes of ancient Egypt. Foot soldiers trained for the army from boyhood. They had to live in barracks where discipline was very tough. Upper-class youths usually joined the chariot corps, which was organized separately. Successful battle commanders received "Gold of Bravery" flies, like those above, as rewards for attacking the enemy again and again.

BATTLE TACTICS
Archers fired from moving chariots. They advanced on the enemy's foot soldiers and then doubled back and attacked from behind.

Discover more in Making Things

MIXED RELIGION
Although Greek pharaohs worshipped their own gods, relief carvings on the temple of Edfu show Ptolemy III and Ptolemy XII with Egyptian deities.

DRINKING CUPS
This Nubian pottery was made on a wheel and decorated with painted and stamped designs.

CHANGING GOVERNMENT
After the Nubian invasion, ancient Egypt was overrun repeatedly by foreigners. For more than a thousand years, power in the ancient world shifted from one conqueror to another.

PERSIANS IN POWER
The Persians introduced camels into ancient Egypt. These could move across the desert from one oasis to another.

• FOREIGN AFFAIRS •

Collapse of an Empire

Many of the pharaohs who came after Ramesses III were not strong rulers. Their subjects began to disobey the laws, and robbers plundered the tombs. Meanwhile, other countries in the ancient world were growing stronger. Foreign conquerors overran the Egyptian empire and invaded the country itself—first the Nubians, then the Assyrians and later the Persians. Alexander the Great brought his army to help the Egyptians expel the Persians. Ptolemy, one of Alexander's generals, founded a dynasty whose rulers spoke Greek and worshipped Greek gods. The Romans took over from the Greeks. Christianity spread through the Roman Empire and came to ancient Egypt. When the Arabs invaded in the seventh century AD, Fustat became their first capital, Islam became the state religion and Arabic became the official language.

GOLD COIN
The Greeks brought coins into Egypt. Pharaoh Ptolemy I is portrayed on this gold piece.

332

ASSYRIANS IN POWER

The well-organized Assyrian army, equipped with iron weapons, swept through ancient Egypt. The Assyrians appointed Egyptian governors to run the country.

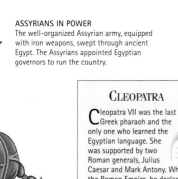

CLEOPATRA

Cleopatra VII was the last Greek pharaoh and the only one who learned the Egyptian language. She was supported by two Roman generals, Julius Caesar and Mark Antony. When Augustus gained power over the Roman Empire, he declared war on Antony and Cleopatra and defeated them in 31 BC. Augustus arrived in Alexandria and demanded Cleopatra's surrender. She was too proud to give in and committed suicide.

ROMAN PORTRAIT

Portraits on coffins became more lifelike during the Roman period. Artists mixed paint with melted beeswax to brighten the colors.

ROMANS IN POWER

Emperor Augustus gained power in 30 BC. The Romans sent gold from the desert mines back to Rome.

GREEKS IN POWER

332 BC, Alexander the Great took possession of ancient Egypt. Later, the Egyptian city of Alexandria became the leading city in the Greek world.

DID YOU KNOW?

Early Christian hermits made their homes in some of the royal tombs at Thebes. They lived in the chapels or offering rooms, not in the actual burial chambers.

SOLDIERS' FOOTWEAR

Archaeologists have found Roman shoes, coins and military equipment at an army outpost in Nubia.

HEAD OF A PHARAOH
In 1816, Giovanni Belzoni arranged for this huge bust of Ramesses II to travel from Thebes to the British Museum in London.

• FOREIGN AFFAIRS •

Discovering Ancient Egypt

How can you discover ancient Egypt? You can visit the pharaohs' treasures in the world's great museums. You can read travelers' tales recorded by writers of the past, such as the Greek historian Herodotus, and you can learn from Egyptologists. When Napoleon Bonaparte's army invaded Egypt in 1798, the French discovered many of its ancient treasures. Since then, Egyptologists have studied monuments, painted friezes, objects from the tombs and things people threw away that the dry climate has preserved. They have deciphered records of daily events and other writing that survives on stone and papyrus. If you ever visit Egypt, you will be able to see the people who now live beside the Nile. They still use some of the old farming methods, and tools have changed little since ancient times. But their crops no longer depend on the time of inundation, or flooding, because the Aswan Dam now controls Egypt's lifeline.

SAVING THE SPHINX
The great Sphinx at Giza is showing its age. From time to time expert restorers have erected scaffolding to make repairs.

334

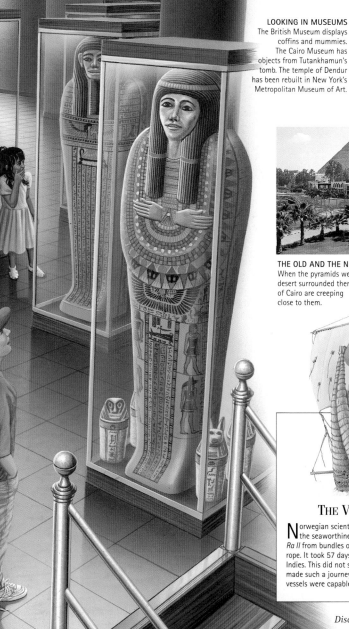

LOOKING IN MUSEUMS
The British Museum displays coffins and mummies. The Cairo Museum has objects from Tutankhamun's tomb. The temple of Dendur has been rebuilt in New York's Metropolitan Museum of Art.

DID YOU KNOW?
To save them from the rising waters of Lake Nasser, the temples from Philae Island were taken piece by piece to Agilkia Island and rebuilt.

THE OLD AND THE NEW
When the pyramids were built at Giza, desert surrounded them. Now, the suburbs of Cairo are creeping close to them.

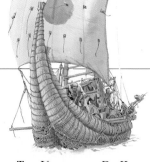

THE VOYAGE OF *RA II*

Norwegian scientist Thor Heyerdahl wanted to test the seaworthiness of reed boats. In 1970, he built *Ra II* from bundles of papyrus lashed together with rope. It took 57 days to sail from Morocco to the West Indies. This did not show that the ancient Egyptians made such a journey, but it did prove that papyrus vessels were capable of surviving long sea voyages.

Discover more in Set in Stone

Dynasties of Ancient Egypt

Egyptologists have pieced together the sequence of the kings of ancient Egypt from fragments of inscribed stone and papyrus. Generally, a dynasty lasted for the time one family or group of pharaohs was in power. There were three very successful periods. During the Old Kingdom, the first pyramid at Saqqara and the Great Pyramid at Giza were built. In the Middle Kingdom, trade expanded and arts, crafts and temple building flourished. The Hyksos were expelled at the beginning of the New Kingdom and the pharaohs of this time established an empire. Listed here are some of the important kings of ancient Egypt and the approximate dates of their reigns.

2920–2575 BC	**ARCHAIC PERIOD**
2920–2770	**1st Dynasty**
2770–2649	**2nd Dynasty**
2649–2575	**3rd Dynasty**
	2630–2611 Djoser
	2611–2603 Sekhemkhet

Polished clay bowl

2575–2134 BC	**OLD KINGDOM**
2575–2465	**4th Dynasty**
	2551–2528 Khufu
	2520–2494 Khephren
	2490–2472 Menkaure
2465–2323	**5th Dynasty**
2323–2150	**6th Dynasty**
	2289–2255 Pepy I
	2246–2152 Pepy II
2150–2134	**7th-8th Dynasties**

Golden head of a falcon

2134–2040 BC	**1ST INTERMEDIATE PERIOD**
	9th–10th Dynasties
	11th (Theban) Dynasty

Golden serpent

2040–1640 BC	**MIDDLE KINGDOM**
2040–1991	**11th Dynasty (all Egypt)**
	2061–2010 Mentuhotpe

Faience hippopotamus

1991–1783	**12th Dynasty**
	1991–1962 Amenemhet I
	1971–1926 Senwosret I
	1929–1892 Amenemhet II
	1897–1878 Senwosret II
	1878–1841 Senwosret III
	1844–1797 Amenemhet III
	1799–1787 Amenemhet IV
	1787–1783 Nefrusobk
1783–1640	**13th-14th Dynasties**

Canopic jars

Bust of Akhenaten

40–1550 BC | **2ND INTERMEDIATE PERIOD**

15th (Hyksos) Dynasty

16th–17th Dynasties

50–1070 BC | **NEW KINGDOM**

50–1307 | **18th Dynasty**

1550–1525	Ahmose
1525–1504	Amenophis I
1504–1492	Tuthmosis I
1492–1479	Tuthmosis II
1479–1425	Tuthmosis III
1473–1458	Hatshepsut
1427–1401	Amenophis II
1401–1391	Tuthmosis IV
1391–1353	Amenophis III
1353–1335	Amenophis IV/Akhenaten
1335–1333	Smenkhkare
1333–1323	Tutankhamun
1323–1319	Aya
1319–1307	Haremhab

07–1196 | **19th Dynasty**

1307–1306	Ramesses I
1306–1290	Sethos I
1290–1224	Ramesses II
1224–1214	Merneptah
1214–1204	Sethos II
1204–1198	Siptah
1198–1196	Twosre

The goddess Selket

96–1070 | **20th Dynasty**

1196–1194	Sethnakhte
1194–1094	Ramesses 111–XI

70–712 BC | **3RD INTERMEDIATE PERIOD**

21st–24th Dynasties

712–332 BC | **LATE PERIOD**

712–657 | **25th Dynasty**
(Nubia and Egypt)

712–698	Shabaka
690–664	Taharqa

664–525 | **26th Dynasty**

610–595	Necho II
570–526	Amasis

525–404 | **27th Dynasty (Persian)**

521–486	Darius I
486–466	Xerxes I

404–399 | **28th Dynasty**

399–380 | **29th Dynasty**

380–343 | **30th Dynasty**

380–362	Nectanebo I

343–332 | **31st Dynasty (Persian)**

332 BC–AD 395 | **GRECO-ROMAN PERIOD**

332–323	Alexander III the Great
304–284	Ptolemy I
246–221	Ptolemy III
51–30	Cleopatra VII Q
30 BC–AD 14	Augustus

Ancient China

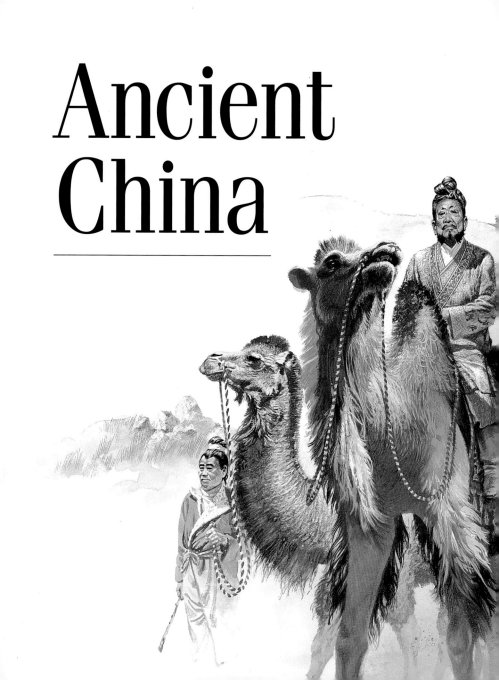

How can Yin and Yang create perfect happiness?

Why is it bad to wear white as a Chinese child?

Who are the Terra-cotta Warriors?

Contents

The Middle Kingdom

The civilization of ancient China began about 8,000 years ago, when people settled in areas of the northeast and beside three great rivers—the Yellow River in the north, the Wei River in the northwest and the Yangzi River in the south. They worked the soil with wood and stone tools, grew millet and rice, and raised pigs and dogs. For centuries, the ancient Chinese were enclosed by mountains, deserts and sea, and had little contact with the rest of the world. They developed their own way of life, and called their country "The Middle Kingdom" because they thought it was the center of the universe. Some Chinese leaders were buried in huge underground tombs. Soldiers, convicts and laborers built a wall to keep out northern invaders, and traders from the west traveled to China along the Silk Road.

TAKLAMAKAN DESERT

Wei Ri

HIMALAYAN
MOUNTAINS

N

S

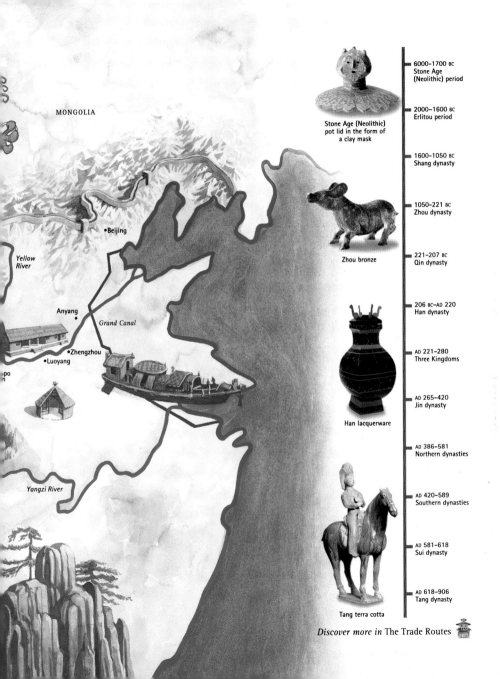

MONGOLIA

•Beijing

Yellow
River

Anyang•

Grand Canal

•Zhengzhou
•Luoyang

Yangzi River

6000–1700 BC
Stone Age
(Neolithic) period

2000–1600 BC
Erlitou period

1600–1050 BC
Shang dynasty

1050–221 BC
Zhou dynasty

221–207 BC
Qin dynasty

206 BC–AD 220
Han dynasty

AD 221–280
Three Kingdoms

AD 265–420
Jin dynasty

AD 386–581
Northern dynasties

AD 420–589
Southern dynasties

AD 581–618
Sui dynasty

AD 618–906
Tang dynasty

Stone Age (Neolithic)
pot lid in the form of
a clay mask

Zhou bronze

Han lacquerware

Tang terra cotta

Discover more in The Trade Routes

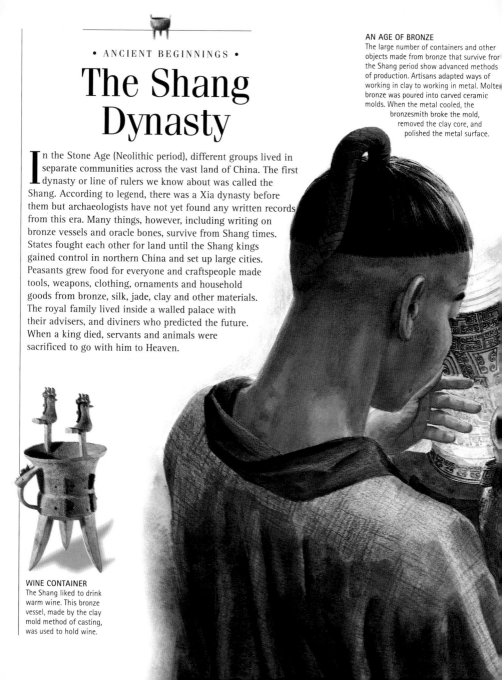

• ANCIENT BEGINNINGS •

The Shang Dynasty

In the Stone Age (Neolithic period), different groups lived in separate communities across the vast land of China. The first dynasty or line of rulers we know about was called the Shang. According to legend, there was a Xia dynasty before them but archaeologists have not yet found any written records from this era. Many things, however, including writing on bronze vessels and oracle bones, survive from Shang times. States fought each other for land until the Shang kings gained control in northern China and set up large cities. Peasants grew food for everyone and craftspeople made tools, weapons, clothing, ornaments and household goods from bronze, silk, jade, clay and other materials. The royal family lived inside a walled palace with their advisers, and diviners who predicted the future. When a king died, servants and animals were sacrificed to go with him to Heaven.

WINE CONTAINER
The Shang liked to drink warm wine. This bronze vessel, made by the clay mold method of casting, was used to hold wine.

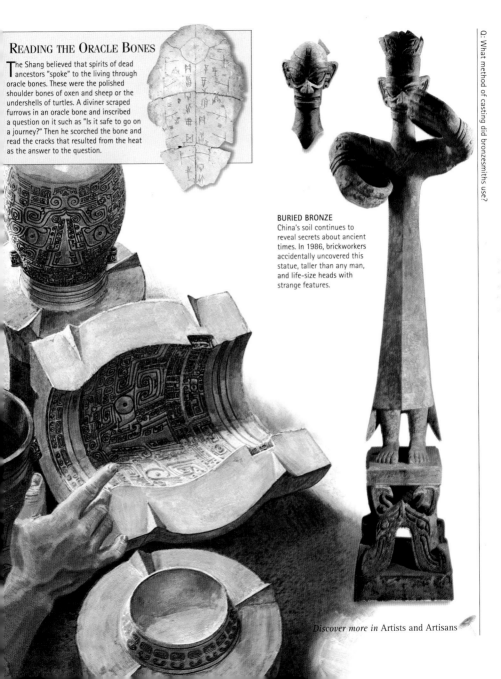

READING THE ORACLE BONES

The Shang believed that spirits of dead ancestors "spoke" to the living through oracle bones. These were the polished shoulder bones of oxen and sheep or the undershells of turtles. A diviner scraped furrows in an oracle bone and inscribed a question on it such as "Is it safe to go on a journey?" Then he scorched the bone and read the cracks that resulted from the heat as the answer to the question.

BURIED BRONZE

China's soil continues to reveal secrets about ancient times. In 1986, brickworkers accidentally uncovered this statue, taller than any man, and life-size heads with strange features.

Discover more in Artists and Artisans

The Qin Dynasty

Qin Shi Huangdi

After more than 800 years of wars and local battles, the powerful Qin people conquered China. The king thanked his ancestors for his success and decided to drop the title wang, which meant "king." He renamed himself Shi (meaning "first") Huangdi (meaning "emperor and divine ruler"). The First Emperor was very important because he unified ancient China by making strict laws, taxing everyone in the country and introducing one script for writing. He commanded his subjects to build roads and canals, and to join existing walls into one long defensive wall. Qin Shi Huangdi did not agree with the teachings of Confucius and other scholars, and ordered their books to be burned. The First Emperor paid magicians, called alchemists, for potions to help him live forever. After his death, his dynasty soon collapsed.

STANDARD COINAGE
Early bronze coins were cast in some unusual shapes. The First Emperor introduced a standard system of money throughout China.

STANDARD WEIGHTS
Qin Shi Huangdi also standardized weights and measures. These included the bronze and terra-cotta cups used for measuring liquids and grains, and the bronze and iron weights that balanced scales.

DID YOU KNOW?
Qin Shi Huangdi, who wanted to live forever, has survived in one way. Qin, pronounced to sound like "chin," gave us the word "China." In a way, the First Emperor's name will never die.

TIGER IN TWO PIECES
An army commander had one part of this model tiger. Messages from the emperor arrived in the second piece to prove that the battle orders were not forged.

PROTECTING THE EMPEROR
In March 1974, well-diggers discovered a silent army guarding the tomb of Qin Shi Huangdi. Pit One contained more than 3,000 life-size foot soldiers and teams of chariot horses.

THE TERRA-COTTA WARRIORS

Qin Shi Huangdi's military companions for eternity march in three pits to the east of his tomb. A fourth pit is empty—the work unfinished at the end of the dynasty in 207 BC. The heads and bodies of the foot soldiers, charioteers and archers (like this one) were made in molds, but no two faces are the same. Some are bearded, others are clean-shaven. Eyes, noses, lips and ears are in many different shapes.

The Han Dynasty

Liu Bang, a government official, gained power and founded the Han dynasty, which lasted for more than 400 years. Han emperors strengthened the Qin system of government and extended ancient China's boundaries. They developed a civil service, based on the teachings of Confucius, to run the empire and keep records in a central place. Scholars who wanted to become government officials had to study very hard. The government organized the salt and iron mines, and state factories began mass-producing objects—from iron and steel farming tools to silk cloth and paper. Han emperors began to control the eastern end of the Silk Road that linked Asia and Europe. Buddhism, one of the most important foreign influences, started to spread throughout ancient China. The Han dynasty finally collapsed after a succession of weak child emperors and droughts and floods.

GILDED BRONZE LEOPARDS
These graceful animals with garnet eyes and inlaid silver spots came from Princess Dou Wan's tomb. They were used as weights.

WATER HIGHWAYS
During the Han era, the people were ordered to build canals to link the cities. These inland waterways made trading, collecting taxes and distributing food during famines much easier. Some families lived on houseboats. Babies often had bamboo floats tied to them until they learned to swim.

AMUSEMENTS FOR THE AFTERLIFE

Acrobats were popular entertainers in ancient China. This tray of tumbling pottery figures was made to amuse the dead in the afterlife.

GOVERNMENT TRANSPORT

Important government officials traveled by horse and carriage. Small models of these were made for their tombs so they would not have to walk in the afterlife.

THE GRAND CANAL

In the Sui period, Emperor Yang decided to link the Yangzi and Yellow rivers by joining new and existing canals. This waterway of 1,550 miles (2,500 km) carried grain and soldiers across his empire. The Grand Canal took more than 30 years to build, and all men 15–50 years of age worked on the project. Every family living nearby also had to send an old man, a woman and a child to the labor force.

THE IMPORTANCE OF HORSES
Horses were a sign of great wealth in ancient China. Lifelike pottery models were often buried in the tombs of noblemen.

CITY MARKETS
Chang'an was the capital of the Han, Sui and Tang empires. By the Tang era, the busy western market was crammed with warehouses. Foreign merchants brought goods from central and western Asia, India and beyond.

• ANCIENT BEGINNINGS •

The Tang Dynasty

The Tang dynasty ruled ancient China for 300 years. This was a time when art, craft, music and literature further developed, and people called it a golden age. Boundaries expanded again as Tang armies fought successfully against the Koreans in the north, the Vietnamese in the south, and the Tibetans and Turks in the west. The Chinese traded with people from these lands and learned much more about the world beyond China. Buddhism, Confucianism and Daoism were still very important, but traders brought other religious ideas to China along the Silk Road. Visiting craftspeople to the city of Chang'an taught local artisans different ways of making things. Clothing showed the influence of foreign fashions. The wealthy developed a taste for imported foods. They ate dumplings in 24 flavors, tasty sauces, and ice cream made from chilled milk, rice and camphor. Tea, made from the leaves of bushes grown in the warm south, reached northern China's markets during the Tang period, and rich people also enjoyed this new drink.

PATTERNS FROM PERSIA
Chinese artisans copied traditional Persian decoration. The mounted hunter on the side of this jug is copied from a Persian design.

SKILLED ARTISTS
This is part of a painting on silk called *The Eight Noble Officials*. The figures in their flowing garments show the lively style used by Tang artists.

> ### DID YOU KNOW?
> Chang'an was laid out in a square. The four walls had three gates in each wall. Each gate had three gateways—the emperor alone used the central one.

FOREIGN WAYS

During the Tang period, gold and silver became more highly prized than before, and rivaled jade and bronze in value. Persian metalworkers, who fled from their own country and came to live in Chang'an, taught the Chinese more delicate methods of using these precious metals. Tang jewelers began to beat them into thin sheets and to make objects from threads of metal.

STROLLING PLAYERS

Goods for sale were loaded onto the backs of camels and into wagons pulled by oxen. Foreign traders, Chinese merchants and the local crowd enjoyed performances by street acrobats and storytellers.

FOREIGN TONGUES

Among the noisy chatter in the market place, the people of Chang'an heard travelers speaking languages from other parts of the world.

ANCIENT DRAGON
Animal charms were
supposed to protect
their owners against
evil influences. This
jade dragon was
modeled during
the Shang period.

PAN GU'S WORLD
The mythical Pan Gu turned his watery
world into stone and then shaped the
moon, sun and stars, and floated them in
space. He asked the phoenix (also known
as the vermilion bird), the Chinese
unicorn (or qilin), the dragon and
the tortoise to help him.

• HEAVEN AND EARTH •

Myths and Symbols

Ancient Chinese myths explain the world's beginning. One story tells of Pan Gu whose body parts became nature after he died. His breath, for example, changed into the wind and clouds, his hair and beard into stars. In another legend, the goddess Nu Gua modeled some human figures from mud. They became rich nobles. Growing tired of such hard work, Nu Gua scattered drops of mud to form poor people. Real and imaginary animals were also featured in ancient Chinese folklore. The blue dragon was linked with spring, the east and rain; the red phoenix bird with summer, the south and drought. The white tiger of the west represented autumn, strength and courage. The black tortoise was associated with winter, the north, long life and wisdom. It was also said that animals were invited to a banquet in Heaven, in which 12—the rat, ox, tiger, rabbit, dragon, snake, horse, goat, monkey, rooster, dog and boar—were chosen to represent the 12-year cycle. Chinese people still link these animals with the year of their birth.

CARRIER BIRD
Mythical animals were often modeled in bronze. This bird, studded with turquoise, has a jade bi disk in its beak. The side cups may have held cosmetics.

YIN AND YANG

Before the world began, there was chaos, shaped like a hen's egg. Pan Gu separated this egg into Yin and Yang—two parts of the same whole. Yin formed the heavier Earth; Yang formed the lighter sky. From then on, Yin stood for all the female, wet, dark things of nature; while Yang represented everything that was male, dry and bright. There could be no perfect happiness until there was an equal balance between Yin and Yang.

GATHERERS OF CLOUDS
Dragons were the rain spirits of ancient China. A Tang jeweler crafted this golden beast with claws outstretched.

MURAL FROM THE HAN ERA
Tigers were popular in Chinese mythology. They were painted on the walls of houses to ward off harm.

DID YOU KNOW?
In ancient Chinese art, horses were symbols of speed and high rank, water buffalo stood for the peasant's life of toil, and flying birds represented freedom.

Discover more in A Time to Celebrate

353

As jade seemed to last forever, the ancient Chinese believed this "stone of Heaven" prevented the body from decaying after death. In the Han dynasty, some royal persons were buried in suits made from tiny pieces of jade.

• HEAVEN AND EARTH •

Sons of Heaven

People believed that kings or emperors received heavenly approval to rule. This was called the mandate of Heaven. The mandate was the idea that the country's leader was the Son of Heaven and obtained power from his celestial or heavenly forefathers. Rulers took part in special ceremonies to ask their ancestors to make sure that rain fell at the right time, that it was safe to go on a journey, that hunting was successful and that many other daily events turned out well. An emperor's subjects expected him to be wise, hard-working, unselfish, good and a brilliant military leader. People rebelled against a bad or weak ruler who did not care about their wellbeing, and believed the heavenly spirits showed their displeasure with him through earthquakes, droughts, famine or floods. The mandate was then taken away and given to someone else.

SYMBOLS OF POWER
A number of jade blades have been discovered from Shang times. They were possibly used during ritual ceremonies and perhaps showed the owner's rank or position in society.

"DAUGHTER" OF HEAVEN

The phoenix bird symbolized an empress. Wu Zetian came to power after the death of her husband Gaozong, a weak emperor during the Tang dynasty. Although other women ruled ancient China at various times, Empress Wu Zetian was the only one to claim she had the mandate of Heaven. She chose her advisers from officials who had passed exams instead of favoring people from rich families.

THE EMPEROR'S COURT
In the Tang dynasty, Sons of Heaven ruled the greatest empire in the whole of the ancient world. They displayed their outstanding wealth at court. Gold glittered and silk shone when the emperors exchanged gifts with ambassadors from foreign lands.

COFFIN COVERING
Coffins often nested one inside the other. Sometimes a silk cloak covered the innermost one—perhaps to clothe the dead person on his or her flight to Heaven.

RIDING IN STATE
A model state coach found near the First Emperor's tomb was drawn by four horses harnessed in gold. It had gold furnishings, and doors and windows that opened.

In Life and Death

The ancient Chinese worshipped their ancestors and looked to them for advice on how to manage their daily lives. The ruler spent many hours communicating with his royal forefathers during special ceremonies. They "spoke" to him through oracle bones and other rituals, and advised him on how to run the country. People believed that life continued after death and that they would need their worldly goods when they joined their ancestors in Heaven. The poor went to their graves in cheap coffins with very few possessions. At first, the tombs of the rich contained human sacrifices. In later times, artisans began making copies of servants and attendants in clay, wood or bronze. Burial pits contained many chambers with walls and ceilings decorated to look like the rooms of real houses. Soldiers and mythological creatures guarded the entrance tunnels.

TANG TOMB GUARD
This glazed pottery official shared his duties as a tomb guard with another official, two spirits, two Buddhist guards, two horses, two camels and three grooms.

FIT FOR A PRINCESS
Princess Yongtai died during the Tang dynasty. Her tomb furnishings reflected her high position in life. Some of the walls were covered with exquisite drawings. Jars of wine, food and other household goods made sure she would want for nothing.

A TUNEFUL AFTERLIFE

Musicians often played bells for ancient Chinese lords at solemn rituals or to entertain visitors. This perfectly preserved set of 64 bronze bells mounted on a wooden rack was found in a Zhou tomb. When struck with a wooden stick, each bell produced two notes. The bells in the second row chimed the melody; the larger ones at the bottom provided the accompaniment.

Discover more in The Shang Dynasty

A TRIO OF LEADERS
The great thinker Confucius (left) was known throughout China as "The Master." Laozi (right), a legendary figure linked with the beginnings of Daoism, was called "The Old Philosopher." Buddha (center), named "The Enlightened One," gave up worldly pleasures to achieve nirvana, or perfect peace.

• HEAVEN AND EARTH •

Three Ways of Thinking

Three ways of thinking—Confucianism, Daoism and Buddhism—influenced the ancient Chinese. Each one might help with a different part of life. Confucius outlined a code of proper behavior, arguing that if families were strong and united, the country would also be strong and united. He praised strict government. The followers of Daoism did not agree. They said that everyone should live by the laws of nature and should not be governed by too many regulations made by people. Daoists thought that there would be fewer wars and crimes when people stopped wanting things they could not obtain honestly. Buddhism taught believers that they could be reincarnated (born again) many times, and that performing good deeds in this life meant better chances in the next one. Besides these ways of thinking, foreigners brought Christianity, Judaism and the beliefs of Islam into ancient China.

DID YOU KNOW?
Daoists worshipped a small group of "immortals," or disciples, who were supposed to possess magical powers such as becoming invisible, turning objects into gold and raising the dead.

TEMPLE STATUES
Some Buddhist statues in the cave temples were made from a clay mixture called stucco, which lasted well. However, the surface painting was often damaged.

ACCORDING TO THE MASTER

Ren, represented by this Chinese written character, was the basis of Confucian teaching. Ren is often translated as virtue or goodness. Confucius himself explained ren as "to love all men." The sayings of Confucius were collected in a book called *The Analects*. The Master said that society would be orderly if prince, subject, father and son kept to their proper places.

CAVES OF A THOUSAND BUDDHAS
Buddhist monks hollowed temples out of caves in remote places. They decorated the walls with detailed images of Buddha and his disciples. Some caves contained portraits painted on paper banners such as this one, or on silk cloth.

Discover more in Writing and Printing

Order in Society

Ancient Chinese society was divided into four main classes. The scholar–gentry class was the highest and most esteemed. Scholars were respected above everyone else because they could read and write. Peasants were the next most important class because the country depended on them to produce food. Artisans (people who worked with their hands) used their skills to make things that everyone needed, such as weapons, tools and cooking utensils. The lowest class were merchants. They made nothing, yet often grew rich from trading goods. Laws governed the lifestyle of people in all classes. The size and decoration of officials' houses depended on their rank. An official of the third rank could build a house with five pillars in a row. Officials of the highest importance could add a gate that was three pillars wide.

Scholar

Peasant

Artisan

Merchant

IN ORDER OF IMPORTANCE
Scholars, peasants, artisans and merchants formed the basic social order of ancient China. Soldiers who made a career of being in the army were not highly regarded and did not belong to a class of their own.

A MAGISTRATE'S DUTIES
A district magistrate was a low ranking official in the many-layered government bureaucracy, or organization. He enforced law and order; collected taxes; counted people; registered births, deaths, marriages and property; inspected schools; supervised building programs; and judged court cases.

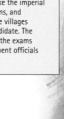

BECOMING A PUBLIC SERVANT

Scholars who were interested in the government trained for the imperial civil service. Han Emperor Wu started a university where students learned the teachings of Confucius. These men usually came from wealthy land-owning families, but anyone could take the imperial civil service exams, and sometimes whole villages sponsored a candidate. The few who passed the exams became government officials and magistrates.

A VERDICT OF GUILTY
When the court believed
that evidence collected by
investigators or statements
from witnesses were
enough to prove a person's
guilt, the accused was
encouraged to confess.

DID YOU KNOW?
Government officials, called censors,
had the task of investigating cases of
injustice or poor government. They also
informed the emperor if they thought
he was failing in his duties.

Discover more in The Han Dynasty

MODEL GRANARY
Harvested grain was stored in a granary, the most important building on an ancient Chinese farm. This knee-high tomb model of a granary is made from glazed terra cotta.

EARLY HOEING
This stone carving shows the mythical father of Chinese agriculture, Shen Nong, digging with a two-pronged stick.

• LIVING IN ANCIENT CHINA •

The Peasant and the Land

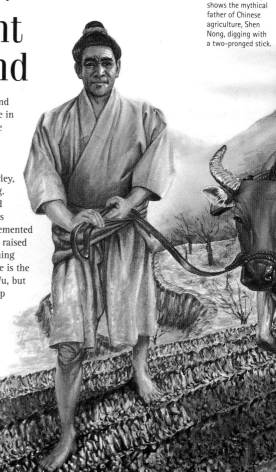

Peasant farmers cultivated small plots, and supplied food to the army and to people in the cities. Farmers made the most of the space by cutting terraces into the hill slopes. In spring, there was an important ceremony when the emperor went to the fields to plow the first furrow. Farmers in the north grew barley, wheat and millet to eat, and hemp for clothing. Those in the south planted rice in the soft mud of the flooded paddy fields. Vegetables, such as snow-peas, and lychees and other fruits supplemented these main crops. In some regions, the women raised silkworms. Every able-bodied person in a farming family worked from dawn to dusk. "Agriculture is the foundation of the world," said Han Emperor Wu, but peasants also had to serve in the army and help with government projects such as building walls, canals and dykes.

DEEP FURROWS
Farmers plowed after rain. Hard iron plows, which turned damp soil more easily, replaced softer ones made from wood or bronze. Peasants kept few large animals because they did not want to waste their precious land growing fodder for livestock.

PROBLEMS WITH WATER

In some regions, peasants pedaled irrigation machines to raise water from canals and streams so that their growing crops had regular moisture. The Yellow River spread rich, fertile silt across the valley in good years. But sometimes this great waterway dried to a trickle. At other times, it flooded or changed its course and destroyed whole villages.

DID YOU KNOW?

Everyone had to pay taxes in ancient China. Farmers often paid their taxes in the form of grain or in time spent working for the government.

Discover more in Discovering Ancient China

Family Life

PAINTED JAR
From earliest times,
the ancient Chinese
shaped, decorated
and fired clay objects.
This lidded earthenware vessel
was made by a Han potter.

Generations of Chinese families lived together in the same house. Grandfathers, fathers and uncles were considered more important than grandmothers, mothers and aunts, and the birth of a boy was celebrated more than the birth of a girl. The first household rule was that sons and daughters should obey their parents at all times. Children were taught that they must care for their mothers and fathers in sickness and old age. Grandmothers became very important when their husbands died because old people were greatly respected. This was how many women from royal families obtained power in ancient China. When a girl married, she left home and had to obey her husband and parents-in-law. Poor families sometimes sold their daughters to be servants of the rich.

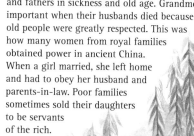

DID YOU KNOW?

Ancient Chinese names reflected the importance of the family. The family name was always written or spoken first and the personal name came last. This is still the custom today.

LACQUERED TABLEWARE
Wealthy people used lacquered wooden tableware. Gold, silver and jade banquet vessels were introduced during the Tang dynasty.

FAMILY GATHERINGS
During festivals, a more relaxed atmosphere replaced the strict daily routines of home. Families might go together to visit the graves of ancestors and fathers would spend time with their wives and children.

MINIATURE MANSIONS
Clay models from Han tombs tell us that the well-to-do lived in large houses built around spacious courtyards. These mansions had gatehouses and watch towers.

ANCESTOR WORSHIP

Long before Confucius drew up his code of behavior and encouraged ancestor worship, the ancient Chinese believed that the living could "talk" to their forefathers in Heaven. In most family homes, there was an altar for making offerings to the dead, and bronze ritual food vessels were often engraved with family achievements and honors.
In return for this respect, people expected the spirits to protect and look after them.

Emperor's hat (mian)

Lacquered gauze cage hat

Water chestnut kerchief hat

Warrior's helmet

Gauze turban (fu tou)

PROPER HEADGEAR
A man's hat completed his outfit, and he would not be seen in public without one. Hat fashions changed through the ages, but they always showed the wearer's occupation and status.

DECORATIVE COMB
Women's long hair was arranged in topknots, held in place by hairpins and other ornaments. The pattern on this comb was hammered out by a jeweler.

• LIVING IN ANCIENT CHINA •

Clothing and Jewelry

Clothing was a mark of class in ancient China. Fabric textures, colors and decoration, jewelry, headgear and footwear all told something about the wearer's rank and position in society. High-ranking officials dressed in the finest silk for public outings and celebrations, and in less expensive clothes at home. Peasants wore a long, shirtlike garment, made of undyed hemp fiber, which altered little until modern times. During some dynasties, the scholar–gentry class wore jade, gold, silver and brass jewelry, while everyone else had copper and iron accessories. Fashions for the wealthy changed as the years passed. Tang noblewomen, for example, favored the hundred-bird feather skirt, but this was later banned to prevent rare birds from becoming extinct. Tang poems praised women's elaborate make-up. One poem claimed that layers of carefully applied face powder and rouge created "a vision of loveliness."

TANG FASHION
The scholar–gentry class dressed in flowing, silk clothes. Fashionable women wore a long skirt and jacket, topped by a short-sleeved upper garment.

UPTURNED TOES
Some silk brocade shoes, made for the nobility, have survived in the tombs. This fashionable pair belonged to Lady Xin, Marquise of Dai.

DU BEADS

craftspeople who made
se beads copied designs
m Egypt and the Middle
t. Layers of colored glass
med a decoration
wn as the eyes.

E MAGIC OF MIRRORS

m Han times, polished bronze mirrors were
ss-produced. Their beautifully patterned
ks represented harmony within the universe.
aller mirrors, thought to ward off evil spirits,
ng from cords at the waist.

FOLLOWERS OF FASHION
Tang princesses wore the
latest styles in gowns,
shoes and hairdressing.
Some design ideas, like
the flowers on this dress
material, came from
countries outside China.

THE IMPORTANCE OF COLOR

Cloth in ancient China was colored with
vegetable dyes, and the color of clothing
indicated importance. This color coding
changed as one dynasty succeeded another.
From Sui times onward, only emperors were
allowed to wear yellow. Ordinary people had to
dress in blue and black. In AD 674, the
government made stricter laws to stop people
from hiding colored clothing underneath their
outer garments. White was for mourning, and
children could not wear white while their
parents were alive.

TANG FASHION
Men wore loose robes.
The wide sleeves were
weighted so they hung
down without flapping.

A Time to Celebrate

FOOD FOR A BANQUET
Food for noblemen's feasts was roasted, fried, steamed, stewed, sun-dried or pickled. Cooks seasoned some dishes with salt, and plum and soy sauces. They sweetened others with sugar and honey, or added ginger, garlic and cinnamon. The Chinese ate with chopsticks when most ancient people still used their fingers.

Festival days provided a rest from hard daily work. The Spring Festival, which welcomed the new year, lasted for several days. People lit lanterns and exploded bamboo firecrackers. They ate specially prepared vegetable dishes, drank spiced wine, watched street entertainment, and took part in ceremonies to cast out demons. Many festivals dated from Han and Tang times, but Cold Food Day reminded people of Jie Zhi Tui, who served Prince Chong Er loyally during the Spring and Autumn period. When Jie Zhi Tui died in a fire, Chong Er declared an annual festival in his memory. On Cold Food Day, kitchen fires were put out and no cooking was done. Children in the emperor's palace competed to kindle new fire by twirling sticks on wooden boards for a prize of three rolls of silk cloth and a lacquered bowl. Some ancient festivals are still important in China today.

LAST DROPS
Ladles, shaped from lacquered wood, were used to scoop wine from deep storage jars. The ancient Chinese made wine from rice.

MUSICAL ACCOMPANIMENT
Musicians performed at banquets and solemn ceremonies. In this group, three players pluck zithers with 25 strings. The other two blow bamboo mouth organs.

TANG TABLEWARE
In Tang times, metalsmiths began to hammer out sheets of gold and silver to make cups and bowls. Banquet tables were set with beautifully shaped vessels decorated with graceful designs.

DRAGON BOAT FESTIVAL

The Dragon Boat Festival, held on the fifth day of the fifth month, is still celebrated today. Rowing boats, decorated at prow and stern to look like dragons, race on lakes and rivers as part of the celebrations. Other craft took to the water in ancient times. Paintings show people enjoying the day in pleasure boats with an upper and lower deck.

Discover more in In Life and Death

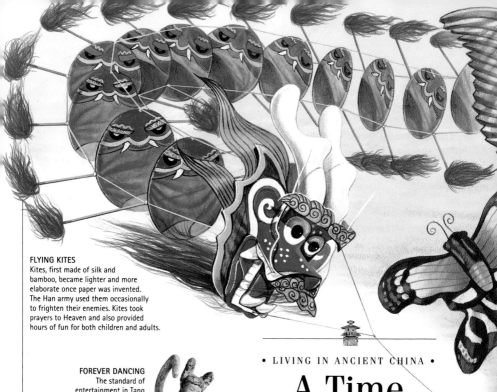

FLYING KITES
Kites, first made of silk and bamboo, became lighter and more elaborate once paper was invented. The Han army used them occasionally to frighten their enemies. Kites took prayers to Heaven and also provided hours of fun for both children and adults.

FOREVER DANCING
The standard of entertainment in Tang times was very high. Graceful figures of dancers, made from clay, were found in burial chambers at Chang'an.

STRUMMING THE STRINGS
Evidence from Tang paintings and tomb models suggests that noblewomen often performed at court. They sometimes played an instrument called the pipa, which looked like a lute.

A Time to Relax

The ancient Chinese loved entertainment. People attended theater and magic shows, acrobatics and martial arts displays, dancing, opera and musical performances. Among the instrume in the orchestra were metal and stone chimes, bamboo flutes and silk-stringed zithers. Confucius thought that harmonious sounds to soothe the soul were nearly as important as food for the body. Both r and women played games such as polo and a form of football, using a ball made from skin. Tang Emperor Xuanzong organized tug-of-wa competitions on the festive Pure Brightness Day. An onlooker record that the noise of beating drums and cheering crowds was deafening. When spring warmed the air and peach trees blossomed, the ladies at court returned to the palace gardens to swing among the budding willows. They competed against one another to reach the dizziest heights.

THE NOBLE GAME OF POLO

Polo, played from horseback, became a favorite game at the Tang court. The invention of the stirrup gave competitors much better control of their horses. Matches were held in the palace grounds at Chang'an where Emperor Zhongzong's team of four players once beat a team of ten from Tibet. Emperor Xuanzong enjoyed polo so much that he neglected his imperial duties.

HOME ENTERTAINMENT
Noblemen invited acrobats, dancers, musicians and other entertainers into their homes to amuse their guests. Performances often lasted for hours.

DID YOU KNOW?

A special Music Bureau was set up during the Han dynasty. Its staff collected official and popular songs and musical compositions.

CAMEL TRANSPORT
Groups of professional musicians traveled from place to place to perform at banquets and ceremonies. Sometimes they traveled on camels, which could carry heavy loads.

Discover more in The Tang Dynasty

Medical Practice

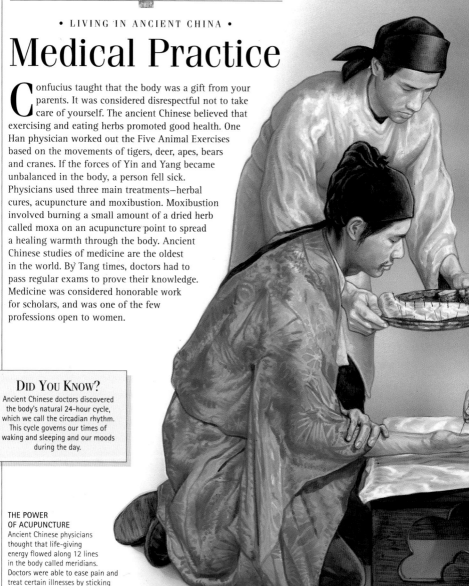

Confucius taught that the body was a gift from your parents. It was considered disrespectful not to take care of yourself. The ancient Chinese believed that exercising and eating herbs promoted good health. One Han physician worked out the Five Animal Exercises based on the movements of tigers, deer, apes, bears and cranes. If the forces of Yin and Yang became unbalanced in the body, a person fell sick. Physicians used three main treatments—herbal cures, acupuncture and moxibustion. Moxibustion involved burning a small amount of a dried herb called moxa on an acupuncture point to spread a healing warmth through the body. Ancient Chinese studies of medicine are the oldest in the world. By Tang times, doctors had to pass regular exams to prove their knowledge. Medicine was considered honorable work for scholars, and was one of the few professions open to women.

DID YOU KNOW?

Ancient Chinese doctors discovered the body's natural 24-hour cycle, which we call the circadian rhythm. This cycle governs our times of waking and sleeping and our moods during the day.

THE POWER OF ACUPUNCTURE

Ancient Chinese physicians thought that life-giving energy flowed along 12 lines in the body called meridians. Doctors were able to ease pain and treat certain illnesses by sticking acupuncture needles just below the skin at points along these meridians.

TO LIVE FOREVER

Followers of the Daoist religion believed they could find an elixir of life made from herbs and extracts of metals. This powerful mixture would allow people to live forever. Alchemists searched for a magic process that would turn base metals into gold fluid. Many elixirs, however, contained small quantities of deadly poisons such as mercury and arsenic.

Star anise

Chinese parsley (coriander)

Garlic

Ginseng

AN ANCIENT WORKOUT
These fragments are part of a chart painted on silk. It was found in a Han tomb and presented more than 40 exercises to keep the body in good shape.

GUARDIANS OF THE HOURS
The ancient Chinese believed that spirits guarded the hours of the day. This Han tile shows the Guardian of Midnight on duty from 11 pm to 1 am.

Discover more in Family Life

The Trade Routes

WESTERN WINE
This pottery figure of a westerner who lived in Chang'an's foreign community was found in a Tang tomb.

KASHGAR MARKETS
In Han times, Chinese camel caravans stopped at Kashgar in central Asia. From there, Middle Eastern merchants, who wanted a share of the profits, took the expensive goods to distant places such as ancient Rome, which bordered the Mediterranean Sea.

Trade began when neighboring peoples wanted luxury goods made in China. At first, there was more raiding than trading, but soon the Chinese began to exchange silk for the Ferghana horses of central Asia. Overland, the Silk Road eventually extended from Chang'an in China to the Mediterranean lands. This route was only possible because the ancient Chinese had camels for transport. These beasts each carried 440 lb (200 kg) of cargo around the "white dragon dunes" of the terrible Taklamakan Desert. They could smell underground water, and warned their riders of deadly, suffocating sandstorms by huddling together, snarling and burying their mouths in the sand.

From at least as early as Han times, sea routes took Chinese traders to Vietnam and later to Korea and Japan. Inside China, merchants transported goods such as grain and salt along the canals and roads that linked the large cities.

VITAL FOR SURVIVAL

The Chinese needed horses to help fight off mounted nomads roaming their borders. Horses were difficult to breed in northern China so the Chinese imported them.

...VES OF LUXURY

...ade goods made people's lives ...ore comfortable. Merchants ...changed Chinese silk, lacquerware, ... and spices for gold, silver, glass, ...ol, pearls, furs and other luxuries.

THE SILK ROAD

In 139 BC, Emperor Wu sent his minister Zhang Qian to Ferghana in central Asia to buy horses. On the way, he was captured by the Xiongnu. When Zhang Qian returned to China ten years later, he brought valuable information about countries to the west. The first Chinese merchants set out into central Asia in 114 BC. This trade route became known as the Silk Road. Within a few years camel caravans traveled frequently along it.

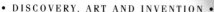

Reading the Heavens

The ancient Chinese believed the night sky could tell them what was about to happen on Earth. Court astronomers studied the stars to improve their methods of making predictions to the emperor. People expected the Son of Heaven to reign in harmony with the universe, and to be able to foretell celestial events. His failure to do this might mean that he was unfit to rule. By 1300 BC, astronomers were recording eclipses of the sun and moon and movements of comets. Later, star catalogs listed individual stars with great precision. Writings on astronomy discovered in one Han tomb showed detailed knowledge of heavenly bodies such as the planets Venus, Jupiter, Mercury, Mars and Saturn. Only a select few were permitted to read the Heavens. There were severe punishments for ordinary people who tried to own astronomical instruments or chart the stars.

STUDYING THE STARS
Every night, astronomers went out to the city walls to observe and record the stars. They had maps of the sky and instruments to help them. Astronomers believed that the appearance of a comet indicated forthcoming disaster.

ASTRONOMICAL MAP
The ancient Chinese grouped the stars into 28 houses. On this box lid, the names of the houses circle the written symbol for the Great Bear star.

STARS IN STONE
This engraving shows five star groups and the animals or persons linked with them. The hare, at the top left, symbolizes the constellation Fang.

CHARTING THE STARS

The Chinese began reading the Heavens a long time ago. Eclipses of the sun and the moon are recorded on Shang oracle bones, and Shang astronomers marked the changing seasons by the stars. This ancient Chinese star chart was based on the work of three astronomers who lived in the fourth century BC. It was the custom to record groups of stars as circles linked with lines.

MEASURING THE STARS
The ancient Chinese used an armillary sphere to measure the stars. It is a collection of bronze rings, each marked as a gauge of measurement. The rings represent imaginary lines round the Earth, such as the equator.

Discover more in Myths and Symbols

FIRST EDITION
Probably the earliest complete printed book to survive is a Buddhist text called *The Diamond Sutra*, printed in AD 868. It is in the form of a scroll nearly 20 ft (6 m) long.

**SIGNE
WITH A SEA**
Seals, usual
impressed into red ink past
frequently replaced signature
The seals had from one to dozer
of characters, and were carve
or molded from bronze, silve
stone, horn, wood or jad

DROP BY DROP
A calligrapher had special equipment, such as this pottery water dropper in the shape of a duck.

• DISCOVERY, ART AND INVENTION •

Writing and Printing

JADE BRUSH REST
Brush rests, often designed in animal forms, stopped wet brushes from rolling when the calligrapher was not using them.

Writing began to develop very early in ancient China. Early inscriptions were written on oracle bones and then on bronze ritual vessels. Bamboo strips, wooden tablets and pieces of silk were also used as writing materials. In about AD 105, government official Cai Lun suggested that pulping bark, roots, rags and old fishnets would improve the quality of paper, which had been invented several centuries before. This process was one of the most important discoveries of all time. In the Tang era, impressions from seals and the making of ink rubbings of engraved stone tablets led to the idea of printing. Text was written on fine paper and pasted, front side down, onto a wooden block. The printer cut away the background to leave raised characters. He then inked the surface of the block and pressed paper sheets against it to produce an image that was the right way up.

KEEPING THE RECORDS
Before people had paper, they wrote on bamboo, cutting the characters into the thin strips of wood from top to bottom. Many government records survive in this form.

CHINESE CHARACTERS

Chinese writing uses symbols for words and phrases and is read vertically (down and up) rather than horizontally (side to side). Some characters have up to 26 brushstrokes, which must be drawn in the correct order. Qin Shi Huangdi ordered a standard form of writing so that imperial commands could be read throughout the country. This script has not changed much until recent times.

口
mouth

日
sun

月
moon

王
ruler, king

HE ART OF CALLIGRAPHY

ncient Chinese calligraphers wrote
ith brushes made from animal
airs, tied together with
ne silk threads and
ued in bamboo tubes.
alligraphers mixed
eir ink by rubbing
solid ink stick
ith drops of water
n an ink stone.

Artists and Artisans

Painting was an important art form that developed alongside calligraphy. Artists had to perfect their brushstrokes, use a variety of colors, produce well-balanced compositions and represent their subject matter accurately. From the fourth century AD, painters were often recognized by name, but artisans who worked in teams usually remained anonymous. They made bronze, jade, clay and other materials into beautiful objects for religious rituals and household purposes, and fashioned thousands of tomb models of almost everything to do with daily life. Once the process of iron casting was developed, governments set up iron foundries to mass-produce agricultural tools and military weapons. Other state factories turned out lacquerware and silk cloth.

In the ninth century AD, an Arab author, Jahiz of Basra, commented that the Turks were the greatest soldiers, the Persians the best kings and the Chinese the most gifted of all craftspeople.

THE ART OF LACQU
Sap from the lacquer tree is t
oldest industrial plastic knov
to humans. Wood, bamboo
cloth utensils, coated with ma
thin layers of lacquer, c
withstand the heat of cookir
The ancient Chinese color
lacquer black, red, brow
yellow, gold and gree

JADE DRAGON
The ancient Chinese valued jade, the "Stone of Heaven," above all other materials. Dragons, which were believed to have special powers, appeared frequently in their art.

TURNING THE PAGE
Books with bound pages took the place of scrolls in ancient China during the Tang dynasty. By this time, more women were learning to read.

WINE VESSEL
Birds with long plumes began appearing as decoration on bronze containers made in the middle Western Zhou period.

380

PATTERNED BOXES
Han potters were the first to glaze with lead. By the Tang period, lead-glazed ceramics were more boldly decorated. Potters often chose bright colors.

PEOPLE OF JADE
Hard jade stone was difficult to work. It was shaped with bamboo drills tipped with bronze, rubbed smooth with abrasive rock sand, and buffed on wood and leather polishing wheels.

BRONZE STATUE
Bronze, a mixture of copper, tin and sometimes lead, lasts forever. Silk and wood often rots to dust.

LACQUER ON THE LINE
Wealthy Chinese babies were fed using lacquer spoons and lacquer bowls. Wealthy Chinese people were buried in lacquer coffins. Artisans on production lines turned out many thousands of costly lacquer goods. This wine cup, surviving from the Han dynasty, is inscribed with the names of eight artisans and five supervisory officials who helped to produce the vessel.

The Finest Silk

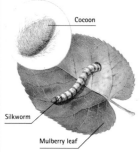

Cocoon

Silkworm

Mulberry leaf

LIFE CYCLE
Female silkworm moths lay hundreds of eggs. After hatching, the caterpillars eat steadily, shedding their tight skins four times as they grow. In four to five weeks they spin cocoons to protect them while they pupate into moths.

The filament from one silkworm's cocoon is several hundred feet long and it takes many thousands of filaments to produce enough thread for a dress length of fabric. The ancient Chinese wrote and painted on this expensive material, and made bed quilts and bags from it. Emperors presented beautifully woven and embroidered pieces to neighboring countries. Women in ancient China developed the silk industry. They began making silk cloth in Stone Age (Neolithic) times, but the rest of the world did not discover their production secrets until many centuries later. Han traders, who carried the precious fabric along the Silk Road to the west, sold a roll of silk for the same price as 795 lb (360 kg) of rice. In Rome, the writer Pliny complained that women's desire to wear silk was ruining the Roman Empire.

EMBROIDERED SILK
In this piece, colored silk and gold-covered threads delicately pattern a cream background. Flower and bird designs were much loved from Tang times onward.

STRANGE BUT TRUE
Silk was very precious. Pliny thought silk came from "the hair of the sea-sheep." Silkworms raised for their thread today eat only mulberry leaves and must be protected from noise, vibrations and strong smells.

PAINTING ON SILK
This painting of four scholars was done on the finest silk, which provided a smooth surface to paint on. But first it had to be coated with a substance called alum to prevent the inks from soaking into the fabric.

WEAVING SILK

The finest quality silk had six or seven smooth filaments twisted to form the thread. Coarser weaves had more. The weavers (shown below) fastened the ends of the vertical warp threads to belts around their waists and held them tight with bars under their feet. They were then able to pass the horizontal weft threads over and under the warp threads.

SILK PRODUCTION

The ancient Chinese gathered mulberry leaves to feed the caterpillars they raised to spin their silky coverings. Women rinsed the cocoons in hot water to loosen the filaments of thread before the adult moths could bite their way out.

Discover more in Clothing and Jewelry

Arched bridge
Engineer Li Chun designed an arched bridge in AD 610. It was stronger and took less stone to build than other bridges.

Horsepower
The Han invented trace (or breast) harnesses for horses. These replaced choking throat straps and greatly increased the animals' pulling power.

• DISCOVERY, ART AND INVENTION •

New Ideas

The ancient Chinese were always looking for practical ways to solve problems. Farmers hung bags of killer ants in their orchards to eat the insect pests that would otherwise have destroyed the mandarin orange crop. Ancient Chinese inventors were way ahead of the rest of the world. They developed wheelbarrows, for example, about 1,300 years before Europeans copied the idea. Their inventions of paper, printing, the compass and gunpowder have probably had more impact on the world than anything else that has ever been invented. The earliest compass, called a "south-pointing fish," consisted of a wooden fish containing a piece of metal floating in a bowl of water. Gunpowder, made from saltpeter, sulfur and charcoal, was used by alchemists and physicians long before it was used for weapons. Other notable inventions included matches, the game of chess and mechanical clocks.

Rudder
Ancient Chinese ships had rudders, essential for steering properly, by the first century AD. European navigators began to use rudders around AD 1180.

TES

e ancient Chinese were probably
e first people to make something
at flew. The earliest kites
ay have become airborne
the fifth century BC.

Fishing reel
The fishing reel was developed
from a battle device designed
to retrieve a javelin after it
had been thrown at an enemy.

UMBRELLA
Oiled paper umbrellas, manufactured from
mulberry bark, protected people against rain and
sun. Emperors used red and yellow umbrellas,
while ordinary people carried blue ones.

TEA SHREDDER
When tea drinking became popular
in Tang times, an inventor devised
a tool for shredding tea leaves.
This was quicker than chopping
them by hand.

Wheelbarrow
Wheelbarrows appeared
around the first century AD.
This type, called a wooden
ox or a gliding horse, could
be pushed or pulled.

EARTHQUAKE DETECTOR

Han Astronomer-Royal, Zhang Heng, designed
a detector to locate areas hit by earthquakes.
Eight dragons' heads were equally spaced around
a bronze vessel containing a pendulum. When an
earthquake made the pendulum move, a ball from
a dragon's mouth fell noisily into a waiting toad's
mouth, and indicated the general direction of the
tremor. Then the government could send help
quickly before food riots broke out.

Discover more in A Time to Relax

YUAN DYNASTY (1279-13
When the Mongols conquered China
founded the Yuan dynasty, they brought g
wealth to the country through internatic
trade. Kublai Khan was the most fam
Mongolian empe

**SONG DYNASTY
(960–1279)**
The Song expanded foreign
trade and built sea-going
junks with sails. Navigators
consulted star charts and
adapted the "south-pointing
fish" compass for use at sea.

SEATED IN PEACE
Yuan sculptors introduced a
more elegant figure style.
This bronze statue of
Bodhisattva Guanyin,
made in 1339, was a
Buddhists monk's
offering in his search
for enlightenment.

DRAGON POWER
Dragons continued
to be a popular
theme in Song art.
This jade plaque,
depicting two of the
mythical creatures
intertwined, shows
how advanced jade work
had become.

• DISCOVERY, ART AND INVENTION •

Into the Modern World

When we refer to ancient China, we are describing a period of time, rather than the rise and fall of a civilization. Ancient China did not disappear like ancient Egypt or the Roman Empire. Imperial Chinese rule survived until the twentieth century. When the Tang fell from power in AD 906, north and south China were briefly divided during the period of the Five Dynasties and Ten Kingdoms. The Song dynasty reunited the country. Invader from Mongolia set up the Yuan dynasty, only to be driven out by the Ming after less than 100 years. In 1644, conquering tribes from northern Manchu established China's final dynasty—the Qing. Each dynasty used achievements from the past but also developed in its own way. New styles of art and craft, for example, flourished in the Ming era, but Qin Shi Huangdi's system of writing changed little until present times.

MING DYNASTY (1368-1644)
Ming emperors were constantly threatened by the Mongols, who fought to regain their lost empire. The Ming rebuilt the Qin dynasty's Long Wall into the Great Wall of China.

QING DYNASTY (1644-1912)
China's empire grew weaker toward the end of the Qing dynasty. In 1908, three-year-old Puyi became the last emperor. He was forced to abdicate (give up his throne) on February 12, 1912.

Q: How did the last emperor end his days?

CHINA FROM CHINA
In the eighteenth century, scenes of the waterfront in Canton, such as this one on a Qing porcelain punchbowl, were painted for European traders to take home as souvenirs.

PRICELESS VASE
The Ming dynasty is famous for ceramics. Ming blue-and-white porcelain dishes, bowls and containers of all types became prized throughout the world.

THE PEOPLE'S REPUBLIC OF CHINA

The republic that replaced imperial government in China was threatened by Japanese invasion in 1937. It joined briefly with the communist party to fight against the Japanese. Unsettled times in China ended when the communist party, led by Mao Zedong (above left), gained power in 1949. Mao ruled China until his death in September 1976.

STRANGE BUT TRUE
The last emperor was treated like a living god until he was six. He became a gardener and citizen of the People's Republic of China and died at the age of sixty-two.

Discover more in Order in Society

387

Discovering Ancient China

RICE FARMING TODAY
In the paddy fields of present-day China, the farmers go about their work in much the same way as they did in ancient times.

W hen people die, the objects they leave behind tell us much about how they lived. Such evidence becomes more difficult to find as time goes on because all materials do not last well. As centuries slip by, natural disasters such as floods and earthquakes destroy villages, towns and even cities. Abandoned settlements often become buried and forgotten. We are fortunate that the early Chinese made things of bronze, clay and jade, and wrote on bone and shell. We are lucky, too, that they placed their dead in sealed underground tombs with treasure stores of beautiful artwork and household possessions. Their passion for keeping detailed records of government affairs also helps us to find out about dynasties from long ago. Discovering ancient China is an ongoing process and thousands of sites across the country are still being examined. Little by little, archaeologists are digging up China's past.

CHARIOTS OF BRONZE
Archaeologists document every find with absolute precision. These chariots unearthed in a chamber on the western side of the First Emperor's tomb in 1980 have been restored to near-perfect condition.

REMARKABLE REMAINS
Lady Xin, Marquise of Dai died sometime after 168 BC. X-rays of her well-preserved remains found the cause of death was a heart attack. She had just eaten a large piece of melon and 138 seeds were discovered inside her.

PRESERVED ON PAPER
The dry atmosphere in the Buddhist cave temples at Dunhuang on the edge of the desert preserved sculptures, silk paintings and paper pages, like this one.

ANCIENT VILLAGES

Banpo, near the modern city of Xi'an in northern China, was first settled at least 6,500 years ago. Archaeologists have learned much from excavating it and other Stone Age (Neolithic) sites. Houses had square, oblong or round frameworks of poles, plastered with mud and thatched with reeds. Potters produced well-shaped vessels without the help of a potter's wheel. They decorated them with bold designs and fired them in a kiln.

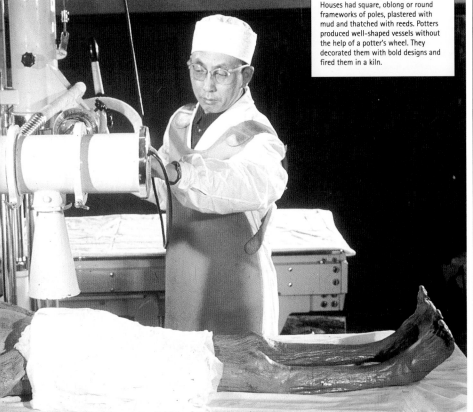

Dynasties of Ancient China

Many dynasties in ancient China lasted for hundreds of years. They took the family name of the first ruler. The emperor had more than one wife and there was usually an eldest son to take his place when he died. If there was no suitable son, another male heir was chosen from the family. It was very rare for a woman to hold power—Empress Wu Zetian was a notable exception. The most common way for a dynasty to end was when a strong opposing group seized government by force. The following list includes some of the most important rulers of ancient China. All dates before the Zhou dynasty are approximate.

Shang bronze

6000–1700 BC	**STONE AGE (NEOLITHIC) PERIOD**

Northeastern China

5000 BC	**Xinglongwa**
4500 BC	**Chahai**
4500–4000 BC	**Zhaobaogou**
3500–2500 BC	**Hongshan**

Stone Age (Neolithic) pottery

North central China

5000–3000 BC	**Central Yangshao**
3000–1500 BC	**Gansu Yangshao**

Eastern China

4500–2500 BC	**Dawenkou**
2500–1700 BC	**Longshan**

Southeastern China

5000–4500 BC	**Hemudu**
5000–4000 BC	**Majiabang**
4000–3000 BC	**Songze**
3000–2000 BC	**Liangzhu**

South central China

4000–3300 BC	**Daxi**
2500–2000 BC	**Shijiahe**

South China

3000–2000 BC	**Shixia**
2000–1600 BC	**ERLITOU PERIOD**
1600–1050 BC	**SHANG DYNASTY**
	1600–1400 BC Erligang period
	1400–1050 BC Anyang period
1050–771 BC	**WESTERN ZHOU DYNASTY**
770–221 BC	**EASTERN ZHOU DYNASTY**
	770–475 BC Spring and Autumn period
	475–221 BC Warring States period
221–207 BC	**QIN DYNASTY**
	221–210 BC Qin Shi Huangdi reigns
	209–206 BC Second emperor of Qin reigns

Zhou bronze

6 BC–AD **9**	**WESTERN HAN DYNASTY**
	202–195 BC Gaodi reigns
	141–87 BC Wudi reigns
9–25	**XIN DYNASTY**
25–220	**EASTERN HAN DYNASTY**
	AD 25–57 Guangwudi reigns
	AD 58–75 Mingdi reigns
	AD 88–105 Hedi reigns
221–280	**THREE KINGDOMS**
	220–265 Wei
	221–263 Shu
	222–280 Wu

Han pottery

Qin terra cotta

AD **420–589**	**SOUTHERN DYNASTIES**
	420–479 Liu Song
	479–502 Southern Qi
	502–557 Liang
	557–589 Chen
AD **581–618**	**SUI DYNASTY**
AD **618–906**	**TANG DYNASTY**
	618–626 Gaozu reigns
	626–649 Taizong reigns
	649–683 Gaozong reigns
	690–705 Empress Wu Zetian reigns
	712–756 Xuanzong reigns
	888–904 Zhaozong reigns
	904–907 Aizong reigns

265–316	**WESTERN JIN DYNASTY**
317–420	**EASTERN JIN DYNASTY**
386–581	**NORTHERN DYNASTIES**
	386–535 Northern Wei
	534–550 Eastern Wei
	535–557 Western Wei
	550–577 Northern Qi
	557–581 Northern Zhou

Tang pottery

Ancient
Greece

Why did people cross the river Styx?

In which city-state would you join the army at the age of seven?

What could a dinner guest expect to eat at a banquet in ancient Greece?

Contents

• THE GREEK WORLD •

• LIVING IN ANCIENT GREECE •

ITALY

MACEDONIA

Mount Olympus ▲

IONIAN SEA

Thermopylae ●

Delphi ●

Thebes ●

Marathon ●

Corinth ●

• THE GREEK WORLD •

A Seafaring People

T he ancient Greek world took shape between 3200 and 1100 BC. This was during the Bronze Age when people first began melting copper with tin to make bronze. Island communities occupied the Cyclades and Crete, and the Mycenaean civilization lived on mountainous mainland Greece. From the beginning, the ancient Greeks farmed the narrow valleys and coastal plains, wherever there was good soil and a river or a freshwater spring. Through the centuries, they cut trees from the mountain slopes for firewood and to build ships. They used ships more than any other means of transport, sailing the seas to trade, to go to war and to settle new places. Ancient Greek legends told of brave seafarers who survived dangerous voyages. Greece's boundaries expanded greatly from around 700 BC onward. Art and ideas developed strongly, and during the Hellenistic Age, the Greek way of life spread to many other countries.

● Olympia

Sparta ●

MEDITERRANEAN SEA

- Troy

EGEAN SEA

ASIA MINOR

- Ephesus

YCLADES

RHODES

nossos •
RETE

AGES OF ANCIENT GREECE

CYCLADIC CIVILIZATION
3200–1100 BC
A Cycladic figurine of a harp player made from marble.

MINOAN CIVILIZATION
3200–1100 BC

MYCENAEAN CIVILIZATION
1600–1100 BC
A Mycenaean pottery cup decorated with a pattern of cuttlefish.

DARK AGE AND GEOMETRIC PERIOD
1100–700 BC

ARCHAIC PERIOD
700–480 BC
This Archaic bronze griffin's head was once attached to a large cooking pot.

CLASSICAL PERIOD
480–323 BC
A Classical red-figure vase showing Odysseus meeting a swineherd.

HELLENISTIC AGE
323–31 BC
A Hellenistic terracotta figure of the god Eros.

Discover more in End of an Empire

FOLDED ARMS
Many marble figurines found on the Cyclades have the right arm folded below the left. The reason for this is unknown.

BULL LEAPING
Bulls were sacred animals to the Minoans. Young men and women leapt over them during religious ceremonies.

Early Settlements

The first Greek civilization was on a circle of small islands in the Aegean Sea called the Cyclades. The islanders grew grain, grapes and olives and raised animals for milk and meat. They produced clay pots and sculpted figurines out of marble, using blades and chisels made from hard volcanic rock and bronze. After many centuries, a disastrous earthquake followed by a huge volcanic explosion buried some Cycladic communities. On the large island of Crete to the south, however, other settlers were flourishing. Archaeologists have named these people after Minos, a legendary Cretan king. The Minoans built towns around large palaces and traded with many foreign countries. They invented a form of picture writing and then began to use symbols in a script archaeologists call Linear A. Mycenaeans on the nearby mainland later took over the island of Crete and became the most powerful people in the Aegean.

WALL PAINTING OF THE FLEET
At Akrotiri, on the Cycladic island of Thera, artists decorated the walls of many houses. This picture, painted around 1500 BC, is one of the best preserved.

SAILING AWAY
The painting tells the story of a sea voyage. The flagship (top center) is carrying the leader of the expedition. Dolphins frolic around the boats in the calm water.

MINOAN GODDESS
It is thought that the Minoans put goddesses above gods in their religion. This goddess, perhaps a protector of the household, holds a pair of sacred snakes.

CUTTING EDGES
The double-headed axe was a sacred symbol in Minoan religion. It appeared on palace walls, pots and tombs.

KNOSSOS FRESCOES
The palace at Knossos was the largest palace on Crete. Built of stone and mud brick, it sprawled across a huge area. Wall paintings such as this have been restored to look as they once did.

THE MINOTAUR

Legend says that King Minos kept the horrible Minotaur—half bull, half man—in a labyrinth (maze) in the center of his palace. Every year, 14 young Athenians were sent to Crete to be fed to this monster. When brave Theseus arrived from Athens, King Minos' daughter Ariadne fell in love with him. She gave Theseus a ball of thread and he tied one end at the entrance to the labyrinth. Then he went to find the Minotaur, unraveling thread to show him the way out. The vase painting (above) shows Theseus killing the monster.

RETURNING HOME
The travelers are wearing Mycenaean armor and carrying Mycenaean weapons, so they may be returning to Mycenae on mainland Greece after visiting the Cyclades.

Discover more in Clay and Metal

399

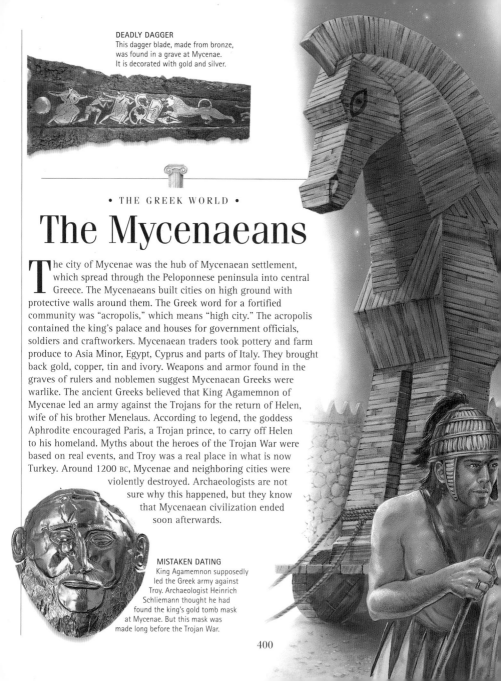

DEADLY DAGGER
This dagger blade, made from bronze, was found in a grave at Mycenae. It is decorated with gold and silver.

• THE GREEK WORLD •

The Mycenaeans

The city of Mycenae was the hub of Mycenaean settlement, which spread through the Peloponnese peninsula into central Greece. The Mycenaeans built cities on high ground with protective walls around them. The Greek word for a fortified community was "acropolis," which means "high city." The acropolis contained the king's palace and houses for government officials, soldiers and craftworkers. Mycenaean traders took pottery and farm produce to Asia Minor, Egypt, Cyprus and parts of Italy. They brought back gold, copper, tin and ivory. Weapons and armor found in the graves of rulers and noblemen suggest Mycenaean Greeks were warlike. The ancient Greeks believed that King Agamemnon of Mycenae led an army against the Trojans for the return of Helen, wife of his brother Menelaus. According to legend, the goddess Aphrodite encouraged Paris, a Trojan prince, to carry off Helen to his homeland. Myths about the heroes of the Trojan War were based on real events, and Troy was a real place in what is now Turkey. Around 1200 BC, Mycenae and neighboring cities were violently destroyed. Archaeologists are not sure why this happened, but they know that Mycenaean civilization ended soon afterwards.

MISTAKEN DATING
King Agamemnon supposedly led the Greek army against Troy. Archaeologist Heinrich Schliemann thought he had found the king's gold tomb mask at Mycenae. But this mask was made long before the Trojan War.

GIFT HORSE

After fighting the Trojans for ten years, the Greeks left a huge wooden horse outside Troy's walls and sailed away. The curious Trojans wheeled the statue into the city. That night, Greek soldiers hiding inside the hollow horse crept out and opened the gates. Their army, which had returned silently, entered Troy and defeated the Trojans.

SEARCHING FOR TROY

As a small boy, Heinrich Schliemann read about the Trojan War in Homer's *Iliad* and he vowed to find Troy. In 1870, relying on details from Homer's poem, Schliemann began to unearth a city at Hissarlik in Turkey. He discovered jewelry that he believed had belonged to Troy's King Priam. Nine cities lying on top of one another have since been found at Hissarlik. The city in the sixth layer is most likely to have been Troy.

Sophie Schliemann wearing jewelry excavated by her husband.

FRESCO AT MYCENAE

Frescoes are pictures painted on the walls of buildings while the plaster is still wet. These donkey-headed demons survive from the thirteenth century BC.

DID YOU KNOW?

Homer, a poet who lived in the eighth century BC, retold well-known Mycenaean legends in the *Iliad* and *Odyssey*. These long poems were possibly not written down until after Homer died.

MARINE THEME

The tentacles of an octopus writhe around this jar. Sea creatures, an important source of food, were popular in designs on Mycenaean pottery.

LIONS ON GUARD

The Lion Gate, named for the animals carved above it, was the main entrance to Mycenae. It was built about 1250 BC when the enormous stone city walls were increased in size.

Discover more in Stories of Daring Deeds

Settling New Lands

MINTED TO MATCH
Early silver coins from some southern Italian colonies had a raised design on one side and the reverse design on the other. These coins were very difficult to make.

When the Mycenaean civilization ended, troubled times fell on ancient Greece. This period, now called the Dark Age, lasted about 400 years. It is thought that people forgot how to write because no writing has been found from this era. When food became scarce some people left their homeland to find new places to settle. They migrated to the coast of Asia Minor, stopping off at the Aegean Islands on the way. By the eighth century BC, Greece had recovered, but the growing population was overcrowding the mainland. Greek colonies spread around the Black Sea to the northeast, and west to southern Italy, and as far as France and Spain. Most colonies became farming communities, but a few were set up as trading posts. Successful settlements sent supplies back to Greece to relieve shortages there. The Black Sea ports exchanged grain and timber for wine, olive oil and honey.

GRIFFIN ON GUARD
Greek craftworkers sometimes learned new skills from their neighbors. This griffin's head was made by bronze hollow casting, a method used in Asia Minor.

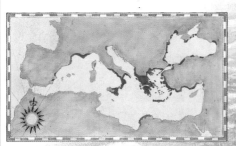

A RING OF COLONIES
The red on this map shows where the Greeks settled around the edges of the seas. Settlers looked for a natural harbor, good soil and a climate similar to their homeland.

RICHES FROM THE COLONIES
This scepter is covered in gold, wrapped in gold wire, and crowned with gold acanthus leaves. It was made in Taranto, a wealthy colony in southern Italy.

TIES WITH HOME
Settlers built new cities to look like those they had left behind. There was a central meeting place and a temple on the highest ground for their special god or goddess.

ON THE RUN
Nike was the winged
goddess of victory. This
bronze statue might once
have adorned the rim of a
bowl in southern Italy.

AFE HARBOR
olonies were usually established
eside the sea. Settlements
eveloped their own systems of
overnment but kept in touch with
eir homeland. Ships sailed to and
o carrying supplies and news.

THE ORACLE OF DELPHI

Leaders of colonizing expeditions
always consulted a priest in
Apollo's temple at Delphi about where
to settle. The buildings at the sacred
site of Delphi on the steep slopes of
Mount Parnassus were a center of
ancient Greek religious life. The vase
below shows a procession on its way
to visit Apollo's shrine. People believed
Apollo answered questions about the
future through his priestesses who
spoke while in trances. Priests
explained these answers, known as
oracles. Oracles
could often be
understood in
more than one
way. So, whatever
happened, the
hearer would think
that the oracle had
come true.

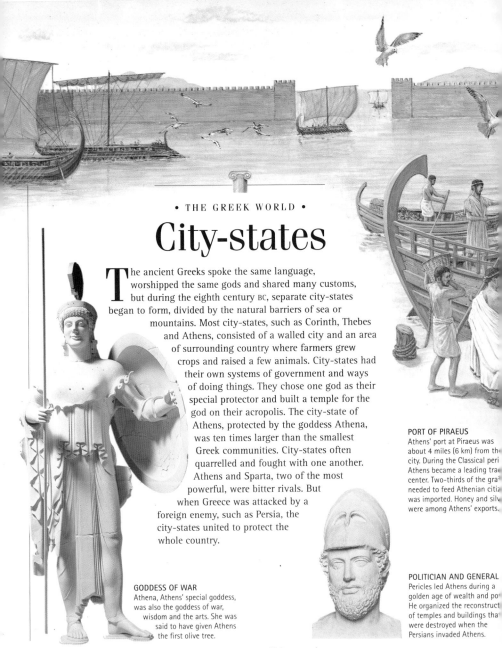

City-states

The ancient Greeks spoke the same language, worshipped the same gods and shared many customs, but during the eighth century BC, separate city-states began to form, divided by the natural barriers of sea or mountains. Most city-states, such as Corinth, Thebes and Athens, consisted of a walled city and an area of surrounding country where farmers grew crops and raised a few animals. City-states had their own systems of government and ways of doing things. They chose one god as their special protector and built a temple for the god on their acropolis. The city-state of Athens, protected by the goddess Athena, was ten times larger than the smallest Greek communities. City-states often quarrelled and fought with one another. Athens and Sparta, two of the most powerful, were bitter rivals. But when Greece was attacked by a foreign enemy, such as Persia, the city-states united to protect the whole country.

PORT OF PIRAEUS
Athens' port at Piraeus was about 4 miles (6 km) from the city. During the Classical peri Athens became a leading tra center. Two-thirds of the gra needed to feed Athenian citiz was imported. Honey and silv were among Athens' exports.

GODDESS OF WAR
Athena, Athens' special goddess, was also the goddess of war, wisdom and the arts. She was said to have given Athens the first olive tree.

POLITICIAN AND GENERAL
Pericles led Athens during a golden age of wealth and po He organized the reconstruct of temples and buildings tha were destroyed when the Persians invaded Athens.

LIVING IN SPARTA

The city-state of Sparta was enclosed by a mountain range, so it did not need protective walls. It was the only city-state to keep a permanent army, and Spartan soldiers were recognized as the best in ancient Greece. All Spartan children belonged to the state and boys began their tough military training at the age of seven. Even when they married, Spartan men still lived in army barracks.

TEST OF TIME
This painting from the nineteenth century shows how well some of the temples on Athens' acropolis survived through the centuries.

Discover more in Building in Stone

WATER CLOCK

In court, speakers were timed by a water clock. Their time was up when all the water from the upper pot had run into the lower one.

SELECTING A JURY

Citizens fit their names into the slots of an allotment machine. A fragment of one is shown here. Colored balls dropped beside the rows of names to show the jurors for that day.

• THE GREEK WORLD •

Government and the Law

In early times, groups of rich landowners ran the city-states, but sometimes one leader, called a tyrant, seized power. Tyrants usually ruled fairly, but some were cruel and unjust. Athens introduced a system of government called democracy. Many other city-states developed the same system. We know most about the way Athens was organized from surviving evidence. In Athens, democracy allowed every citizen to have a say in state affairs. But only men who were born in the city-state and were not slaves could become citizens. A council of 500 citizens, drawn annually in a lottery, suggested new laws and policies. Citizens voted at the assembly to accept, change or reject these suggestions. Juries of more than 200 citizens tried most Athenian law cases. Jurors were also chosen by lot. There were no lawyers, and only citizens could speak in court.

TRIAL BY JURY

After a trial, jury members cast their verdicts with bronze discs. They used tokens with solid centers to show the accused was innocent, and tokens with hollow centers to show the accused was guilty.

PUBLIC SPEAKER

Oratory is the art of making public speeches. Aeschines, the famous Athenian orator, started a school in Rhodes for speech makers.

POINTS OF VIEW
Any Athenian citizen, rich or
poor, could explain his point
of view to the assembly. At
least 6,000 citizens had to be
present before a meeting could
begin. Once all opinions about
a matter had been heard, the
whole assembly voted on it.

LOSING THE VOTE

The Athenians had a way of getting rid of
politicians they did not trust. Once a year,
assembly members could vote against the ones they
disliked. Citizens wrote the names of unpopular politicians
on pieces of pottery called ostraca. A man who received
more than 6,000 votes had to leave Athens for ten years. This
method of exiling people was called ostracism. Arissteides and
Kimon, named on these fragments, were both ostracized.

On Mount Olympus

The top of Mount Olympus, the highest mountain on mainland Greece, is often hidden by clouds. The ancient Greeks imagined that gods and goddesses, who looked and behaved like humans, lived on the mountain. However, these supernatural beings drank nectar and ate ambrosia, which made them immortal—they could not grow old or die. The people believed the gods and goddesses controlled events in life and nature, and had power to shape the future. Apollo made the sun rise and set. Hermes cared for travelers and led souls to the Underworld. Zeus, the king of the gods, roared with thunder and threw lightning bolts when he was angry. Poseidon whipped up storms at sea or caused earthquakes. Every village had its own guardian god or goddess although some were not important enough to live on Mount Olympus. The Greeks built temples for their deities and organized animal sacrifices, processions, plays and games to please them.

THINGS FROM THE PAST
Sometimes, archaeologists find it difficult to identify the things they find. This statue is thought to be the goddess Hera.

Zeus
God of the sky and thunder.

Head of Zeus

Poseidon
God of earthquakes, the sea, horses and bulls.

Hestia
Goddess of the family and the hearth.

Hermes
Messenger of the gods and protector of travelers.

Aphrodite
Goddess of love and beauty.

Athena
Goddess of wisdom, art and war.

Ares
God of war.

THE GODDESS ARTEMIS
This terracotta figurine of Artemis dates from the Hellenistic Age. It was made in Myrina, a Greek city in Asia Minor.

THE GODDESS APHRODITE
According to one legend, beautiful Aphrodite was born from sea foam. Here, her attendants are helping her to rise from the water.

PAN, A NATURE SPIRIT

Country people worshipped nature spirits. Pan, protector of shepherds and their flocks, was the best known of these lesser supernatural beings. Pan was a satyr—half man, half goat. Although his upper body was human, his face was goatlike and he had horns. Pan wandered across lonely mountain tops playing his panpipes—a wind instrument made from seven hollow reeds, graded in length and bound together.

GODDESS WITHOUT A NAME
The early Mycenaeans worshipped some deities that later Greeks placed on Mount Olympus. Mycenaean sites also contained terracotta images of unknown gods and goddesses, such as this one.

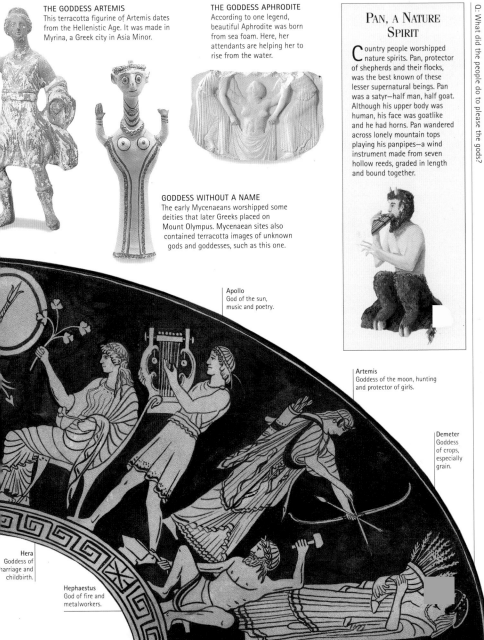

Apollo
God of the sun, music and poetry.

Artemis
Goddess of the moon, hunting and protector of girls.

Demeter
Goddess of crops, especially grain.

Hera
Goddess of marriage and childbirth.

Hephaestus
God of fire and metalworkers.

Stories of Daring Deeds

CLASH OF WARRIORS
In this scene from the Trojan War, Menelaus from Greece and Hector from Troy fight over the body of Euphorbus, whom Menelaus has just slain.

The heroes of Greek legends were usually successful in war or adventure. They inspired the ancient Greeks to be brave, strong and clever. Poets told stories about the heroes in long poems that people learned by heart. Pottery, sculpture and other Greek art often showed heroic deeds. Heroes were helped by the gods and goddesses, and sometimes they even had an immortal as a parent. Zeus was the father of Heracles, the ancient Greek superman. Clothed in a lion's skin, Heracles faced great dangers. The story of the Trojan War had many heroes, such as the Greek warrior Achilles. Strange creatures in Greek mythology included Pegasus, the beautiful white-winged horse (above left), and the ugly Furies with dogs' heads, bats' wings and snakes for hair. Medusa, one of three hideous sisters, also had snakes for hair. Anyone who looked at her was turned to stone. The god Hermes showed Perseus, another son of Zeus, how to kill this monster.

GARDEN OF THE HESPERIDES
King Eurystheus ordered Heracles to steal apples of eternal youth from a garden tended by nymphs called the Hesperides. A serpent with 100 heads guarded the apple tree. Heracles killed this monster and picked the golden apples. The gods made Heracles immortal for successfully completing the 12 tasks that Eurystheus had set him.

HALF MAN, HALF HORSE
Centaurs were shaped like a horse but had a man's head and upper body. Legend says that the healing god Asclepius learned about medicine from a centaur.

THE 12 LABORS OF HERACLES
Heracles, called Hercules in Roman myths, had to complete 12 difficult tasks for the legendary King Eurystheus. In one labor, Heracles had to fetch three-headed Cerberus from the Underworld.

THE WANDERINGS OF ODYSSEUS

Odysseus (called Ulysses in Roman mythology) was the king of Ithaca. He fought for ten years at Troy and took ten more years to sail home. On the way, he outwitted the one-eyed giant Cyclops, persuaded the enchantress Circe to help him, and visited the Underworld. Odysseus resisted the sirens who lured sailors to their death, and escaped from the sea nymph Calypso, pictured here. He eventually reached Ithaca, where his wife Penelope and son Telemachus were still waiting patiently for him to return.

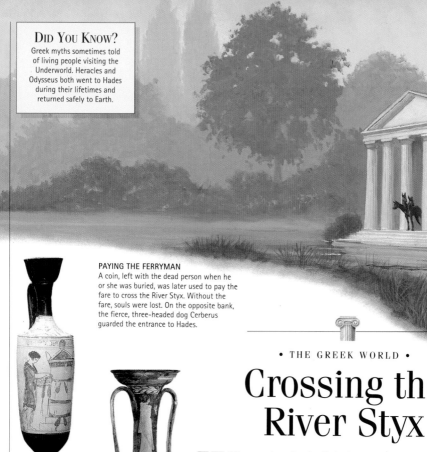

PAYING THE FERRYMAN
A coin, left with the dead person when he
or she was buried, was later used to pay the
fare to cross the River Styx. Without the
fare, souls were lost. On the opposite bank,
the fierce, three-headed dog Cerberus
guarded the entrance to Hades.

OIL FOR THE DEAD
Women took offerings
of perfumed oil to the
tombs of their dead
relatives. Tying ribbons
around the tombstone
was another way of
showing respect for
the dead.

WATER VESSEL
A jug, like this one depicting
women in mourning, was
placed on the tombs of
people who died unmarried.

• THE GREEK WORLD •

Crossing the River Styx

When ancient Greeks died, they believed their souls traveled across the River Styx to the Underworld. This underground kingdom, called Hades, was ruled by the god Hades. A dead person lay at home for a day so that sad relatives and friends could say their farewells. Grieving women cut their hair short. The next morning, before dawn, a funeral procession took the body to the burial ground. There was music and weeping and wailing. Food, drink and personal belongings were placed in the tomb to give comfort in the afterlife. Once across the River Styx, which divided the living world from the Underworld, all souls faced three judges. Those who had been good on Earth were sent to everlasting happiness in the Elysian Fields. Wrongdoers had to endure endless punishments in Tartarus. Large crowds of souls who were neither good nor bad were condemned to wander forever on the dreary Plain of Asphodel.

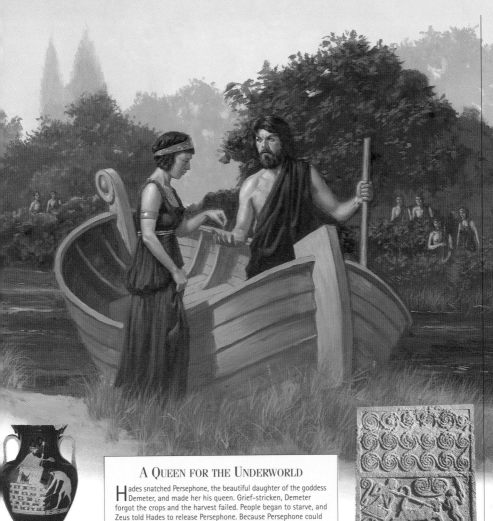

FUNERAL PYRE
Human ashes found in graves from the Dark Age, and scenes on vases made much later, show that the Greeks cremated their dead as well as burying them.

A QUEEN FOR THE UNDERWORLD

Hades snatched Persephone, the beautiful daughter of the goddess Demeter, and made her his queen. Grief-stricken, Demeter forgot the crops and the harvest failed. People began to starve, and Zeus told Hades to release Persephone. Because Persephone could not stay on Earth if she had eaten in the Underworld, Hades tricked her into swallowing a pomegranate seed. Thereafter, Persephone was allowed to spend only two-thirds of the year above ground. When she returned to Hades, Demeter plunged the Earth into winter.

MYCENAEAN GRAVESTONE
Large stone slabs called stelae marked important people's graves. They were often elaborately carved. This one was found above a royal tomb in Mycenae.

Discover more in Going to War

CHILDREN'S TOYS
Spinning tops and dolls were made
from terra-cotta—a mixture of
clay and sand. The writing
on the pottery baby's
bottle says, "drink,
don't drop!"

• LIVING IN ANCIENT GREECE •

In the Home

Houses had stone foundations, mud-brick walls and roofs of pottery tiles. Small, high windows kept out heat and burglars. Doors and shutters were made of wood. The central open courtyard contained an altar where the family worshipped their gods. Some courtyards also had a well, but water was usually fetched from public fountains. Women ran the household with the help of slaves. In ancient Greece, women had to obey their fathers, husbands, brothers or sons. A father could abandon his newborn child. He might do this if the baby was sickly, but healthy infant girls were also abandoned sometimes. Most women married at 15, while men married at 30 or more. The father chose his daughter's husband and gave the bridegroom money or valuables to save for his wife in case he wanted a divorce or died before she did. Sometimes, a bride met her husband for the first time on the day she was married.

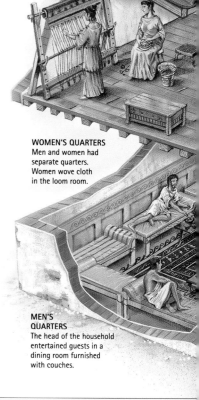

WOMEN'S QUARTERS
Men and women had separate quarters. Women wove cloth in the loom room.

MEN'S QUARTERS
The head of the household entertained guests in a dining room furnished with couches.

LOOM DUTIES
The women of the family produced a great amount of cloth for household furnishings and clothing. Girls learned how to spin and weave from their mothers.

GETTING MARRIED

On her wedding day, a bride bathed in sacred spring water and dressed in white. That night, the bridegroom and his friends came to fetch her. The very wealthy had horse-drawn chariots, while others had carts or walked. At the bridegroom's home, the couple knelt at the altar and his family showered them with nuts, dried fruit and sweets before they went to the bridal chamber. The next day, the two families celebrated at the husband's house.

DID YOU KNOW?

Children stopped being infants at the age of three. In Athens, at a spring festival, they were given miniature wine jugs, such as this, to mark the end of their babyhood.

THE KITCHEN
...es baked bread in a
...ery oven and cooked food on
...pen fire. The smoke escaped
...ugh a hole in the roof.

OUT TO PLAY
Girls played
...nucklebones using
...he ankle joints of
...ts or sheep. But
...day before they
...arried, girls had
...o leave all their
...aythings at the
...ple of Artemis.

HOME FURNISHINGS
Couches, tables, stools,
chairs, beds and storage chests
were made from wood and bronze.
Oil-burning lamps provided light. Wealthy
people had bathrooms in their homes and
plenty of slaves to fill the tubs with water.

Discover more in Eating and Drinking

• LIVING IN ANCIENT GREECE •

Writing and Education

AT SCHOOL
Boys learned reading, writing and arithmetic. Students wrote with pointed sticks on wooden tablets covered in soft wax. Mistakes could be rubbed out easily. Athletics and dancing were also important lessons.

Education in ancient Greece was not free. Only the sons of wealthy citizens could afford to go to school, where they attended classes from about the age of seven. The sons of poorer citizens learned their father's trade. At 18, youths were trained to fight so they were prepared to go to war when necessary. Some girls were taught to read and write at home, but lessons in housework were considered much more important. One writer even said that sending a girl to school would be like "giving extra poison to a dangerous snake!" In Sparta, education was much tougher than elsewhere in Greece. When they were seven, Spartan boys went to board in army barracks. They were given so little to eat that they had to steal food. This was supposed to teach them to be cunning soldiers. Spartan girls attended gymnastics, dancing, music and singing lessons.

INSPIRED BY MUSES
This stone carving shows three of the nine Muses, the goddesses of arts and science. They were thought to inspire poets, playwrights, musicians, dancers and astronomers.

416

GREEK ALPHABET
The word alphabet comes from two Greek letters—alpha and beta. There were 24 letters in the ancient Greek alphabet.

APOLLO AND THE MUSE
Apollo, god of music and poetry, is shown here talking to a Muse. Apollo worked closely with the nine goddesses.

THE DEVELOPMENT OF WRITING

The Linear B script on these clay tablets was adapted by the Mycenaeans from Minoan Linear A script. It was forgotten during the Dark Age. Greeks began to write again in the eighth century BC. They borrowed an alphabet from the Near Eastern Phoenicians and altered it slightly. Phoenicians wrote from right to left but the Greeks eventually reversed this direction. Early Greeks wrote only in capitals with no spaces between words and no punctuation.

MUSIC LESSON
Boys learned to play the lyre and pipes from a teacher called a kitharistes. This teacher also taught poetry and students had to memorize very long poems.

PIPES AND LYRES
Music lessons, like this one painted on a water vase, were often attended by several students. It seems that dogs and pet cheetahs were also welcome.

Discover more in Discovering Ancient Greece

Dressing for the Climate

Summers in ancient Greece were hot and dry. Winters were wet with chilly north winds. The poet Hesiod said that winter gales were cold enough to "skin an ox." People wore clothing made from rectangular pieces of material wrapped around the body in soft folds. They covered their summer garments with warm cloaks in winter. Statues and paintings on vases show that fashions changed slowly. The women of the household spun sheep's fleeces into fine woolen thread and flax fibers into linen. They dyed the yarns in bright colors and sometimes wove a contrasting color or a pattern into the edge of the fabric. Imported silk cloth was very costly. Cotton was introduced into Greece after Alexander the Great reached India. The ancient Greeks went barefoot or wore sandals, shoes or boots. Wealthy men and women owned fine jewelry. Slaves in the mines and stone quarries, who were at the poorer end of the social scale, wore only loincloths.

DID YOU KNOW?

Suntans were unfashionable in ancient Greece. In summer, said Hesiod, "the sun scorches head and knees." Both men and women wore broad-brimmed hats to protect their faces when they went outdoors.

SHOULDER PINS
These bronze dress pins came from mainland Boeotia. They measure about 18 in (45 cm) in length.

MIRROR IMAGE
Containers of sweet-smelling oils often pictured lifelike figures on a background of white clay. Greek women cared greatly about face make-up and hairstyles.

WOMEN'S WEAR
Women wore a long tunic made of wool or linen, fastened on the shoulders with brooches or large pins and tied at the waist and sometimes at the hips. Wraps ranged from thin shawls to heavy traveling cloaks.

SPINNING AND WEAVING

Women and girls spent many hours weaving fabric from local sheep's wool. They made clothing for the household, wall hangings, and covers and cushions for wooden furniture. Raw wool had to be spun into thread before it could be woven into material. The woman pictured on the vase (left) is spinning with tools called a distaff and a spindle. In her left hand she holds the distaff wrapped in unspun wool. She pulls out the wool with her right hand and twists it slowly to form thread. The thread is fed onto the spindle, which is weighted to keep it steady.

JEWELRY FROM RHODES

This chest ornament belonged to a rich man. He wore it by pinning the rosettes to his shoulders. The seven gold plaques show winged goddesses and lions. Pomegranates hang beneath them.

MENSWEAR

Men and boys usually wore thigh-length tunics; old men favored longer hemlines. Winter cloaks were often draped to leave one shoulder bare.

ELEGANT FOLDS

Women wore two main tunic styles—folded over at the top (below left) and usually made from wool, or fastened on the shoulders in several places (below right) and usually made from linen.

Making a Living

Many ancient Greeks worked hard. Fishermen and country folk provided food, such as sea creatures, olive oil, wine, grain, fruit, vegetables and honey, for people who lived in the towns. Farmers kept sheep, goats, pigs, poultry, donkeys and cattle, but only people in northern mainland Greece had enough pasture to breed horses. City people earned their living from making or selling things, such as leather goods, furniture, pots, tools, weapons and jewelry. It was unusual for well-born women to work outside the home, but some were priestesses in the temples. Low-born women became midwives, shopkeepers, dancers or musicians. Citizens who could afford to buy slaves or hire laborers looked down on those who had to work for a living. Rich men owned farms or ran silver mines and lived off the profits. They took part in public affairs, fought in the army when required, and paid high state taxes.

HARD AS IRON
During the Greek Dark Age, people learned to forge iron. They began to make sickles and other farm tools out of this strong metal.

DAILY NEEDS
Women ground barley to make porridge, gruel and bread. They baked loaves in ovens. Wheat flour was more expensive and bought only by the wealthy.

METAL WORKING
Metal ores were heated in furnaces. Workers took the red-hot ore from the fire with tongs. They hammered it into shape while it was soft.

MAKING WINE

Grapes were picked when they ripened in September. A few bunches were set aside for eating. The rest were tipped into huge vats and stamped on to squeeze out the juice. This was poured into jars and left to ferment into wine.

SERVED BY SLAVES

Slaves did much of the work in ancient Greece. They were often prisoners of war or pirates' captives and were bought and sold in the markets. Many were educated. The children of slaves also became slaves. Sometimes slave traders reared abandoned babies to sell later. Household servants were usually treated well. A few even saved their tiny wages and bought their freedom. But slaves in the silver mines and stone quarries worked in terrible conditions.

OLIVE HARVEST

The scene on this vase shows men beating the ripe olives from the lower branches of olive trees with poles. Boys climbed the trees to reach the higher fruit.

Discover more in Sickness and Health

Meeting Place

Every city had a central open space for meetings and markets. It was called the "agora." Men frequently did the shopping in ancient Greece and slaves carried the purchases. The agora must have been a noisy place, with customers jostling for attention at the covered stalls. Donkeys trotted to market bearing large loads of country produce. Strong smells mingled with fragrant, spicy odors. Merchants displayed foreign goods and stocked up with new cargoes to take to their next port. Some traders sold slaves and the poor offered themselves for hire as laborers. Citizens met friends under the shady colonnades to discuss business deals, politics and new ideas. Women came to fill their water pots. Altars and statues honoring gods, local athletes and politicians were erected in many marketplaces. The center of Athens, surrounded by important government buildings, impressed all who visited the city.

DID YOU KNOW?

Sparta did not issue coins until three centuries after the other city-states. The Spartans used iron rods for money until the fourth century BC.

TEXTILE TRADE
The Athenians on this pot are hurrying across the agora with bolts of cloth. Shops sold textiles made in foreign lands.

GOODS AND SERVICES
The marketplace provided all kinds of goods and services. Fishmongers sold fish, kept cool on marble slabs. Cobblers made sandals to fit the wearer's feet. Barbers trimmed hair and beards.

CENTRAL SPACE
In the agora at Athens, the city council met in one large building. Another big building housed public records and state documents. Merchants traded from stalls in the open square or from shops in the long, colonnaded buildings.

FETCHING WATER
Water gushed from the lion-headed spouts of Athens' public fountains. Poor women and slaves visited them daily to collect the household supply.

SACRIFICIAL CALF
This calf was purchased as a sacrifice. The carcass was later cooked and divided among poor citizens who could not afford to buy meat.

ROAD TO THE ACROPOLIS
A long road climbed uphill from the agora to Athens' acropolis. On festival days it was crammed with people taking part in religious processions.

THE ORIGIN OF COINS

The Lydians of Asia Minor first made coins from electrum, a natural mixture of gold and silver. Neighboring Greek colonies soon copied them, stamping the coins with their own emblems. Ephesus, for example, had a bee and a stag. Coins from some city-states, mostly made from silver, were accepted throughout the Greek world. There was no fixed exchange rate and people bargained over prices.

AT HIS EASE
This bronze figurine of a reclining man was made in the sixth century BC. He is a guest at a drinking party called a symposium.

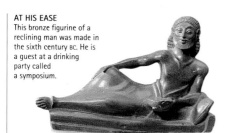

A GOOD CATCH
Wall paintings at Akrotiri on the island of Thera tell us about Bronze Age life. Greeks have always depended on food from the sea.

• LIVING IN ANCIENT GREECE •

Eating and Drinking

In early times, the Greeks worked extremely hard to produce enough food. They did not always succeed, and there were famines during the Dark Age. Grape vines grew well on terraced hillsides and olive trees thrived in poor soil. But there was always the problem of needing more flat, fertile land for wheat and barley. Things improved once grain could be imported from the colonies or from Egypt. People ate greens—cabbage, lettuce, spinach and dandelion leaves—and root vegetables such as radishes, carrots and onions. Eggs, goat's milk cheese, almonds, figs and other fruit were also available. Squid, sea urchins, fish and shellfish were plentiful and provided protein. Meat was a rare treat reserved for the rich who could afford roast goat, sheep or pig, and for those who hunted wild deer, hares and boars. The Greeks sweetened their cakes and pastries with honey. Seasonings included garlic and herbs such as mint and marjoram.

STRINGED INSTRUMENTS
The kithara was a complicated form of lyre, usually played by professional musicians. They plucked the strings with a small instrument called a plectrum.

FEMALE COMPANIONS
Foreign or low-born women, called hetairai, amused the guests at symposiums. They were unmarried, beautiful, clever and trained in the art of conversation, making music and dancing.

DID YOU KNOW?

Greeks today use some vegetables that were unknown in ancient times. Peppers (capsicums), eggplants (aubergines), potatoes and tomatoes reached Europe in the 1500s, after the Spanish invasion of the Americas.

PARTY FOOD

At first, the food at symposiums was simple. By the third century BC, fashionable dinners began with roasted songbirds, artichoke hearts, mushrooms, grasshoppers, snails, fish roe and other snacks. Tuna fish, stuffed with herbs, might follow. Cooks often flavored the meat course with cheese and aniseed, and the feast finished with honeyed sweetmeats, nuts and seeds. Bread was part of every meal. The Greeks used olive oil instead of butter.

Discover more in End of an Empire

THE OLYMPIC OATH
During the Olympic ceremonies, a wild boar was sacrificed to Zeus. Athletes swore on this beast that they were freeborn Greeks and would not cheat.

TEST OF STRENGTH
The winner in upright wrestling had to throw an opponent three times and push the back of his shoulders to the ground.

WOMEN'S GAMES
Women had their own games every four years to honor Hera, wife of Zeus. They ran foot races, which were divided into three different age groups.

FIRST PRIZE
Winners at the Panathenaic Games in Athens received jars of olive oil. The jar was decorated with a scene from the winner's sport.

• LIVING IN ANCIENT GREECE •

Festival Games

People believed that when the gods were happy they sent good fortune. They arranged festivals of music, poetry, drama and athletic games to please the gods. The four main festivals were the Olympic, the Pythian, the Isthmian and the Nemean. The five-day Olympic Games, honoring Zeus, took place in August every four years from 776 BC to AD 393. Wars between city-states ceased for a month beforehand so that thousands of priests, competitors and spectators from all over Greece could travel to Olympia in peace. The valley looked like a huge fairground. The visitors put up tents and there were food stalls and entertainers. Women did not compete in the games, but unmarried women, foreigners and slaves could watch from the stadium. Ancient Olympic events included running, wrestling, boxing, and chariot and horse races. Jockeys rode bareback, and horses frequently finished without their riders!

DRESSED IN OIL
Athletes competed naked in all events except the chariot race. They oiled their body thoroughly beforehand and afterwards cleaned their skin with bronze scrapers.

OLYMPIC CONTESTS

The pentathlon tested all-round athletes in five skills. These were running, long jumping, wrestling, javelin throwing and discus throwing. The discus was a heavy circular plate of stone or bronze. The athlete rubbed it with sand for a good grip.

CROWNING THE WINNER

Athletes trained hard for the games. Winners received woolen ribbons, jars of olive oil, palm branches and wreaths. The crowning wreaths were made from olive leaves at the Olympics, laurel leaves at the Pythian Games, pine needles at the Isthmian Games and parsley sprigs at the Nemean Games. Crowds went wild when successful athletes returned home. Some city-states gave them presents and great banquets. Statues of sporting superstars (above) stood beside those of the gods in the temples and open spaces at Olympia.

Discover more in City-states

427

Sickness and Health

The ancient Greeks admired physical fitness. Citizens gathered at the public "gymnasium" to practice sports, bathe and discuss philosophy. Studies of skeletons show that many women died at about 35 years of age and men at about 44. According to written records, some philosophers lived very long lives. For centuries, ancient Greeks believed that the gods sent accidents and illnesses as punishments. Asclepius, the god of healing (above left), usually carried a snake coiled around his staff. Sick people made sacrifices at his temples, then stayed overnight, hoping to be cured as they slept. The temple priests were Greece's first doctors. They treated their patients with a mixture of magic and herbs, special diets, rest or exercise. Towards the end of the fifth century BC, Greek physicians, led by Hippocrates, began to develop more successful methods of healing. They tried to find out how the body worked and what caused diseases. They also noted the effects of different medicines.

JASON THE PHYSICIAN
A man's tombstone often showed what he did for a living. Here, an Athenian doctor called Jason examines a boy with a very swollen belly.

DOCTOR'S APPOINTMENT
From the time of Hippocrates onward, blood was thought to contain disease. Doctors sometimes removed a little from a patient's arm. Physicians prescribed herbal remedies for many complaints and consulted scrolls of recorded medical information.

Eyebright
The ancient Greeks soaked eyebright in hot water to make an eyewash.

Hyssop
Hippocrates prescribed hyssop for coughs, bronchitis and other chest infections.

Mullein
Mullein, also an ancient treatment for coughs, is still used today.

Motherwort
Motherwort was thought to ease the pain of heart disease and childbirth.

Smooth sow thistle
This oddly named herb was supposed to relieve stomach complaints and scorpion bites.

Q: What remedies did the temple priests prescribe?

THE FATHER OF MEDICINE

Hippocrates, the founder of scientific medicine, practiced and taught on the island of Cos. He said doctors could not understand the parts of the body until they understood the whole system. Hippocrates carefully observed patients' symptoms before making a diagnosis. In many countries, newly qualified doctors swear the Hippocratic oath, promising to care for the sick as well as they can.

SURGICAL INSTRUMENTS
Greek doctors did not perform many operations. Surgery, without anaesthetics and with simple instruments (right), was very painful. Patients often died from shock or infection.

DID YOU KNOW?
The ancient world had no protection against epidemic diseases. Between 430 and 429 BC, a terrible plague swept through Athens. The great Pericles was one of its victims.

WINE JUG
Some inventive potters began modeling parts of clay vessels. This jug, in the shape of a head, was made in the fourth century BC.

GIFTS FOR THE GODS
Small bronze horses, such as this mare with her foal, were made during the Geometric period. Worshippers left these statues in temples for the gods.

SHAPED IN TERRA-COTTA
Many Greek cities produced terra-cotta figurines, each developing its own style. This sphinx, dating from the late fifth to early fourth century BC, came from southern Italy.

• ARTS AND SCIENCE •

Clay and Metal

The ancient Greeks loved beauty. They wanted buildings and useful objects to be balanced and graceful. Designers calculated how to fill spaces with the perfect amount of decoration. Early potters fashioned elegant vessels to be used as drinking cups, water jugs, storage jars for wine and olive oil and for other practical purposes. From then onward, most potters used the same shapes. Large numbers of craftworkers made sandals, furniture, cooking vessels and other items for everyday use.

In Athens, workshops lining the agora's south and west sides produced pottery, bronze and marble goods, and terra-cotta figurines. Athenian potters also lived and worked beside the cemetery. Metalworkers had quarters near the temple of Hephaestus—their special god. They used bronze for armor and household articles, and harder iron for tools and weapons that required sharp edges. Coins and jewelry were often made from gold and silver. Many craftworkers, especially those who produced weapons and armor, became very rich.

SILVER CHAIN
As the centuries passed, jewelers became more skilled at working silver and gold. This silver necklace was probably made in northern Greece between 420 and 400 BC.

DID YOU KNOW?

By the fifth century BC, 20,000 slaves were laboring in the silver mines near Athens. They worked shifts of ten hours in narrow tunnels lit by oil lamps.

POTTERY WORKSHOP
The use of the wheel gave potters both hands free to shape the clay. They prepared slip, a special liquid clay, for decoration and then fired the vessels in a kiln. Sometimes touches of color were added to the pots after firing.

THE BLACK AND THE RED

Thousands of pots survive from all over ancient Greece. From about 550 to 300 BC, Athenian "figure" ware was more popular than any other pottery. The vessels showed scenes from the lives of gods and heroes, as well as everyday subjects. Vase painters first developed the black-figure technique—drawing black figures on a red clay background (below left). Later, the red-figure technique (below right) became fashionable. Painters covered the red clay surface with a black slip background and left the figures outlined in red.

RUINS AT CORINTH
The temple of Apollo at Corinth was built of limestone in about 540 BC. At the time it was coated in white plaster to look like marble.

• ARTS AND SCIENCE •

Building in Stone

Few houses or early public buildings of ancient Greece survive. They were made of timber and sun-dried mud bricks, and have crumbled and rotted away. Later structures of marble and limestone have survived earthquakes, fires, wars and weather. Throughout the Greek world, priests and priestesses cared for sacred sites. The best known site was the Acropolis, the hill overlooking Athens, where there was a collection of temples, altars, statues and memorial stones, as well as the state treasury. The Parthenon, a temple decorated with enormous sculptures, housed a statue of Athena. Pheidias, the sculptor in charge, made it from wood, gold and ivory. Like many other temples, the Parthenon was huge, rectangular in shape and surrounded by grooved (fluted) columns. The pieces of each column were heaved into place with winches and pulleys. Architects used clever tricks in their buildings such as making columns lean slightly inwards so that they would appear straight up and down from a distance.

PARTHENON RUINS
The Parthenon was built in Doric style between 447 and 432 BC. Plastic copies of the original marble-relief friezes are being used to restore the building.

HORSEMEN IN PROCESSION
A marble-relief frieze almost 525 ft (160 m) long ran like a ribbon under the roof around the Parthenon's four outer walls. It showed the Panathenaic festival procession.

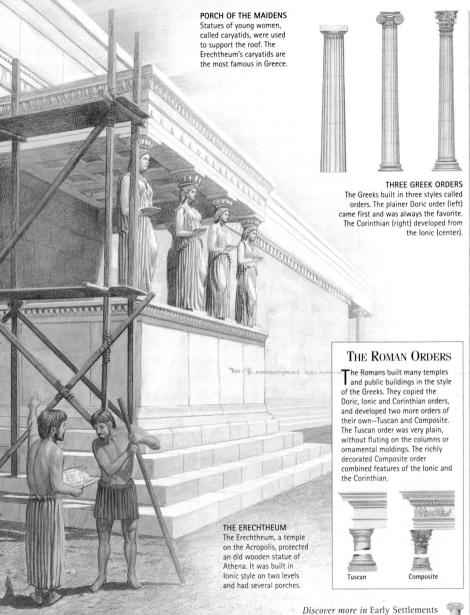

PORCH OF THE MAIDENS
Statues of young women, called caryatids, were used to support the roof. The Erechtheum's caryatids are the most famous in Greece.

THREE GREEK ORDERS
The Greeks built in three styles called orders. The plainer Doric order (left) came first and was always the favorite. The Corinthian (right) developed from the Ionic (center).

THE ROMAN ORDERS

The Romans built many temples and public buildings in the style of the Greeks. They copied the Doric, Ionic and Corinthian orders, and developed two more orders of their own—Tuscan and Composite. The Tuscan order was very plain, without fluting on the columns or ornamental moldings. The richly decorated Composite order combined features of the Ionic and the Corinthian.

Tuscan   Composite

THE ERECHTHEUM
The Erechtheum, a temple on the Acropolis, protected an old wooden statue of Athena. It was built in Ionic style on two levels and had several porches.

Discover more in Early Settlements

• FOREIGN AFFAIRS •

Going to War

The ancient Greek city-states fought each other over land and trade. Sparta had a full-time army, but other city-states trained freeborn men to fight and called them up in times of war. In Athens, men aged between 20 and 50 had to defend their state whenever necessary. Greeks who could afford horses usually joined the cavalry, but most served as foot soldiers called hoplites. Poorer citizens who were unable to buy their own weapons and armor rowed the warships. When the Persians invaded Greece, some city-states banded together against the foreigners. The Persian Wars lasted from 490 to 449 BC, and in 480 BC the Persian army destroyed Athens. In 447 BC the Greeks rebuilt the city. Sparta fought Athens for 27 years in the Peloponnesian War from 431 to 404 BC. Both sides were supported by other city-states. Sparta eventually won, but every city-state that took part in the conflict was weakened by loss of lives and money.

BATTLE SCENES
War was a common theme in vase art. These paintings provide useful information about the way warriors dressed for battle and the weapons they used.

DID YOU KNOW?
The Athenian navy had long, narrow, timber warships called triremes. In battle, a trireme was powered by 170 oarsmen. They tried to sink enemy ships by ramming them with the bronze prow.

BODY SHIELDS

Hoplites carried shields, made of bronze or leather, to protect them from neck to thigh. The symbols on the soldiers' shields represented their family or city.

FIGHTING THE AMAZONS

The Amazons were legendary women warriors who were believed to have helped Troy in the Trojan War. This marble frieze shows the Greeks and Amazons fighting.

PROTECTIVE CLOTHING

Soldiers wore bronze or leather breast and back plates joined on the shoulders and at the sides. Helmets, which came in various styles, protected their head and face.

▮CTORY AT SEA

▮ter the Persians destroyed Athens, the ▮reek fleet trapped the Persian fleet in a ▮arrow channel of water between the island ▮ Salamis and the Greek mainland. There, ▮e Greek triremes rammed the larger Persian ▮arships and forced them to retreat.

FIRST HISTORIANS

Herodotus, who wrote a history of the Persian Wars, has been called the father of history. He traveled widely to get information for his books. Another ancient Greek called Thucydides wrote a history of the Peloponnesian War. Both Herodotus and Thucydides tried to write factual accounts of what had happened. They interviewed many people who had fought in the wars.

• FOREIGN AFFAIRS •

The Macedonians

Macedonia in northeastern Greece was not a democratic city-state. It was ruled by kings who claimed to be descended from Macedon, a son of Zeus. For centuries, Macedonia was weak and frequently overrun by invaders. When Philip II came to power, he began to improve Macedonia's fortunes. By 338 BC, he controlled all of Greece and afterwards declared war on Persia. When Philip II was murdered in 336 BC, he was succeeded by his 20-year-old son Alexander, who eventually conquered Persia in 333 BC. Alexander (left) had blond hair and eyes of different colors—gray-blue and dark brown. He loved reading the *Iliad* and modeled himself on two great heroes, Achilles and Heracles. Alexander fought many successful battles and seized kingdoms throughout the eastern world. He earned the title "the Great" and married a Persian princess. When he died in 323 BC, his vast empire stretched as far as India.

HUNTING FOR SPORT
Alexander, whether hunting or fighting, was a popular subject with sculptors. This is part of a carving on the tomb of a king of Sidon in ancient Phoenicia.

ON THE MOVE
The Macedonian army trudged long distances through deserts and over mountains. Besides thousands of cavalry and foot soldiers, there were servants, grooms, women, children, pack animals and wagons. Indian elephants became part of the Macedonian army after Alexander's death.

Q: Which Greek heroes did Alexander the Great most admire?

FIGHTING IN FORMATION

Philip II trained his hoplites to fight in a formation called a phalanx. In battle, the front ranks extended their long spears. The men behind rested their spears on the row in front to form a barrier against arrows. A phalanx's weakest point was the right side, where the men were only half protected by their shields. Flute music helped the marching hoplites to stay in step.

DID YOU KNOW?

When Alexander was eight or nine, he tamed a pedigree stallion that had defeated his father's horse trainers. Alexander rode this horse, Bucephalus, into almost all his major battles.

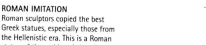

ROMAN IMITATION
Roman sculptors copied the best
Greek statues, especially those from
the Hellenistic era. This is a Roman
statue of the goddess Aphrodite,
known to the Romans
as Venus.

The Hellenistic World

The ancient Greeks called their country "Hellas" and themselves "Hellenes." After Alexander the Great died, the three generals Antigonus, Seleucus and Ptolemy divided up his empire between them. Antigonus ruled Macedonia and the rest of Greece and founded the Antigonid royal family. Seleucus took Asia Minor, Persia, and other eastern countries and began the Seleucid line of rulers. Ptolemy governed Egypt as the first ruler of the Ptolemaic dynasty. This was the beginning of the Hellenistic Age when Greek customs and ideas spread far beyond the boundaries of Greece. Architects building new cities throughout the Hellenistic world used Greek styles. These settlements adopted Greek law and language and the people attended Greek entertainment in theaters and stadiums. Hellenistic artists were interested in realism. Portraits on coins throughout the empire began to represent people's faces rather than making everyone look like gods or heroes. Sculptors chose a wider range of subjects and showed childhood, old age and suffering in a realistic way.

LIFELIKE IN STONE
Hellenistic sculptures showed
movement and feeling. This Altar
of Zeus at Pergamum in modern
Turkey features warring
gods and giants.

THE PTOLEMIES OF EGYPT

The Ptolemy dynasty governed Egypt from 323 to 30 BC. In Alexandria, Ptolemy I (left) built a huge library that had laboratories, observatories and a zoo. This city soon became the most important center of learning in the ancient world. The Ptolemy dynasty taught the Egyptians many Greek ways, but Cleopatra VII, the last ruler of the dynasty, was the only one who spoke Egyptian as well as Greek. After her death, the Romans took over Egypt.

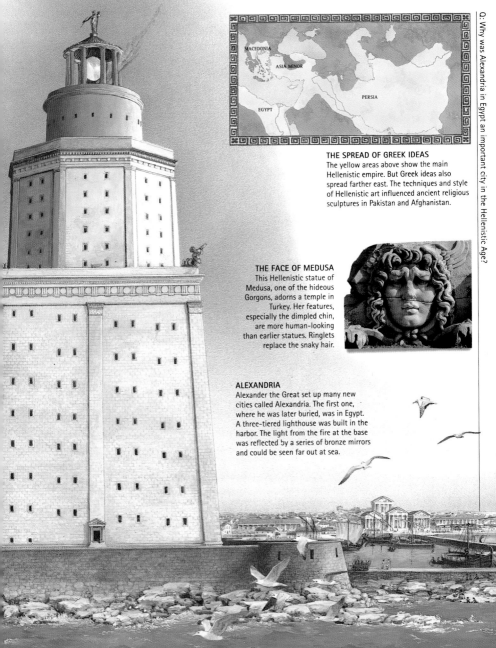

MACEDONIA

ASIA MINOR

PERSIA

EGYPT

THE SPREAD OF GREEK IDEAS
The yellow areas above show the main Hellenistic empire. But Greek ideas also spread farther east. The techniques and style of Hellenistic art influenced ancient religious sculptures in Pakistan and Afghanistan.

THE FACE OF MEDUSA
This Hellenistic statue of Medusa, one of the hideous Gorgons, adorns a temple in Turkey. Her features, especially the dimpled chin, are more human-looking than earlier statues. Ringlets replace the snaky hair.

ALEXANDRIA
Alexander the Great set up many new cities called Alexandria. The first one, where he was later buried, was in Egypt. A three-tiered lighthouse was built in the harbor. The light from the fire at the base was reflected by a series of bronze mirrors and could be seen far out at sea.

End of an Empire

The division of Alexander the Great's empire by his generals weakened Greece's hold on the ancient world. From 509 BC, a new power—Rome—had been growing in Italy. Roman rule spread gradually across the Hellenistic world. In 275 BC, the Romans captured the Greek colonies in southern Italy and Sicily. Between 148 and 146 BC, Macedonia and all of southern Greece became part of the Roman Empire. To the east, the Hellenistic empire slipped away as Rome acquired one kingdom after another. Finally, the Roman emperor Augustus defeated the Egyptians in 31 BC and demanded Cleopatra VII's surrender. The following year Cleopatra killed herself and Egypt became the last province in the Greek empire to fall into Roman hands. Although Greece no longer existed as a political and military power, Greek literature, art and architecture became models for the Romans who also adopted Greek gods and heroes. Many Roman boys were educated in Athens before Athenian schools closed down in the sixth century AD.

TWO NAMES—ONE GOD
The Romans renamed most of the Greek gods and heroes when they adopted Greek mythology as their own. Hephaestus, the Greek god of fire, became Vulcan.

GODDESS JUNO
This Roman bust of Juno, wife of Jupiter, was copied from a Greek original. In Greek mythology, Juno and Jupiter were known as Hera and Zeus.

DID YOU KNOW?

Jesus and his disciples, who lived in the Roman Empire after the Hellenistic Age ended, spoke a language called Aramaic. Their teachings were written in ancient Greek as the New Testament.

THE GREEK ORTHODOX CHURCH

Byzantium was a Greek colony from mid-600 BC until the Romans occupied it in mid-100 BC. The Roman emperor Constantine renamed the city Constantinople. This city became the center of eastern Christendom and Christians living there founded the Greek Orthodox Church. Religious pictures, called icons (above), are sacred to Greek Orthodox worshippers. Early religious artists used ancient Greek styles. Constantinople is now Istanbul, the capital of Turkey.

TEMPLE OF ZEUS

The Roman architect Cossutius began the Temple of Zeus in Athens in 174 BC. This Corinthian-style building, which took until AD 132 to finish, was the largest Hellenistic shrine on mainland Greece.

PORTRAIT ON WOOD

Like the Egyptians, Roman citizens living in Egypt mummified their dead. Roman artists copied Greek styles and painted lifelike portraits, such as this, on the mummy cases

Discover more in Writing and Education

441

SAILORS' LANDMARK
The temple of Poseidon at Cape Sounion, south of Athens, was a landmark to Athenian sailors from the fifth century BC. Its ruins can still be seen from the sea.

• FOREIGN AFFAIRS •

Discovering Ancient Greece

SHIPWRECKED EROS
This bronze statue of Eros was found off the coast of Tunisia in modern times. It was part of a cargo of Greek art bound for ancient Roman villas in North Africa.

When Pericles planned to rebuild Athens in the fifth century BC, he said, "Future ages will wonder at us, as the present age wonders at us now." His words have come true, for people still admire ancient Greece. They can study Greek art and architecture and visit museums to see what has been collected from ancient sites. Archaeologists continue their quest for information about ancient Greece. Clues about Greek civilization are still being found in shipwrecks as well as from beneath the ground. By the 1980s, experts knew enough about triremes to build one and sail it on the Aegean Sea. Most Greek writing is lost forever, but some stone inscriptions and copies of manuscripts survive, and we can still learn to speak and write ancient Greek. Echoes of ancient Greece also linger in languages and systems of government. Perhaps you live in a democracy or speak words that come from ancient Greek such as "theater," "orchestra," "gymnasium" and "Olympic."

ARCHITECTURE LIVES ON
Since the eighteenth century, architects designing public buildings have often used Greek styles. The front porch of this bank in Philadelphia, Pennsylvania, has Greek columns.

IN THE BRITISH MUSEUM
Part of the Parthenon was destroyed in 1687 when gunpowder accidentally exploded. Lord Elgin, an art collector, later bought some of the damaged sculptures for Britain.

CLUES FROM THE SEA

The ancient Greeks made hazardous sea voyages, with few navigational aids. Many ships sank. With the help of modern scientific equipment, divers can find some of these shipwrecks and retrieve cargoes that have lain undisturbed for centuries under the water.

THE RENAISSANCE

The Renaissance was a period in history that began in Italy in the fourteenth century AD and spread throughout Europe. During this time, writers, sculptors, architects and painters rediscovered ancient Greek artists and scholars and turned to them for inspiration. This Renaissance painting by Alessandro Allori shows the sea nymph Ino rescuing Odysseus after he was shipwrecked. The great Italian Renaissance artists Michelangelo, Raphael and Leonardo da Vinci were influenced by Greek art.

DID YOU KNOW?

In shipwrecks divers have found pottery jars that once held olive oil or wine. One sunken cargo included the remains of more than 10,000 almonds, probably grown on the island of Cyprus.

Portraits from Ancient Greece

Greek art from the Bronze Age onward portrayed gods, heroes and important people. They were featured on wall and vase paintings and in sculpture. Statues adorned temples and homes, marked graves, and were erected in public places. Few of the early statues made from wood survive, but many portraits in clay, stone and bronze have lasted. Although the bigger bronze figures were often melted down so that the metal could be reused, some large bronze statues have been recovered from shipwrecks.

Late Bronze Age
1600–1100 BC

MYCENAEAN SCULPTURE
This painted plaster statue was found on the acropoli at Mycenae. The artist has highlighted the cheeks and chin of this woman who is a goddess or a sphinx.

BRONZE AGE
3200–1100 BC

Early Bronze Age
3200–2000 BC

SCULPTURE FROM THE CYCLADES
Many marble figures, carved in the same style as this one, were made on the Cyclades islands in the early Bronze Age. Here, a musician is plucking a harplike instrument.

DARK AGE AND GEOMETRIC PERIOD
1100–700 BC

PATTERNED CENTAUR
Few works of art survive from the centuries known as the Dark Age, when it seems the Greeks had to struggle just to stay alive. This pottery centaur displays strong geometric patterns.

Middle Bronze Age
2000–1600 BC

MASTER OF THE ANIMALS
Figures appeared in Minoan jewelry made between 1700 and 1500 BC. This intricate golden pendant shows a Cretan god clasping two geese by their necks. His feet are shown pointing the same way.

FIGURES IN SILHOUETTE
During the last century of the Dark Age, human figures were painted within the patterned bands on pots. They were shown in silhouette and, like these, were often taking part in a funeral.

444

ARCHAIC PERIOD
700–480 BC

EGYPTIAN INFLUENCE
In the Archaic period, sculptors began making statues of young men and women from stone or bronze. Archaic artists learned figure sculpture from the Egyptians, but used more relaxed poses.

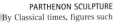

CLASSICAL PERIOD
30–323 BC

PARTHENON SCULPTURE
By Classical times, figures such as this one of Dionysus, or Heracles, had become more like real human beings. The sculptor Pheidias designed the sculptures that decorated the Parthenon, but he did not carve them all.

HEAD OF APOLLO
Classical sculptors modeled faces in a more lifelike way, but they still did not make them completely realistic. This is one of the few surviving bronze sculptures from this period.

HELLENISTIC AGE
323–31 BC

VENUS DE MILO
Hellenistic sculptors portrayed the human form and clothing with great realism and attention to detail. This statue of Aphrodite (Venus) was found on the island of Milos in the Cyclades.

PREPARING TO KILL
Many Hellenistic statues showed people's actions and feelings at dramatic moments. In this one, cast in bronze in about 100 BC, a young hunter tenses his body before he spears an animal.

Ancient Rome

What did Romans believe about death?

Why did the emperor Hadrian have a wall built right across Britain?

What was it like to live in Rome?

Contents

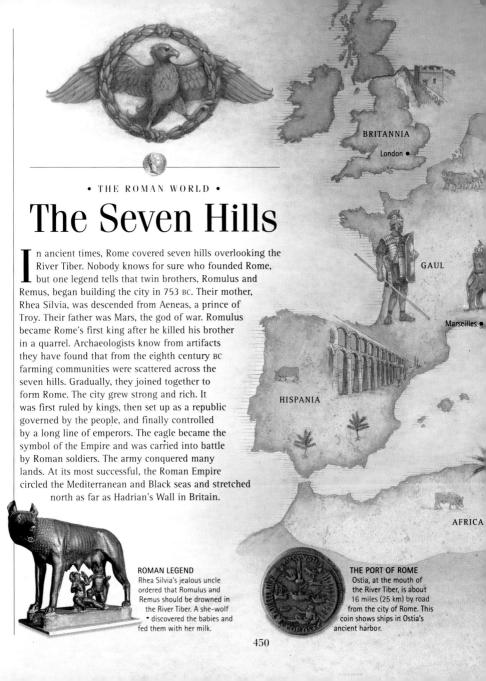

The Seven Hills

I n ancient times, Rome covered seven hills overlooking the River Tiber. Nobody knows for sure who founded Rome, but one legend tells that twin brothers, Romulus and Remus, began building the city in 753 BC. Their mother, Rhea Silvia, was descended from Aeneas, a prince of Troy. Their father was Mars, the god of war. Romulus became Rome's first king after he killed his brother in a quarrel. Archaeologists know from artifacts they have found that from the eighth century BC farming communities were scattered across the seven hills. Gradually, they joined together to form Rome. The city grew strong and rich. It was first ruled by kings, then set up as a republic governed by the people, and finally controlled by a long line of emperors. The eagle became the symbol of the Empire and was carried into battle by Roman soldiers. The army conquered many lands. At its most successful, the Roman Empire circled the Mediterranean and Black seas and stretched north as far as Hadrian's Wall in Britain.

BRITANNIA

London •

GAUL

Marseilles •

HISPANIA

AFRICA

ROMAN LEGEND
Rhea Silvia's jealous uncle ordered that Romulus and Remus should be drowned in the River Tiber. A she-wolf • discovered the babies and fed them with her milk.

THE PORT OF ROME
Ostia, at the mouth of the River Tiber, is about 16 miles (25 km) by road from the city of Rome. This coin shows ships in Ostia's ancient harbor.

450

753 BC	753–510 BC	510–27 BC	27 BC	AD 395	AD 476	AD 1453
The legendary date for the founding of Rome.	Rome is ruled by a series of kings.	The Roman Republic.	The Roman Empire is established under Augustus, the first emperor.	The Empire is split into West and East.	The Western Empire collapses.	The last city of the Eastern (Byzantine) Roman Empire is captured.

GERMANIA

DACIA

BLACK SEA

Ravenna

Constantinople

MACEDONIA

ASIA

ADRIATIC SEA

Rome

Ostia

Pompeii

ACHAEA

SYRIA

Athens

CYPRUS

INIA

IONIAN SEA

AEGEAN SEA

Damascus

SICILIA

CRETA

Jerusalem

Carthage

MEDITERRANEAN SEA

JUDAEA

Alexandria

AEGYPTUS

DID YOU KNOW?

"Italos" was the Greek word for bull-calf. Because the earliest Romans used cattle as a form of money, this "land of calves" soon became known as Italy.

Discover more in Empire in Decline

451

MYTHICAL BEAST
The Etruscans adopted some figures from Greek mythology, such as this lion-like monster called a chimaera. It has a goat's head in the middle of its back and a serpent for a tail.

BANQUETING IN STYLE
Large wall paintings in Etruscan tombs, which were sealed off from the air for about 2,500 years, have lasted amazingly well. Scenes, such as this one of people at a sumptuous banquet, show how much the Etruscans enjoyed themselves.

• THE ROMAN WORLD •

The Etruscans

GUARDIAN OF THE GATE
This two-faced god is looking forwards and backwards in time. He may be an earlier Etruscan version of Janus, the Roman god of exits and entrances.

P eople lived in Italy long before Rome was built. There were Latins, Samnites, Umbrians, Sabines, Greeks and the most powerful of all—the Etruscans. They occupied Etruria, which spread to the north and south of Rome. Etruria had a good climate, rich soil for farming, rocks containing useful metal ores, and thick forests to provide wood for building houses, temples, boats and bridges. One story suggests that the Etruscans came to Italy from Lydia in Asia Minor when food was scarce in their own country. Archaeologists cannot prove this, but they know a great deal about Etruscan daily life and their well-planned cities from what these early settlers buried in their tombs. They furnished the underground tombs like their homes with many fine bronze and terracotta clay statues. The Etruscans excelled at making music, and raising and riding horses. They were also skilled engineers. At times Etruscan kings ruled Rome, and the Etruscans and Latins later became one group of people.

MODELED IN TERRA-COTTA
In the ancient world, men and women usually dined separately. In Etruria, however, husbands and wives often feasted together, as shown on the lid of this coffin.

DID YOU KNOW?
The Etruscans made colors for their paintings from rocks and minerals. Crushed chalk gave them white, powdered charcoal made black, and oxidized iron granules made red.

TELLING THE FUTURE

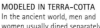

T he Etruscans worshiped many gods. Priests, called augurs, claimed they could tell what the gods wanted from certain natural signs. They read meaning into thunder, lightning, and the flight of birds, and told the future from the insides of dead animals, especially the liver. This replica of a sheep's liver (left) is marked out in sections, each with the name of a different god. It was possibly used by augurs when examining a liver, as shown on the back of the mirror above.

ON GUARD
Etruscan sculptors made many statues from bronze. They often portrayed warriors with huge, crested helmets, which made the soldiers look taller and thinner.

Government

Kings ruled Rome until 510 BC when the citizens expelled the last king, Tarquin the Proud. Rome then became a republic governed by officials who were elected by the people. Each year, the citizens chose two consuls and other government administrators from a group called the Senate. The idea was to prevent any one man from having too much power. Julius Caesar (above left), a brilliant general, had many military successes, which helped him to gain popularity and power in the Republic. In 49 BC, he marched his army to Rome and seized power. A civil war followed in which Caesar defeated his rivals and became the ruler of Rome. This one-man rule worried some senators and Caesar was murdered in 44 BC. His death brought renewed civil war and the collapse of the Republic. Caesar's adopted son Octavian gained control and brought peace to the Roman world. Octavian was renamed Augustus, and in 27 BC he became the first of Rome's many emperors. Some emperors, such as Augustus and Trajan, governed well. Others, such as Domitian and Nero, used their power badly.

DID YOU KNOW?
Roman emperors did not wear crowns like kings. Instead, they wore laurel wreaths on their heads. These had once been given to generals to celebrate victories in battle.

SYMBOL OF LAW
The fasces was a bundle of rods and an axe that symbolized the power of a magistrate. It was carried for him by his young attendant. The Romans took this symbol from the Etruscans.

HONORING AUGUSTUS
The marble Altar of Augustan Peace in Rome is carved on all sides with subjects that reflect the greatness of Rome's first emperor, Augustus. This panel shows members of the imperial family.

FACES IN STONE
Carved pictures in layered stone, called cameos, often portrayed important people. This one shows the emperor Claudius, his wife Agrippina the Younger, and her relatives.

THE SOCIAL ORDER

The people of ancient Rome were either citizens or non-citizens. Citizens were divided into three levels: wealthy patricians (such as the man shown here holding busts of his ancestors); businessmen called equites; and commoners called plebeians. At first, only patricians were allowed to be senators. Later, plebeians gained representation in the Senate, but the emperors took away this power. Non-citizens included women, slaves, foreigners and people who lived in the provinces.

Discover more in Growing Empire

Worshiping the Gods

RELIGION FROM EGYPT
The worship of Isis, the goddess of heaven and earth, first became fashionable in Rome after the Egyptian Queen Cleopatra spent a year there. Later, Isis had followers throughout the Empire.

The ancient Romans worshiped the same gods as the Greeks, but they gave them different names. The Greek god Zeus, for example, was called Jupiter, and the goddess Hera became Juno. People worshiped their gods through prayer, offerings of food and wine, and animal sacrifices. They also believed that natural things, such as trees and rivers, and objects that were made, such as doors, hinges and doorsteps, all had their own divine spirits. The head of the household was responsible for religious rituals in the home. Government officials performed religious duties in temples. Priests, called augurs, examined the insides of dead animals and looked for meaning in the flight of birds, which they believed were messengers from the gods. The expansion of the Empire introduced the Romans to new religions. These were permitted, provided worshipers did not ignore the Roman gods or threaten Roman government. The Jews and the Christians endured centuries of hardship because their beliefs challenged the emperors' authority.

HOUSEHOLD SHRINE
Every Roman house had an altar for the household gods. Here, spirits of the home, the larder and the family's special god (center) stand above a sacred serpent.

WORSHIPING BACCHUS
Followers of Dionysus sacrificed goats to him. This Greek god of wine and the theater was known to the ancient Romans as Bacchus (right).

A PERSIAN GOD
Many Roman soldiers worshipped Mithras, the bull-slayer, god of light and wisdom. Mithras was supposed to have been born holding a sword.

A TEMPLE FOR AUGUSTUS
This temple at Nîmes, in what is now France, was built during the reign of the emperor Augustus. It was dedicated to the worship of the emperor.

THE GOD OF BEGINNINGS

Two-faced Janus, looking backwards and forwards, supposedly understood both the past and the future. Janus was the god of beginnings—such as the first hour of the day. The first month of the year, January, was named after him. This early Roman god of exits and entrances guarded doors, gateways and arches. The doors to his shrine in Rome were always open in times of war, but closed during peace.

HE VESTAL VIRGINS

esta, the goddess of the hearth, as worshiped in every household. x priestesses, called Vestal Virgins, nded her state shrine in Rome's forum. ney had to keep her sacred flame rning continuously. It was a great onor to be chosen as a Vestal Virgin.

457

INTERIOR DECORATION
The rooms in grand homes were built around courtyards and colonnades. Artists covered the inside walls with scenes painted on wet plaster. This picture is in a villa at Pompeii.

JOINING OF HANDS
A Roman bride wore a white tunic, a saffron-colored veil and a wreath made from marjoram. The couple clasped right hands while the marriage contract was read and sealed. Later, the groom carried the bride over the threshold of her new home.

FACES FROM THE PAST
Roman couples often had their likenesses painted on wood or as frescoes on the walls of their houses. This portrait was found in a house at Pompeii.

AT PLAY
Rich women amused themselves by playing music. They often sat on high-backed bronze chairs, which were a sign of wealth. Most Romans sat on wooden stools.

• LIVING IN THE EMPIRE •

Marriage and Home Life

The father was head of the household. He was called the paterfamilias and he had complete power over his family and slaves. Although men legally controlled women, husbands often consulted their wives in private about public matters. Women were not entirely without rights, but they never became full citizens. Widows could own property, and some women became priestesses, shopkeepers, hairdressers, midwives or doctors. Highborn women employed slaves to raise the children, shop, clean and cook. Parents arranged marriages for their children. The wedding date was chosen carefully, and the second half of July was considered a lucky time. A priest would study the insides of a dead animal looking for signs that would indicate good fortune for the couple. Girls married at about 14 and wore a metal wedding ring (above left) on the third finger of their left hand. A marriage was not final until a wife had stayed in her husband's home for a full year.

A SLAVE'S LIFE

Thousands of prisoners from Greece, Egypt, North Africa, Britain and other conquered lands labored as slaves in homes, fields, mines, workshops and on building projects. They were punished severely for work badly done or for running away, and some slaves had to wear identity tags (above) and collars (below). But not all were ill-treated. "Treat your slave with kindness" advised the philosopher Seneca. Many educated Greek slaves became teachers or doctors. Some slaves earned wages and saved to buy their freedom.

TIME TO CELEBRATE
This children's procession was held in Ostia. Processions were a way of celebrating religious festivals, battle victories and weddings.

WRITING ON WAX
This girl holds wooden tablets that are coated on one side with wax and tied together to form a type of notebook. She is writing with a bone stylus.

• LIVING IN THE EMPIRE •

Children and Education

Wealthy couples thought children were a blessing and hoped to have large families. But women often died in childbirth and many babies and small children did not survive because the Romans had little knowledge of hygiene and childhood diseases. Parents sometimes abandoned their sickly infants outside the city's walls. Poor people found children costly to raise and their offspring went to work at an early age without any schooling. There was no free education in ancient Rome, and when children were about six, those from wealthy families started school where fees had to be paid, or began lessons at home. Many families employed educated Greek slaves as tutors. At 11, some boys went to secondary schools where they learned history, geography, geometry, astronomy, music and philosophy. At 14, boys who wanted a career in politics or law began to study public speaking. Girls learned basic reading, writing and mathematics and how to run a home.

A BOY'S LIFE
This carving shows four stages in a boy's childhood: at his mother's breast, accepted by his father, riding in a toy chariot drawn by a donkey, and with his teacher.

A NEW BABY
A father held his newborn infant in his arms to show that he accepted his son or daughter. In a ceremony nine days after being born, the baby received a name and a neck charm, called a bulla, to ward off evil spirits.

LEARNING LESSONS

Schoolboys read their lessons from long papyrus scrolls. Papyrus came from Egypt where the plants grew plentifully beside the River Nile. Paper was made by beating the fibers of the stems.

Q: What did the ancient Romans use to write on a wax surface?

THE TOGA OF MANHOOD

At the age of 14, boys were expected to behave as men. Every March, during the feast of the god Bacchus, special coming-of-age ceremonies for citizens' sons were held in the forum. The boys took off their neck charms (right), and exchanged their childhood clothing for adult togas. Barbers gave them their first shave and they were registered as citizens. The ceremonies ended with the boys making offerings of sacred honey cakes on the altar of Bacchus.

MOTHER AND CHILD

This gravestone in Palmyra, Syria is an early tribute to a mother and child. The theme is repeated in Christian paintings of Mary and Jesus.

SPRING BOUQUET
Here, Primavera, the goddess
of spring, walks through the
countryside gathering flowers.
Her clothing, falling in soft
folds, shows the elegance of
fashionable female dress.

DRESSED FOR BE
Fashions in clothing were slow to change
ancient Roman times. These two people a
dressed in early Empire style. The woman
wearing a short tunic under a full-length o
and has a long robe draped over the t
The man is clad in tunic and to

DID YOU KNOW?

Men went to special shops to have
the hair on their arms removed.
A mixture of bats' blood and
hedgehog ashes was popular for
getting rid of unwanted hair.

• LIVING IN THE EMPIRE •

Togas and Tunics

Togas, the dress of citizens, were first worn by the Etruscans. Although togas were uncomfortable and difficult to keep clean, these garments were popular with the emperors and remained fashionable for centuries. They were semicircular in shape and usually made from a single length of woolen cloth. The wearer held the heavy folds together as he moved, which left only the right arm free. Emperors wore purple togas; senators' togas were white and edged with a purple stripe. Citizens wore wool or linen tunics under their plain white togas, but other men, women and children wore tunics only. These were folded and pinned, and held with a belt. They seldom had stitched seams because sewing with thick bone or bronze needles was so difficult. During imperial times, wealthy people liked clothing made from finely woven Indian cotton and expensive Chinese silk. Jewelers crafted beautiful ornaments, such as gold earrings (above left). Women also wore jewelry made of polished amber from the cold Baltic countries, which was carved in the town of Aquileia in northeast Italy.

462

DELICATE JEWELRY
Jewelers worked with many precious stones and metals imported from the provinces. This necklace is made from emeralds threaded on gold rods and joined with gold links.

IN FASHION
Beards went in and out of fashion through the centuries. In the second century AD, emperors such as Hadrian and Septimius Severus (right) led a trend towards beards.

SWEET PERFUMES
Women used great quantities of sweet-smelling oils. The most expensive came from Arabia and were bought in glass bottles, such as this one, from shops that specialized in perfumes and scented ointments.

HAIRSTYLES OF THE EMPIRE

Slaves spent hours dressing their mistresses' hair. They dyed and perfumed it, curled it with heated tongs, and coiled it into buns. Slaves who did a bad job could be beaten. Sometimes, dark-haired Roman ladies wore wigs made from blonde locks, or had hairpieces fixed in place with hairpins. Hairstyles changed often. Vibia Matidia, the emperor Trajan's niece (far right), and Faustina (right), wife of the emperor Marcus Aurelius, have two different imperial hairstyles.

Discover more in Food and Feasts

463

City Life

SEWER SYSTEMS
Sewers carried running water to private lavatories in wealthy homes and to public lavatories in the streets. In Rome, the waste ran into the River Tiber.

Roman cities were laid out in a square with the main roads crossing at right angles. Buildings included a basilica that contained offices and law courts, temples, shops, workshops, public bathhouses and public lavatories. Aqueducts brought in water, while sewers removed waste. There was often a theater for plays and an amphitheater for other entertainments. Statues and decorated arches and columns were erected to commemorate important events. Cities were crowded and dirty. The rich had spacious town houses but most people lived in cramped conditions. Blocks of flats made of stone and wood were three to five storeys high. Sometimes they collapsed because they were so badly built. Most houses had no plumbing, and residents tipped waste down communal drains. Later, when the city of Rome became extremely overcrowded, the emperor passed a law forbidding people to move wheeled vehicles in daylight, so carts clattered after dark. The poet Martial complained, "there's nowhere a poor man can get any quiet in Rome."

STREET SCENE
Rome's narrow side streets were lined with stalls and shops. At night, the shopkeepers secured their goods behind heavy wooden shutters. Walls were covered with signs and advertisements. Some people carried their wares while others used donkeys to transport their goods.

CITY SHOPPING
There were no supermarkets in Rome. Small shops specialized in particular goods. These relief carvings show a greengrocer (top) and a butcher (bottom).

TAKE-OUT SHOP
Most apartments had no kitchens. Hot food was sold from shops (right) and stalls, and many people cooked in the street using portable stoves.

In Memoriam

Ancient Romans cremated or buried the dead, and believed that their spirits crossed the mythical River Styx to the Underworld (Hades). A coin placed in the dead person's mouth paid the ferryboat fare. Mourners and musicians accompanied the body through the streets to the cemetery outside the city's walls. Many monuments to the dead survive. The one above presents portraits of two freed slaves surrounded by the rods and staffs of the freedom ceremony (left), metalworker's tools (top) and carpenter's tools (right).

Did You Know?

City dwellers greatly feared fires. These were easily started by candles, oil lamps and portable stoves. Rome had seven fire brigades, equipped with hand pumps, buckets, hooks and axes, but they seldom managed to save houses that caught fire.

IMPERIAL GRANDEUR
The emperor Hadrian's villa at Tibur (now called Tivoli), near Rome, was set in a beautifully landscaped garden. Greek sculptures stood beside the many lakes and waterways.

COUNTRY ART
People spent a great deal of money on their country villas. Wall paintings, such as this one, sculptures and mosaics were just as elaborate as the art that decorated town houses.

• LIVING IN THE EMPIRE •

Country Life

Throughout Italy and the provinces, large farming estates produced food for city dwellers and the army. Provinces such as Egypt and North Africa grew grain, Spain was famous for olive oil, Italy made the best wines, and Britain supplied beer and woolen goods. Farm slaves toiled from dawn to dusk, and were frequently whipped by the overseers who supervised them in the fields. Oxen and cattle pulled plows, and donkeys and mules were used for other work. The famous Roman poet Virgil wrote a practical handbook for farmers in the form of a poem called the *Georgics*. In it, he advised them to "enrich the dried-up soil with dung," and told them how to rotate crops and keep corn free of weeds. The crops grown depended largely on soil and climate. In Italy, farmers grew grapes, olives and many vegetables. Pork was a favorite meat, and large numbers of pigs were raised. Sheep and goats were kept for wool and milk. Farmers also raised chickens, ducks, geese and pigeons, and kept bees for their honey.

DID YOU KNOW?
In the fourth century AD, St. Augustine wrote about the number of spirit gods worshiped by country people. At least 12 gods were linked to the growing of crops. Silvanus, for example, was the spirit of the boundary between cultivated fields and woodland.

COUNTRY HOLIDAY
Wealthy Romans owned luxurious houses in the country. Set among the estate's plowed fields, orchards and livestock barns, the "villa urbana" had all the comforts of the owner's town dwelling. The part of the estate that housed the farm staff was called the "villa rustica."

GARDEN FRESCO
Livia, wife of the emperor Augustus, had a villa at Primaporta outside Rome. This fresco on the wall of one of the small rooms created a cool feeling during summer.

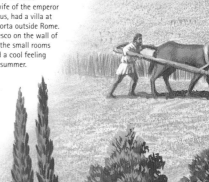

466

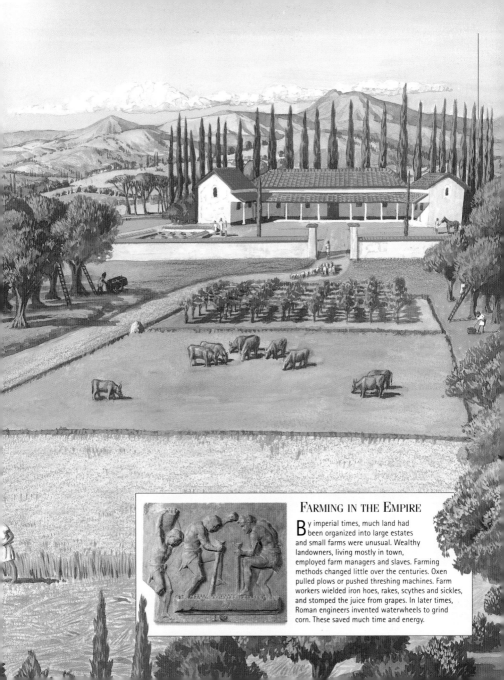

FARMING IN THE EMPIRE

By imperial times, much land had been organized into large estates and small farms were unusual. Wealthy landowners, living mostly in town, employed farm managers and slaves. Farming methods changed little over the centuries. Oxen pulled plows or pushed threshing machines. Farm workers wielded iron hoes, rakes, scythes and sickles, and stomped the juice from grapes. In later times, Roman engineers invented waterwheels to grind corn. These saved much time and energy.

MAKING MOSAICS

Mosaicists used hammers and chisels to cut sandstone and marble into small colored cubes called tesserae. Working over a pattern or picture, they pressed the tesserae and pieces of glass into a bedding of soft mortar. Finally, they cleaned and polished the surface thoroughly.

DID YOU KNOW?

Pottery was traded across the Empire. Scientists can now find out where the clay came from originally by examining tiny pieces under powerful microscopes, or by chemical analysis of fragments.

DANCING FAUN

This bronze faun, a country god, stood in the courtyard of a house in Pompeii. Like so many other Roman statues, it was copied from a Greek sculpture.

RESTING EMPRESS

A full-length marble statue was a common form of portrait for important people during imperial times. This is Agrippina, wife of the emperor Claudius.

• LIVING IN THE EMPIRE •

Craft Skills

In small workshops behind city shops, craftworkers turned metal, clay, stone, glass, wood, animal bones, ivory, cloth, leather and other materials into useful and decorative objects. Potteries in Gaul (present-day France), Italy and North Africa produced great numbers of red pots, which were exported all over the Empire. Fresco painters, stonemasons, mosaicists and carpenters worked at people's houses or on public buildings, which were often commissioned by the emperors. Cameo carving (the art of making pictures from layered stone) and mosaic (the art of making designs from small pieces of colored stone and glass) were both popular during the time of the Empire. Cameo carvers usually worked with sardonyx, a semiprecious stone that formed naturally in layers of different colors. In the centuries before the Empire, generations of Roman craftworkers learned from Etruscan art and from Greek pottery, painting and sculpture. They copied Greek statues and the technique of carving friezes in stone. Craftworkers passed on their skills to their sons.

CARVED GLASS
The dark inner layer of this glass vase was encased in opaque white glass, which was then skillfully cut away to produce the decoration.

DRINKING CUPS
Silver cups survive from the first century BC. This one features a scene from a Greek tragedy by Sophocles.

BLOWING GLASS

Glass-blowing, which was developed in the first century BC, was an important technological breakthrough. Glass-blowing probably began in Syria, but it spread quickly across the West. Glass workshops were hot and uncomfortable, but this new technique allowed skilled craftworkers to make objects in many shapes. Sometimes they combined old molding methods with blowing. Glass became a cheap, everyday material, and most people throughout the Empire could afford some glassware.

DRINKING DOVES
This mosaic, made from thousands of tesserae, decorated the emperor Hadrian's villa. The Roman artist copied a Greek original.

Discover more in The Etruscans

469

Architecture and Engineering

For centuries, Roman architects were content to adapt Greek styles to suit their own tastes. They also adopted the arch from Mesopotamia. Arches spread the weight of a load and were used in bridges, aqueducts and amphitheaters. In the second century BC, the Romans developed concrete, a mixture of volcanic ash, lime and water that set as hard as rock. Concrete made it possible to build large, strong, domed roofs, such as the one on the Pantheon in Rome. Engineers used a tool called a groma to check that buildings and roads were straight. They worked out efficient water systems and invented waterwheels to drive flour mills. Many emperors organized building programs. Augustus boasted that he found Rome a city of brick and left it a city of marble. Vespasian was responsible for building Rome's great amphitheater, the Colosseum, and Trajan ordered the construction of a towering victory column. Hadrian's many building successes included his luxury villa at Tivoli and a frontier wall in northern Britain.

Ax

Stonemason's square

Tongs

Cutting tool

Scaled dividers

BUILDERS' TOOLS
Most Roman tools were made from iron and some had wooden handles. The tools above are similar in shape to modern tools.

OLIVE PRESS
Olive presses operated on a screw thread, a clever device that made turning easier. The oil was crushed from the fruit under the block at the base.

PANTHEON OF GODS
Rome's Pantheon, built mainly during Hadrian's reign, honored all the state gods and still stands. Its domed roof, the largest of its time, represented the heavens; the circular opening stood for the sun. People offered gifts to the gods at altars inside this magnificent building.

METAL FORGE
A metalworker (center) holds the hot metal in tongs while he beats it into shape. More metalworkers' tools can be seen on the right of this relief carving.

WATER SUPPLIES

Imperial cities had fresh water supplies. At one time, 11 great aqueducts fed Rome's bathhouses, fountains and public lavatories. Some of them were linked with springs 30 miles (48 km) away. The water flowed gently from a higher level to a lower one. Arches, in one, two or three tiers, supported the pipes across steep valleys. Tunnels took them through hills. Invading barbarian tribes destroyed many Roman aqueducts, but ruins like the ones at Segovia in Spain (left) show us what they were like.

DID YOU KNOW?
Not all slaves were owned by
individuals. Some belonged to towns.
These slaves led miserable lives,
laboring in chain gangs on road
works and other engineering projects.

Writers and Thinkers

L atin was the official language of business and trade in the empire in the West, while Greek was used in the empire in the East. People in the provinces spoke local tongues. Latin, which had an alphabet of 22 letters, was inscribed on stone or wax tablets, or written in ink (inkpots above left) on papyrus scrolls, pieces of wood, or thin animal skins called vellum. Only educated Romans could read and write. After they conquered ancient Greece, the Romans' admiration for Greek literature and ideas increased. Homer's Greek *Odyssey* inspired Virgil to write the *Aeneid*. Horace, Ovid and Martial also produced fine poetry. Some historians, including Livy, Pliny the Elder and Tacitus, recorded the events of the day. Two Greek philosophies, Epicureanism and Stoicism, attracted Roman followers. Epicureans believed that a person's happiness depended on pleasure and avoiding pain. Stoics believed that happiness came from thinking things through carefully and not surrendering to your feelings.

PTOLEMAIC MAP
Ptolemy, c.AD 90 to c.168, was a mathematician, astronomer and geographer who had a lasting influence on Western scientists. This map of the world, drawn in 1540, is based on his ideas.

EYEWITNESS
Pliny the Younger, a Roman writer, watched with his aunt as Mt. Vesuvius erupted in AD 79. His uncle Pliny the Elder died from the fumes. Pliny the Younger later recorded what he saw. He also traveled widely, and his letters to his friend Tacitus and to the emperor Trajan tell us much about life at this time.

LITERARY MASTERPIECE
Here Virgil sits between the Muses of epic poetry and tragedy. His long poem the *Aeneid* describes the adventures of the Trojan prince Aeneas and how he came to the land of the Latins.

ROMAN NUMERALS
The letters I, V and X represent the numbers 1, 5 and 10. If "I" is before a letter, it is subtracted from it. IV stands for 4. If "I" comes after a letter, it is added to it. VI represents 6.

THE JULIAN CALENDAR

I n 46 BC, Julius Caesar established the calendar that became accepted throughout the Roman Empire. It was devised by the mathematician Sosigenes. In order to make it work, 67 days were added in 45 BC. The new calendar gave the year 365 days, which were divided into 12 months. Seven months had 31 days, four had 30 days and one had 28 days, as in the modern Western calendar. An extra day was added every four years. The Romans worked for seven days. Every eighth day was market day.

DAILY EXERCISE
Some young women exercised daily in the gymnasiums to keep trim. The ancient Romans believed that a healthy body ensured a healthy mind.

DID YOU KNOW?
Although Roman citizens knew nothing about germs and bacteria, they liked to be clean. They used olive oil instead of soap and a wet sponge attached to a stick instead of toilet paper.

THE ORDER OF BATHING
Bathers began in a warm room with a pool filled with tepid water. Next, they moved to the steam room where the water was very hot. Finally, they plunged into cold water to cool off. Slaves attended to bathers' needs, cleaned the baths and stoked the furnaces.

• LIVING IN THE EMPIRE •

Staying Healthy

Diseases such as typhoid, dysentery and tuberculosis killed many people in the Empire's overcrowded cities, but minor ailments could also be serious. Cures were often unsuccessful because doctors did not understand the causes of illness. Treatments depended on basic scientific knowledge combined with folklore and prayer. The Empire's only hospitals were reserved for wounded soldiers. Every town had at least one public bathhouse, and cities often had several. Entrance fees were low for both men and women; children were admitted free. In Rome, besides baths, these huge complexes contained gymnasiums, massage rooms, take-out food shops, libraries, and gardens where citizens met to exchange gossip and news. The emperors Titus, Trajan, Caracalla and Diocletian all sponsored the building of public bathhouses. The Baths of Caracalla could take 1,600 bathers a day. Tooth decay was not a serious problem for ancient Romans because they had no sugar in their diet. Honey was expensive and used sparingly.

BODY OIL
This oil flask and tools, called strigils, once belonged to a Roman athlete. Athletes oiled their bodies before exercising and scraped themselves clean afterwards.

474

HEALING WATER
The baths at Bath in England are filled with water from natural hot springs. These were sacred to the ancient Romans and used for healing purposes.

MEDICAL MATTERS

Many doctors practicing in ancient Rome were Greek slaves or freedmen. They advised their patients on exercise and diet, prescribed herbal remedies that could be bought in herbalists' shops (above) and did basic operations (instruments below). Sick people also appealed to magic and the gods, particularly to Asclepius, the god of healing. Patients who slept in his temples expected him to appear in their dreams with miraculous cures. Surgery was performed without anaesthetics, and women often died during childbirth.

475

FOOD SCRAPS
This mosaic on a dining-room floor cleverly copies the debris from dinner. Guests would discard wishbones, fishbones, shells, lobsters' claws and fruit pits during the meal.

BANQUET BEHAVIOR
Men and women ate together at Roman banquets. Guests dressed in their best clothing, but removed their shoes once inside the host's house. Diners reclined three to a couch and ate mostly with their fingers. They drank wine mixed with water.

BIRDS FOR DINNER
Artists studied animals carefully and reproduced their behavior in accurate detail. In this mosaic, a cat is about to dine on a plump bird.

COOKING POTS
Cooking pots had to be strong because they were used frequently. Large pans (top center) rested on iron grids above hot coals.

SERVING BOY
The children of slaves also became slaves. The luckier ones worked indoors in private homes and helped to prepare food.

COOKING WITHOUT TOMATOES
There were many imported foods at banquets. But tomatoes did not reach Italy until the sixteenth century, after they were discovered in the Americas.

• LIVING IN THE EMPIRE •

Food and Feasts

Romans ate their main meal of the day late in the afternoon. Lower classes had wheat and barley porridge, bread, vegetables, olives and grapes, and made cheap cuts of meat into sausages, rissoles and pies. Emperors arranged for hand-outs of grain and oil to the very poor. If grain deliveries were delayed, riots sometimes broke out in Rome. The emperor Tiberius warned the Senate that stopping the corn dole would mean "the utter ruin of the state." Unlike the poor, wealthy citizens ate extremely well. There was a saying in imperial Rome that the rich fell ill from overeating and the poor from not eating enough. Slaves from the East, who were skilled in preparing exotic dishes, were in great demand as cooks. Hosts spent huge sums of money on food for a banquet, which might last well into the night. The meal had three courses, each consisting of a range of dishes served on pottery, glass or even silver or gold platters.

BANQUET DISHES

R oman cook books listed dishes that required hours of preparation. Fish sauce, made from sprats, fish intestines, olive oil and herbs, was fermented in the sun for three days. Cooks might prepare appetizers of sows' udders or jellyfish stuffed with salted sea urchins.

Main courses could include flamingo with dates, roast parrot, boiled ostrich, and dormice stuffed with pork and pine nuts. Romans liked fruit for dessert, and army generals competed with one another to bring back new fruits from the provinces.

POORLY PROTECTED
These terra-cotta figurines are models of gladiators. Although they are wearing Thracian helmets, the rest of their scanty clothing leaves parts of their body dangerously exposed.

THE NET MAN
Unlike other gladiators, who wore some protective armor, the almost naked retiarius carried only a weighted net, a Neptune's trident used by tuna fishermen and a short sword.

THE ROAR OF THE CROW
Several pairs of gladiators might fight at th same time: Samnite against Samnite (belo\ armed with sword and shield; Thraciu against Samnite (center); and retiariu against Samnite (right). The audienc cheered winners and booed losers loud\

Spectator Sports

E mperors and provincial governors arranged chariot races and violent games to amuse people on the frequent public holidays. People of all classes flocked eagerly to these entertainments. Rome's Circus Maximus was always packed on race days, when chariots thundered round the track to the deafening noise made by 260,000 onlookers. Most charioteers were slaves and the successful ones earned freedom. Convicts and slaves (both men and women) trained as gladiators. Equipped in the styles of warriors, such as Samnites and Thracians, or armed with fishing gear, they fought each other and wild animals, while musicians played bronze horns and water organs. Rome's amphitheater, known as the Colosseum, seated 60,000 spectators. The contestants shouted, "We who are about to die, salute you!" as they filed past the imperial stand during opening parades. Wounded gladiators could appeal for mercy, but jeers from the crowd and the thumbs-down signal from the referee brought death. Victorious gladiators were treated as stars and won their freedom.

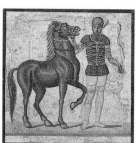

CHARIOTEERS' COLORS
In Rome, there were four important chariot teams—Reds, Blues, Whites and Greens (above). Every team had a large group of fans. Sometimes fights broke out between the different groups.

Did You Know?
One of the games staged in Rome on orders from the emperor Trajan lasted for 117 days. More than 10,000 gladiators took part.

FIRST PAST THE POST
Chariots were drawn by teams of two, three or four horses (left). Charioteers fell frequently during the race. First place went to the winning chariot, either with or without a driver.

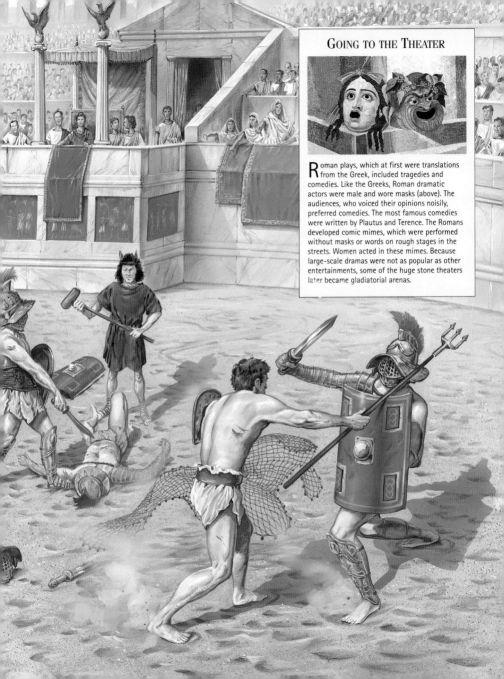

GOING TO THE THEATER

Roman plays, which at first were translations from the Greek, included tragedies and comedies. Like the Greeks, Roman dramatic actors were male and wore masks (above). The audiences, who voiced their opinions noisily, preferred comedies. The most famous comedies were written by Plautus and Terence. The Romans developed comic mimes, which were performed without masks or words on rough stages in the streets. Women acted in these mimes. Because large-scale dramas were not as popular as other entertainments, some of the huge stone theaters later became gladiatorial arenas.

Roads and Travel

HARBOR SCENE
The Romans built deep harbors like this one, which is probably in the Bay of Naples, so that goods and passengers could be loaded and unloaded safely onto ships.

R oman roads, constructed by the army and slave laborers, carried soldiers, messengers, travelers and traders across the Empire. These straight, level highways followed the most direct routes, tunneling through hills and bridging rivers. They formed a network that allowed troops to march quickly to any trouble spot, and connected town cities and ports with Rome, the capital and center of government. Road construction usually began with digging a ditch about 3 ft (1 m) deep. This was filled with sand, and then with small stones and gravel mixed with concrete. Paving stones were laid on top and milestones set in plac to mark distances. Roman roads were crammed with all kinds of people and animals. The rich traveled in carriages with several slaves. They sometimes slept in their vehicles or in tents pitched by the roadside because they were frightened of being robbed at the inns. Some roads lasted for many centuries and many present-day European roads follow old Roman routes.

MEASURED BY MILESTONES
Inscriptions in Roman numbers on low pillars of stone, placed at the edges of roads, told travelers the distances between towns.

DID YOU KNOW?
Road transport was slow and therefore expensive. Horses were faster than bullocks, but the Romans had no harnesses that allowed horses to pull heavy wagons. Horses were ridden or used to pull light carts.

e first Roman road, built in
2 BC, linked Rome to Capua
southern Italy. It was called
Via Appia and later
ended even farther south
present-day Brindisi.

TOURING THE EMPIRE

Many Roman tourists visited the Seven Wonders of the
World, such as the Egyptian pyramids at Giza (left), and
other places in the Empire. There were even some maps and
guidebooks available. A few sightseers left lasting evidence of
their travels in Latin graffiti on stone monuments in Egypt.
Travelers often asked soothsayers to read the future before they
began what might be a dangerous adventure. Robbers ambushed
tourists on Italian roads, and barbarians attacked them in
foreign lands. Sea voyagers ran into pirates and foul weather.

ROADSIDE INNS
During long journeys travelers often
stopped to eat and sleep at inns beside
the road. Surviving accounts of people's
journeys tell us that these inns frequently
served poor food and bad water. The noisy
behavior of some of the guests often
kept others awake.

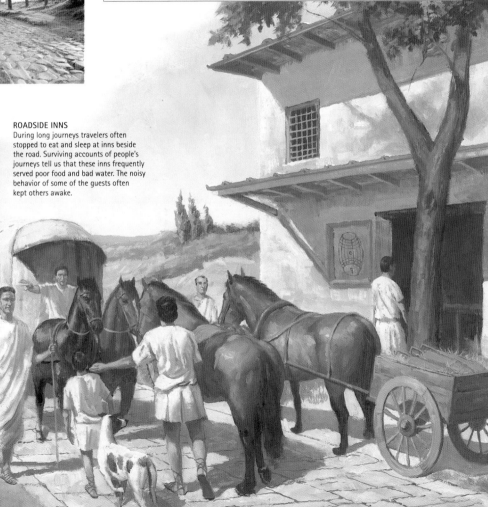

CARGO FOR THE CAPITAL
Large trading vessels, equipped with one main sail and often with elaborately carved prows, unloaded at the port of Ostia at the mouth of the River Tiber. The river was too shallow for these ships to travel farther. Smaller boats then took the goods on to Rome.

• EXPANSION AND EMPIRE •

Traders at Home and Abroad

When the Empire expanded, trade increased within Roman boundaries and with countries beyond. Pliny the Younger boasted that "the merchandise of the whole world" could be bought in Rome. Farmers loaded local produce onto bullock wagons (left), mule carts, donkeys and camels and took it to the towns. Spices, perfumes, silk and cotton came from the mysterious East. Road travel, however, was very expensive and only luxury goods were taken over long distances. Shipping was the preferred method of transport, and grain was the most important cargo. Vessels also carried neatly stacked, tall pottery jars, called amphorae. These were filled with wine; fish sauce; and olive oil for lamps, cooking and oiling people's bodies. Merchants could make large fortunes, but they faced high risks of losing cargoes. Roman sailors, who had no navigational instruments, steered across the Mediterranean Sea by looking at the sun, moon and stars. Their wide-hulled, wooden sailing ships were often plundered by pirates or wrecked by storms.

LOADING THE GRAIN
Some tomb paintings pictured people at work. The captain of this boat, named Farnaces, is taking sacks of grain upriver from Ostia to Rome.

482

THE WILD-ANIMAL TRADE

The ancient Romans enjoyed watching fights between wild beasts, and between wild beasts and gladiators. They imported animals such as tigers (right) from Asia, rhinoceroses from North Africa, wolves from Ireland and bears from Scotland. Towns, such as Leptis Magna in Africa, which specialized in exporting wild animals became very wealthy. During the opening games at the Colosseum in Rome, 9,000 animals died in 100 days. This kind of slaughter wiped out lions in Mesopotamia and elephants in North Africa.

On the March

T he conquests of ancient Rome depended on its well-trained army—a powerful fighting force that marched across much of the known world pushing out the Empire's frontiers. At first, only Roman men of property were allowed to serve as soldiers. Then, at the end of the second century BC, General Gaius Marius reformed the army and let citizens without property join. Many poor townsmen signed on for a period of up to 25 years. Military life was tough and punishments were harsh, but when soldiers retired, they were given money or a small plot of land to farm. By Julius Caesar's time, Rome had a highly efficient, wage-earning, permanent army. It was divided into 60 units, or legions, of foot soldiers called legionaries. Augustus later reduced the number of legions to 28 units. Legionaries were supported by auxiliary soldiers who were recruited mostly from the provinces. They formed infantry units (on foot) or cavalry units (on horseback).

MUSEUM PIECES
Many museums in Europe have collections of Roman military equipment. This cavalryman's ceremonial helmet is on display in London's British Museum.

MILITARY EQUIPMENT
A bronze cheek guard (top), an iron spear and javelin head, and the handle from a helmet were found at a fort in England. They were once used by Roman soldiers.

GUARDING THE EMPEROR
The Praetorian Guard, numbering about 9,000 soldiers, was the only part of the army stationed in Rome. Augustus formed this unit to protect the emperor and Italy.

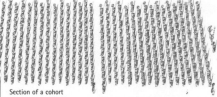

Section of a cohort

THE ROMAN LEGION

A Roman legion consisted of about 5,000 foot soldiers. The legion was divided into nine groups, called cohorts, of equal size, which were led by a tenth, larger cohort. Cohorts were split into six centuries. A century originally contained 100 men, but this number was later reduced to 80 to make the group easier to manage. Centuries were divided into groups of eight soldiers who shared a tent and ate together. Each legion carried a silver eagle into battle. If the eagle fell into enemy hands, the legion was disbanded.

GOING TO WAR
Each group of soldiers, called a century, was commanded by a centurion (front left) and had its own standard-bearer (front center). Legionaries marched with all their equipment and belongings. Besides their weapons, legionaries carried food for three days and tools for making camp, digging canals, laying roads and building bridges.

DID YOU KNOW?
Regular army pay attracted Rome's poorer citizens into the army. A salt allowance, called a salarium, formed part of a soldier's rations. The English word salary, describing payment of wages, comes from this Latin word "salarium."

RED SLIPWARE
Potteries in Gaul, Italy
and Asia Minor produced
huge amounts of red
tableware. These fine
drinking cups show the
uniformity of shapes used
throughout the Empire.

PORTRAIT
OF THE DEAD
When the Romans
invaded Egypt, Egyptian
artists were influenced
by Roman styles. They
began to paint more
lifelike portraits, such as
this, on mummy cases.

• EXPANSION AND EMPIRE •

Growing Empire

HADRIAN'S WALL
The emperor Hadrian ordered the army to
build a wall in Britain at the northernmost
boundary of Roman territory. Hadrian's
Wall, about 75 miles (120 km) long, wound
across the country near the present
Scottish border. The wall, built mainly
between AD 122 and 129, linked 14 forts
and was intended to keep out invaders.
Parts of the wall still stand.

Rome's frontiers began to expand long before
imperial times. By 264 BC, the Romans dominated
the whole of Italy and, after successful wars, ruled
the island of Sardinia, territory in Spain and southwest
Europe, and Carthage in North Africa. Roman government
took hold in Greece in 146 BC, and during the next century, Rome's
boundaries extended to the eastern Mediterranean. In 31 BC, before
he became emperor, Augustus finally conquered Egypt. The emperor
Claudius overran Mauretania and Thrace, and ordered the invasion
of Britain. The emperor Trajan extended the Empire farthest
with the conquests of Dacia and large areas of the Middle
East. This expansion was later abandoned by his
successor Hadrian (above left). Rome divided its
territories into provinces, which were governed by
senators. Several emperors were born
in the provinces. Trajan, for
example, came from Spain and
Septimius Severus from Africa.
Syria and Asia Minor were
among Rome's richest provinces,
for in those days good crops grew
in what is now desert.

MILDENHALL TREASURE
This large dish, found at
Mildenhall in Suffolk, England,
is one of the finest surviving
pieces of silverware from ancient
Roman times. The central motif
represents a sea god. The outer
frieze shows followers of Bacchus.

DID YOU KNOW?
The provinces paid taxes to Rome
but were not governed by Roman
law. Laws, which took into account
local customs, were made for each
province added to the Empire.

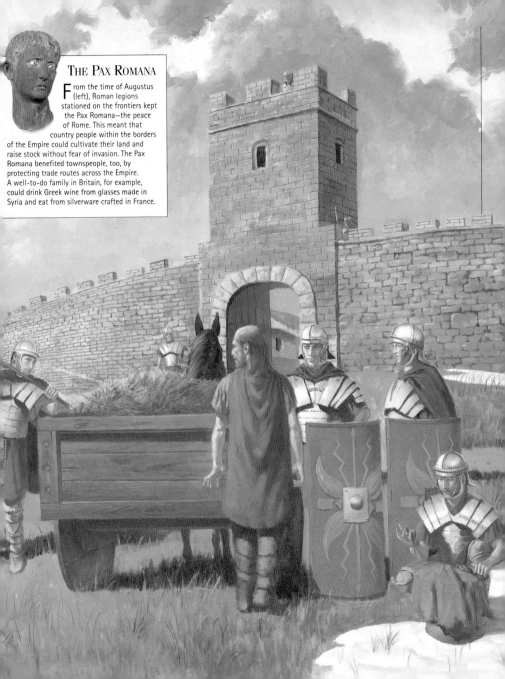

THE PAX ROMANA

From the time of Augustus (left), Roman legions stationed on the frontiers kept the Pax Romana—the peace of Rome. This meant that country people within the borders of the Empire could cultivate their land and raise stock without fear of invasion. The Pax Romana benefited townspeople, too, by protecting trade routes across the Empire. A well-to-do family in Britain, for example, could drink Greek wine from glasses made in Syria and eat from silverware crafted in France.

The Beginnings of Christianity

IN THE ROUND
This mosaic comes from the floor of a Roman villa in Hinton St. Mary in Dorset, England. It is the earliest picture of Jesus Christ in Britain.

J esus, a carpenter by trade, lived in Nazareth, a village in the province of Judaea. When he was 27 or 28 years old, he began preaching to a growing band of followers. These people believed Jesus was chosen by their god to lead them. Some years later, Pontius Pilate, the Roman governor, declared Jesus a rebel against the state and ordered him to be crucified under Roman law. His death inspired the spread of the Christian religion throughout the Empire. Two of Jesus' followers, Peter and Paul, took this new religion to Rome. In its early years, Christianity was very popular with slaves and the poor because it promised everlasting life, regardless of wealth. The Christian belief in one god conflicted with the Roman state religion and with the official view that the emperors were gods. Because Christians would not take part in the ceremonies on special festival days of emperor worship, they were cruelly punished. They were forced to meet secretly.

BEARDED CHRIST
This fourth-century painting is on the ceiling of Commodile Cemetery in Rome. Here, Jesus Christ has a beard as he does in most later Christian art.

GOING UNDERGROUND
When the Romans refused to allow the Christians to bury their dead in official burial places, the Christians dug large passages, called catacombs, beneath uninhabited sections of Rome and Naples, and under some cities in Sicily and North Africa. They placed the shrouded bodies in openings along the walls, which they often decorated with paintings.

CATACOMB PAINTING
Early paintings of Jesus Christ, such as this one on the wall of a catacomb, often showed him as a short-haired, beardless young man in the role of the Good Shepherd.

TOMB INSCRIPTION
This inscription from Rome's catacombs dates from the third century. It is simpler than most tomb inscriptions and may have been carved secretly and in haste.

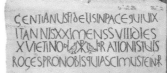

THE EDICT OF MILAN

Christianity soon attracted followers both rich and poor. It spread so rapidly that by 311 the emperor Galerius passed a law allowing Christians to worship openly "on condition they in no way act against the established order." Two years later, the Edict of Milan issued by the emperor Constantine I gave Christians the freedom to inherit and dispose of property, and to elect their own church government. Constantine I (above) was baptised as a Christian shortly before he died in 337. In 380, under Theodosius I, Christianity became the official religion of the Empire.

THE GOOD SHEPHERD
This fine statue of Jesus Christ as the Good Shepherd is remarkable because it was made in secret at a time when Christianity was still generally forbidden.

Empire in Decline

BATTLING THE BARBARIANS
Scenes from battles were often carved on large stone coffins, known as sarcophagi. This one belonged to a general who fought in the Germanic wars in Marcus Aurelius's army.

DID YOU KNOW?
Diocletian sent out officials to count the Empire's population. They brought back detailed information. The census counted everything from people to livestock and olive trees.

T he huge Roman Empire was difficult and expensive to run. From AD 161 to 180, the emperor Marcus Aurelius had to fight many campaigns to protect the boundaries. By the third century AD, the army was stretched too far and taxes were raised to cover the Empire's costs. Farmers who could not afford the taxes abandoned their farms, and cities suffered as the economy slumped and their markets declined. Many emperors were weak, so generals competed for power. Civil wars raged, and barbarians, sensing the Empire's weakness, attacked the frontiers. Diocletian (above left) became emperor in AD 284. He split the Empire in order to make it easier to manage, appointing Maximian to rule the West, and keeping the wealthier East for himself. Diocletian also reorganized the army and the provinces. Soon after Diocletian and Maximian retired in 305, civil wars broke out again until Constantine took power over the whole Empire in 324. He moved the imperial court to Byzantium, which he renamed Constantinople.

TRIUMPHAL ARCH

The Arch of Constantine was built to celebrate a victory at the Battle of the Milvian Bridge near Rome in 312. Many of the arch's best carvings were stolen from second-century monuments.

COLLECTING TAXES

Emperors needed huge sums of money to support the large army and pay for their extravagant lifestyles, expensive works of art, large building programs and the corn dole. This revenue was raised through taxation. People had to pay taxes on land, slaves, crops, roads and goods from shops. Tax officials (above) were required to collect set amounts of money and had to make up any shortfall from their own wages. They were harsh and unfeeling about collecting what was owed.

DIOCLETIAN'S TETRARCHY

These four men, called tetrarchs, represented Diocletian's new system of imperial leadership. It was his idea to have an emperor for East and West and two junior emperors to assist them.

ATTILA THE HUN

The Huns were a barbarian tribe from central Asia. By 447, Attila, king of the Huns, and his large army of warriors had conquered all the countries between the Black Sea and the Mediterranean. They defeated the Roman army in three battles, but did not capture Constantinople and Rome.

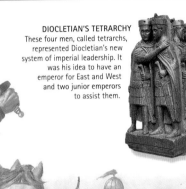

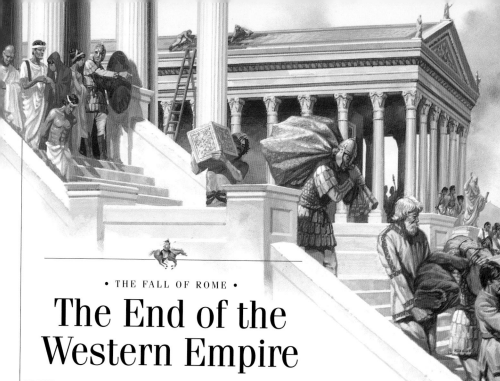

The End of the Western Empire

The Romans, like the ancient Greeks, called all tribes whose language they could not understand "barbarians." The emperor Theodosius I allowed German barbarian tribes to settle in the north of the Empire when they were driven there by the fierce Asiatic Huns. But by the early fifth century, other barbarian people, including the Huns themselves, came looking for land to settle. They fought the imperial armies and the Germans, and soon occupied large areas that had once been Roman territory. After Theodosius I died in 395, the Western and Eastern Roman empires split forever in matters of government, but the West still depended upon the East for money and grain supplies. King Alaric led the Visigoths into Rome in 410 and became the first person to conquer the city in 800 years. The Vandals later plundered Rome in 455. The Eastern Empire refused to help the Western Empire, which finally ended in 476 when the last Western emperor, Romulus Augustulus, was exiled by the barbarians.

SACKING OF ROME
Gaiseric, leader of the barbarian Vandals, sailed to Ostia in AD 455. His soldiers entered Rome and spent 12 days stripping buildings of everything valuable, including the gilded roof tiles of important temples. Gaiseric took the widow of the emperor Valentinian III and her daughters hostage.

BESET BY BARBARIANS
Hordes of barbarians seeking land attacked the frontiers and swarmed into the Western Empire.

VILLAS FOR VANDALS
When the Vandals overran Roman territory in North Africa, local landowners faced slavery or exile. This mosaic from Carthage shows a Vandal in front of a Roman villa.

SAFE HARBOR
Ravenna became the Western Empire's capital in 402. The city was connected to the mainland by a causeway. Ships entering the harbor had to pass between lighthouse towers.

SPREADING CHRISTIANITY

The Christian Church survived when the Western Empire collapsed and Rome became the holiest of Western Christian cities. The first Christian monasteries were in Egypt, but soon monastic communities were formed throughout the former Western Empire. The monks preserved many ancient Roman manuscripts by patiently copying them out by hand. The symbol at right, called a Chi-Rho, was one of the earliest Christian symbols and is found on many Christian objects. It combines the first two letters of Christ's name in Greek.

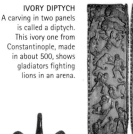

IVORY DIPTYCH
A carving in two panels is called a diptych. This ivory one from Constantinople, made in about 500, shows gladiators fighting lions in an arena.

PANEL PICTURE
The Byzantine Christians produced panel pictures called icons. They were often painted on wood, like this one of St. Gregory.

SIGN OF VICTORY
Before winning the Battle of the Milvian Bridge, Constantine I claimed he saw a flaming cross and the words "In this conquer." The cross became a Christian symbol.

• THE FALL OF ROME •

Eastern Empire

The Empire in the East, which came to be known as the Byzantine Empire, prospered as the Western Empire weakened. The city of Constantinople grew wealthy. Its geographical position between Europe and Asia was good both for trade and for managing its territory. The frontiers of the Byzantine Empire extended west to Greece, south to Egypt, and east to the border with Arabia. Although Greek was the official language of the East, Latin was often still spoken at the emperor's court. During his rule from 527 to 565, Justinian (above left) regained some of the western provinces in Africa, Italy and Spain, but he did not hold them for long. Many other Eastern emperors had long reigns and governed well. The Byzantine Empire was never a great military power and tried to settle difficulties with its neighbors by peaceful means. Its people were Christians, and invaders who threatened the Empire were frequently persuaded to join it instead and become Christians, too.

EASTERN CAPITAL

Like Rome, Constantinople (left) covered seven hills. The emperor Constantine I began a grand building program to beautify Constantinople, which was protected by water on three sides. Later, a wall was built around the city by Theodosius II. Justinian I built Hagia Sophia, which was the largest Christian church of the time.

OTTOMAN TURKS

During the thirteenth century, the Ottoman Empire, which was located in what is now Turkey, began to expand its frontiers. The Ottoman Turks followed the Islamic religion and set out to conquer Christian cities. Gradually they controlled most of the Eastern Roman Empire. The troops of Sultan Mehmet II (above) besieged Constantinople for six weeks before the city finally fell in 1453. This was the end of the Byzantine Empire and Constantinople became known as Istanbul.

GOLDEN GIFT

The empress Theodora, Justinian's wife, was very generous to the Christian Church. Here, attended by her ladies-in-waiting, she presents a golden chalice to the Church of San Vitale in Ravenna, Italy.

NEARBY TOWN
The town of Herculaneum was also buried by the eruption of Mt. Vesuvius. Many people fled through these streets to shelter in the boat storage chambers beside the beach. Their skeletons were discovered in 1982.

DID YOU KNOW?
The people of Pompeii and Herculaneum were completely unprepared for the eruption of Vesuvius in AD 79. They did not even know they were living beside an active volcano until that fateful day in August.

• THE FALL OF ROME •

Discovering Ancient Rome

The Roman Empire was one of the largest the world has ever known. Many of its structures still stand and they tell us about Roman architectural styles, building methods and materials, and town planning. Remains of frontier fortifications, aqueducts and roads show the extent of this great civilization. Surviving art and collections of objects once in daily use help us form a picture of the Roman people. We also know a lot about their hopes, ambitions and feelings from inscriptions on stone and metal, and from the many surviving copies of original histories, poems, plays, scientific works, recipe books, letters and official lists. The Latin alphabet, expanded from 22 to 26 letters, is used today throughout the Western world. Latin shaped the languages of modern Italy, France, Spain, Portugal and Romania. Many English words also come from Latin. Scientists, doctors and lawyers still use Latin phrases, and every known species of plant and animal has a Latin name.

NO PROTECTION
Many of Pompeii's victims died with their arms raised above their heads. People tried to cover their heads with their cloaks as the fiery ash rained down upon them.

GUARD DOG
This plaster model shows how a watchdog outside the House of Vesonius Primus in Pompeii died during the eruption. He was chained up and could not flee from the ash.

496

FROZEN IN TIME

When Mt. Vesuvius in southern Italy erupted in AD 79, Pompeii was buried under 10 ft (3 m) of boiling ash and lapilli in only a few hours. Archaeologists have been uncovering the city since 1748. The remains of streets, shops, villas and gardens were well preserved beneath the ash.

WALL PAINTING

Many paintings in Pompeii were preserved because they were buried. Some fine frescoes survive, such as this one dating from the second century BC in the Villa of the Mysteries.

ABANDONED MEAL

When Mt. Vesuvius erupted, the terrified people ran from their homes. The food being cooked in these pots at the House of Vettii was never served.

BEWARE OF THE DOG

In one Pompeii house, a realistic floor mosaic took the place of a real guard dog. The words "cave canem" mean "beware of the dog."

RECORDING THE PAST

E dward Gibbon was an English historian who thought history should record "the crimes, follies and misfortunes of mankind." When he visited Rome in 1764, he decided to write *The History of the Decline and Fall of the Roman Empire.* This work, which is still read, covered the years from the destruction of the Western Empire to the end of the Eastern Empire. Gibbon read documents from the ancient world, and studied maps, coins and ruins to support his account of ancient Rome. The first volume was published in 1776; the sixth in 1788.

Discover more in Growing Empire

— Tributes to the Emperors —

When Augustus, the first emperor, died in AD 14, the Roman Senate declared him a god. From then onward, people worshiped the emperors, and temples were dedicated to them all over the Empire. These impressive buildings reminded everyone of the emperors' absolute power. Monuments erected to celebrate victories in war also helped to advertise the strength of rulers. Emperors issued coins bearing their images, and commissioned paintings, mosaics and sculptures of themselves and their families. The style and symbolism of imperial portraits often tell us more about the way an emperor wished his subjects to see him rather than giving us a true likeness. Some of the most important emperors are listed below.

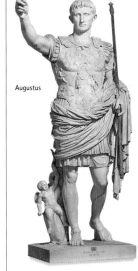

Augustus

Augustus	27 BC–AD 14	Philip	244–249	**DIVISION OF THE**		
Tiberius	14–37	Decius	249–251	**ROMAN EMPIRE** AD 395		
Caligula	37–41	Trebonianus		**WESTERN EMPIRE**		
Claudius I	41–54	Gallus	251–253	Honorius	394–423	
Nero	54–68	Aemilianus	253	Valentinian III	423–455	
Galba	68–69	Valerian I	253–259	Maximus	455	
Otho	69	Gallienus	259–268	Avitus	455–456	
Vitellius	69	Claudius II	268–270	Majorian	457–461	
Vespasian	69–79	Quintillus	270	Severus	461–465	
Titus	79–81	Aurelian	270–275	Anthemius	467–472	
Domitian	81–96	Tacitus	275–276	Olybrius	472	
Nerva	96–98	Florianus	276	Glycerius	473–474	
Trajan	98–117	Probus	276–282	Nepos	474–475	
Hadrian	117–138	Carus	282–283	Romulus		
Antoninus Pius	138–161	Carinus	283–285	Augustulus	475–476	
Marcus Aurelius	161–180	Diocletian	284–305			
Commodus	180–192	Maximian	286–305	**EASTERN EMPIRE**		
Pertinax	193	Constantine		Arcadius	395–408	
Didius Julianus	193	and Licinius	307	Theodosius II	408–450	
Septimius		Constantine I	324–337	Marcian	450–457	
Severus	193–211	Constantine II	337–340	Leo I	457–474	
Caracalla	211–217	Constans	340–350	Zeno	474–491	
Macrinus	217–218	Constantius	340–361	Anastasius I	491–518	
Elagabalus	218–222	Julian	361–363	Justin I	518–527	
Severus		Jovian	363–364	Justinian I	527–565	
Alexander	222–235	Valentinian I	364–375			
Maximinus	235–238	Valens	364–378			
Gordian I	238	Gratian	378–383			
Gordian II	238	Valentinian II	375–392			
Gordian III	238–244	Theodosius I	379–395			

MESSAGES FROM STATUES
Augustus (above) appears as a godlike hero. His breastplate symbolizes military successes and the boy-god Cupid links the emperor with Venus, the goddess of love. Caligula (below) assumes a commanding pose on horseback.

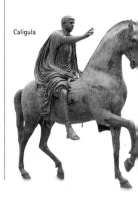

Caligula

PORTRAITS OF EMPERORS

Portrait busts of new emperors were sent to the provinces so their subjects could have some idea of what the ruler looked like. Although many emperors traveled widely, they could not visit every part of the Empire.

EMPEROR ON HORSEBACK

This bronze statue of Marcus Aurelius, designed to inspire his subjects, shows him as a forceful soldier. In his writings, however, he comes across as a quiet, thoughtful student of philosophy.

Marcus Aurelius

Caracalla

Valerian I

Nero

Constantine I

Hadrian

Honorius

Trajan

Romulus Augustulus

IMPERIAL COINAGE

A portrait of the emperor of the time adorned one side of imperial coins, which were made of gold, silver or bronze. Images on coins were an excellent way to make sure that people in the provinces knew the emperor's face.

Native Americans

Why did some Native American babies have so many names?

When did Native American boys become warriors?

What can totem poles tell us?

Contents

Where Did They Come From?

N ative Americans tell wonderful creation stories to explain where they came from. Some say Coyote shaped people from mud; others think Raven called them from a clam shell because he was lonely. Archaeologists, however, suggest people arrived in several groups, beginning at least 15,000 years ago, perhaps much earlier. The first Americans came from Asia and followed herds of grazing animals across a land bridge formed during the Ice Age, when the great glaciers sucked up the shallow seas and left dry land. Later, when the Earth began to warm, this land bridge disappeared and became the Bering Strait. The people trekked slowly southward into North America through a harsh landscape. They were excellent hunters and speared huge animals such as woolly mammoths and long-horned bison. These enormous beasts later died out, and the people were forced to hunt smaller game and collect wild plants for food.

TOOLS
Early Native Americans made knives from stone and bone, which we sharp enough to slic up large animals. They used scrapers t remove the fur and flesh from skins.

SHAMAN'S MASK
Many thousands of years ago, an Arctic woodcarver shaped this mask for a shaman (healer) to wear during the ceremonies that called upon the spirits to bring good health and good hunting.

CREATION STORY

The Northeastern Iroquois describe Skywoman's fall from her home. Water birds guided her to an island that a muskrat was building from mud and a turtle's shell. The island grew to become the Earth, and the world began when Skywoman gave birth to a daughter.

HUNTING AND FIGHTING
Early hunters and warriors flaked stone into deadly tips for their spears and lances. They made heavy war clubs from whale bone.

WALK ACROSS THE STRAIT
A strip of water called the Bering Strait now separates Siberia from Alaska. In the Ice Age, when the sea level dropped, the strait became a land bridge for herds of migrating game and the hunters who pursued them.

Siberia

Alaska

Land bridge

Glaciers

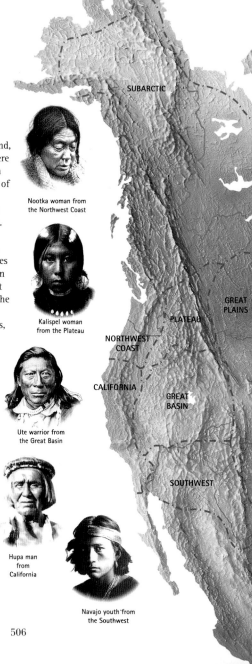

Where Did They Live?

Tribes of Native Americans spread across the land, depending on nature for food and shelter. Where they lived shaped the way they lived, and each group began to develop different customs and ways of doing things. On the rugged Northwest Coast, the "Salmon and Cedar People" built with wood and ate mainly fish, like their neighbors on the high Plateau. The Inuit (Eskimos) hunted polar bears across the treeless tundra and whales in freezing Arctic waters. Caribou provided nearly everything Subarctic families needed, while buffalo sustained the Plains Indians. In California, the mild climate meant that tribes there had plenty to eat, unlike the parched Great Basin where food was scarce. In the Northeast and Great Lakes, people traveled the rivers and cleared forest plots to grow corn and tobacco. Most Southwest and Southeast tribes became farmers and lived in villages.

SUBARCTIC

Nootka woman from the Northwest Coast

Kalispel woman from the Plateau

GREAT PLAINS

PLATEAU

NORTHWEST COAST

CALIFORNIA

GREAT BASIN

SOUTHWEST

Blackfoot medicine man from the Great Plains

Ute warrior from the Great Basin

Hupa man from California

Navajo youth from the Southwest

SOUTHWEST LANDSCAPE
A parade of saguaro cactuses in the dry Southwest provided nourishment when the fruit formed and ripened.

ARCTIC

Nunivagmiut man from
the Arctic

SUBARCTIC

WOODLANDS
Trees grew thickly in many areas, and Native
Americans soon began making canoes, weapons,
food containers and other items from wood.

Cree warrior from
the Subarctic

Sauk warrior from
the Northeast

NORTHEAST

SOUTHEAST

SIGN LANGUAGE

As people separated into groups, the way they spoke slowly
changed. Eventually, there were many different languages.
To overcome the problems of not understanding their neighbors,
the tribes on the Great Plains invented sign language— a clever way
of communicating without words. Gesturing with their hands,
chieftains made peace bargains, hunting parties discussed
the whereabouts of game, and Mandan farmers traded
surplus corn for Sioux buffalo skins.

Hello   Riding a horse   Peace   Friend

Seminole warrior
from the Southeast

Discover more in Village Life

507

• THE PEOPLE •

What Did They Wear?

Native Americans loved decorated ceremonial costumes, but had simple everyday clothes. They dressed to suit the weather on windy plains, in the chill Arctic, in damp rainforests, or in the dry desert, where often they wore very little. Most garments fitted loosely for easy movement—loincloths, shirts, tunics and leggings for men; skirts and dresses for women. In winter, people added shawls, blankets and extra clothing. They made garments from the things around them. Animals provided skins for cloth, sinews for thread and bones for needles. One moose could clothe a person in the Northeast, a single caribou supplied a jacket in the Subarctic, but a man's robe in the Great Basin required a hundred rabbit skins. Where game was scarce, the people wove cloth from plant material, such as nettle fiber and cedar bark.

DRESSED FOR BEST
Both men and women wore ceremonial robes made from soft buckskin, decorated with fringe and porcupine quill embroidery. The brave's full-length war bonnet and coup stick are trimmed with ermine tails and eagle feathers. The feathers show his courage in hand-to-hand fighting.

SNOW GOGGLES
The Inuit (Eskimos) wore goggles shaped from walrus tusk. These protected their eyes from the intense glare of the sun reflected off the snow.

KEEPING WARM
Men, women and children dressed alike against the cold in waterproof pants, hooded parkas, boots and mittens. Caribou and seal skins with the fur turned inside were most popular. Here, the mother chews skin to soften it, while the father carves an ivory ornament with a bow drill.

508

BARK WEAVING
While her grandchild rocks gently in a cradle, this woman weaves a basket from dampened strips of cedar bark. Her clothing is also woven from shredded bark. Cedar bark was one of the main plant fibers used by the people of the Northwest Coast.

FOOTWEAR
Plains women stitched leather moccasins with hard or soft soles. They decorated them with dyed porcupine quills and elaborate beadwork.

CLOTHING MATERIALS

Native Americans made their clothing from the animal and plant materials they found around them.

Areas where animal fur (such as caribou, seal and polar bear) and hide (such as buckskin and buffalo) were used for clothing.

Areas where cotton was mainly used for clothing.

Areas where both animal skins and plant materials were used for clothing.

Discover more in Getting Around

Life as a Child

C hildren spent most of the first year of their lives strapped snugly in a cradleboard, carried everywhere by their mothers. Later, a large family of parents, aunts, uncles and grandparents watched over them and taught them tribal ways. Girls practiced preparing food, sewing, tanning hides, making pottery, basket weaving and embroidering with dyed porcupine quills. Boys learned to make tools and weapons and how to hunt and fight. There was always time for games, such as hurtling down snowy slopes on sleds made from buffalo ribs. When children reached puberty, there were important ceremonies, sometimes with dancing, special clothes and the gift of a new name. Apache girls were showered with yellow tule pollen; Alaskan girls had their faces tattooed. After puberty, girls joined the women in the tribe, but boys had to pass tests of courage, such as wounding or killing an enemy, before they could become true hunters and warriors.

UMBILICAL POUCHES
Some tribes sewed newborn babies' umbilical cords into beaded pouches shaped like lizards or turtles. They believed these creatures would bless their infants with long life.

TARGET PRACTICE
Sioux boys learned how to use small bows and arrows, shooting first at still targets and then at moving jack rabbits.

UNDER THE STARS
Like Native American children everywhere, the boys and girls of the Great Basin gathered to hear stories. The retelling of myths, legends and folktales taught the children tribal history and customs.

DRESSING WARM
St. Lawrence Island children in the Arctic wore parkas made from reindeer skin with the hairy side on the inside. The tops of boys' heads were shaved like the men of the tribe. Girls kept their hair long.

NAMING A BABY

A baby's name was very important, but it was often chosen by relatives and tribal elders, not by the parents. When a Southwest Hopi baby was 20 days old, the father's mother and his sisters visited with many blessings and suggested names. Inuit (Eskimos) gave newborn infants another name every time they cried, so some tiny babies had dozens of names. A Seminole boy was rewarded with a new name when he proved his courage in battle.

Seminole child

CRADLEBOARD
This Cheyenne cradleboard was made of beaded cotton lashed securely to a wooden frame. It fitted comfortably on the mother's back.

• THE PEOPLE •

Choosing a Partner

Some Native American couples began life together with a simple exchange of presents. Then the girl (perhaps no more than 13 years old) moved into her husband's home, or he joined her family. Subarctic partners set up a wigwam together. Marriage ceremonies varied greatly between tribes and regions. The richest of the Northwest Coast Tlingits gave huge wedding feasts and valuable presents. A Plains boy courted his sweetheart with fluted love tunes. In the evening, outside her tepee, the couple hid from curious passers-by with a blanket over their heads, while they chatted to see if they liked each other. In later times, wealthy Plains bridegrooms gave horses to the bride's family. Southwestern Hopis sealed their marriage partnership when the mothers washed their hair together in one bowl.

WISHRAM BRIDE
A Wishram bride on the high Plateau wore wedding finery made from panels of dentalium shells, edged with beads and coins.

READY FOR MARRIAGE
A Hopi girl's elaborate hairstyle indicated to a Hopi boy that she was old enough to be married.

MARRIAGE DOLLS
Menominee newlyweds near the Great Lakes received a pair of dolls as a wedding present. These were "good medicine" for a long and happy marriage.

512

ACROSS THE WATER
This Kwakiutl bride arrived at her future husband's village in the family canoe.

WEDDING BASKETS
Navajo wedding baskets always belonged to the bride. When a man moved into his wife's home, he looked after her goods but never owned them himself.

READY FOR MARRIAGE
A Hopi boy wore several fine bead necklaces on his wedding day.

HOPI WEDDING

A Hopi bride ground corn for three days in her chosen partner's house to show her wifely skills. After the hair-washing ceremony, she stayed there while the groom and his male relatives wove her wedding clothes. Then she walked home in one outfit, carrying the second one in a reed container. Women were buried in their wedding garments so that when they entered the spirit world, they would be properly dressed.

Wedding blanket sash

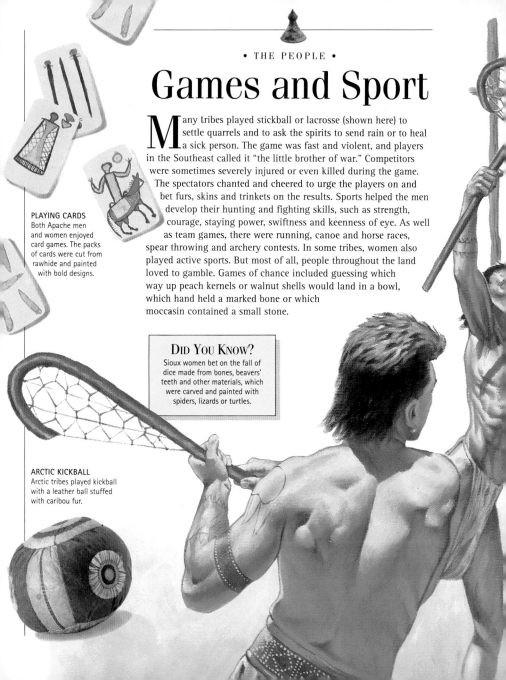

Games and Sport

Many tribes played stickball or lacrosse (shown here) to settle quarrels and to ask the spirits to send rain or to heal a sick person. The game was fast and violent, and players in the Southeast called it "the little brother of war." Competitors were sometimes severely injured or even killed during the game. The spectators chanted and cheered to urge the players on and bet furs, skins and trinkets on the results. Sports helped the men develop their hunting and fighting skills, such as strength, courage, staying power, swiftness and keenness of eye. As well as team games, there were running, canoe and horse races, spear throwing and archery contests. In some tribes, women also played active sports. But most of all, people throughout the land loved to gamble. Games of chance included guessing which way up peach kernels or walnut shells would land in a bowl, which hand held a marked bone or which moccasin contained a small stone.

PLAYING CARDS
Both Apache men and women enjoyed card games. The packs of cards were cut from rawhide and painted with bold designs.

DID YOU KNOW?
Sioux women bet on the fall of dice made from bones, beavers' teeth and other materials, which were carved and painted with spiders, lizards or turtles.

ARCTIC KICKBALL
Arctic tribes played kickball with a leather ball stuffed with caribou fur.

RING AND PIN
In this game, players held the wooden pin and tried to pass the loop of string over the deer-foot bones.

PLAYING PATOL

Patol is a game of chance played by two to four people. They use counters called "horses," stick dice and stones arranged in a circle or rectangle. Players become very skilful at throwing the dice. Patol was popular in the Southwest, where games took place outside in front of interested onlookers.

GAMBLING GAME
The sticks in this game were shuffled under a cover and divided into two bundles. Players then guessed which bundle contained a specially marked stick.

Hoop

Lance

HOOP AND POLE GAME
This was a game of great skill. Competitors had to throw a lance through a hoop as it rolled along the ground. A hit on the center hole scored the most points.

Canoes and Kayaks

P eople built boats for fishing, moving between
hunting grounds, carrying goods and going to war.
Some hollowed out massive tree trunks with fire,
others wove crafts from reeds, or covered wooden frames with
birch bark and sealed the seams with hot black spruce gum.
Low-ended crafts steered best in calm waters. Boats with high
bows and sterns resisted rough waves and were more suitable
for the open ocean. The tall prow of a ceremonial Northwest
Coast canoe (shown here) was carved and painted to reflect
the family's importance. The fearsome bear is a villager
dressed in his winter dance costume. The Californian
Chumash put to sea in canoes built from pine planks.
Inuit (Eskimos) made lightweight, waterproof kayaks,
by stretching oiled animal skins over driftwood frames.
Most kayaks were for one person, though some held two.

LONG BOAT
It took 11 men to launch this
canoe, which was carved from
a single tree trunk.

HUNTING CRAFT
Inuit hunters stalked sea
mammals in their speedy,
silent kayaks.

516

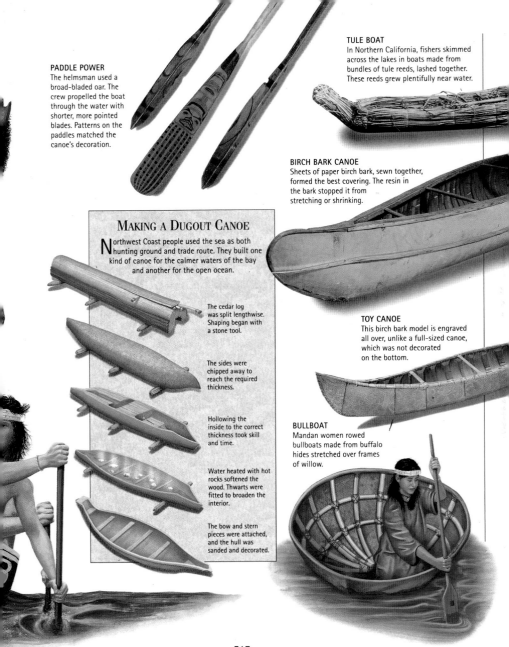

PADDLE POWER
The helmsman used a broad-bladed oar. The crew propelled the boat through the water with shorter, more pointed blades. Patterns on the paddles matched the canoe's decoration.

TULE BOAT
In Northern California, fishers skimmed across the lakes in boats made from bundles of tule reeds, lashed together. These reeds grew plentifully near water.

BIRCH BARK CANOE
Sheets of paper birch bark, sewn together, formed the best covering. The resin in the bark stopped it from stretching or shrinking.

MAKING A DUGOUT CANOE
Northwest Coast people used the sea as both hunting ground and trade route. They built one kind of canoe for the calmer waters of the bay and another for the open ocean.

The cedar log was split lengthwise. Shaping began with a stone tool.

The sides were chipped away to reach the required thickness.

Hollowing the inside to the correct thickness took skill and time.

Water heated with hot rocks softened the wood. Thwarts were fitted to broaden the interior.

The bow and stern pieces were attached, and the hull was sanded and decorated.

TOY CANOE
This birch bark model is engraved all over, unlike a full-sized canoe, which was not decorated on the bottom.

BULLBOAT
Mandan women rowed bullboats made from buffalo hides stretched over frames of willow.

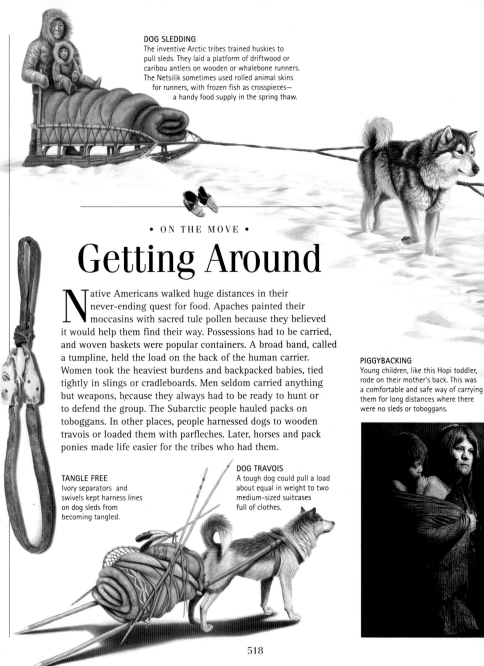

DOG SLEDDING
The inventive Arctic tribes trained huskies to pull sleds. They laid a platform of driftwood or caribou antlers on wooden or whalebone runners. The Netsilik sometimes used rolled animal skins for runners, with frozen fish as crosspieces— a handy food supply in the spring thaw.

• ON THE MOVE •

Getting Around

Native Americans walked huge distances in their never-ending quest for food. Apaches painted their moccasins with sacred tule pollen because they believed it would help them find their way. Possessions had to be carried, and woven baskets were popular containers. A broad band, called a tumpline, held the load on the back of the human carrier. Women took the heaviest burdens and backpacked babies, tied tightly in slings or cradleboards. Men seldom carried anything but weapons, because they always had to be ready to hunt or to defend the group. The Subarctic people hauled packs on toboggans. In other places, people harnessed dogs to wooden travois or loaded them with parfleches. Later, horses and pack ponies made life easier for the tribes who had them.

PIGGYBACKING
Young children, like this Hopi toddler, rode on their mother's back. This was a comfortable and safe way of carrying them for long distances where there were no sleds or toboggans.

TANGLE FREE
Ivory separators and swivels kept harness lines on dog sleds from becoming tangled.

DOG TRAVOIS
A tough dog could pull a load about equal in weight to two medium-sized suitcases full of clothes.

DID YOU KNOW?

One legend explained that baskets walked by themselves until Coyote, the mischievous wolf spirit, said they looked silly. He made women carry them from then on.

SNOWSHOE SHUFFLE

Snowshoes, shaped like bear paws or beaver tails, were the perfect winter footwear for journeying through deep snow drifts. They allowed hunters to keep pace with caribou and other large game without their feet sinking into the soft snow. Subarctic tribes laced bent frames of birchwood with strips of wet caribou hide. When the webbing dried, it was tight and light.

PLAIN AND SIMPLE

The Southeastern Seminoles made their moccasins from a single piece of soft buckskin gathered at the seam.

WALKING WARM

In summer, Eskimos gathered grasses and braided them into socks. They were shaped to fit the foot snugly.

HUSKY HELPERS

Husky dogs pulled sleds in teams. They also used their keen noses to track down seals for the hunters.

Discover more in Tepees

Horses and Ponies

SADDLE BLANKET
Respected horses were as
well-dressed as their owners.
It took many hours to edge
this blanket with beads.

Horses brought speed to the Plains Indians and changed their hunting and fighting ways. They were called "Spirit Dogs" or "Medicine Dogs," but were much stronger and faster pack animals than real dogs. With horses, whole settlements could move freely to follow the buffalo. The hunter–warrior had a special bond with his horse; only he trained and rode it, bathed it in summer, covered it with buffalo skins in winter and tethered it beside his tepee. Horses, extinct since the Ice Age, were reintroduced by the Spanish in the sixteenth century and spread gradually from tribe to tribe. The Plains Comanches excelled in taming wild horses. Comanche women joined the men on antelope hunts, and the children rode solo by the age of five. Plateau tribes bred Appaloosa horses and Indian ponies and became expert horse traders.

WHIPS AND QUIRTS
Whips or quirts urged horses
on. This quirt is made
from an elk antler
and a leather strip.

STRANGERS ON HORSEBACK
This Navajo cave painting of a Spanish friar and his companions has survived from the sixteenth century.

SADDLE UP
Men usually rode bareback. Women used saddles and stirrups. Crow women spent many hours trimming theirs with beads.

Saddle pommel

SADDLE BAG
Bags were made from deerskin, stitched up the sides. They were tied to the pommel of the saddle.

Stirrup

BATTLEDRESS

Plains warriors depended on their war horses. The animals' reactions in battle could mean the difference between life and death. They had to endure the noise, move quickly, turn sharply and respond instantly to their riders' commands. Warriors shared battle honors with their mounts and painted them with the symbols they used on their own bodies (below). The horses wore eagle feathers and scalp locks, and their manes and tails were often trimmed and dyed.

Plains pony

Appaloosa horse

War party leader

Enemy killed in hand combat

Hail

Mourning marks

TAILPIECE
A crupper passed from the back of the saddle under the horse's tail to keep the saddle from slipping.

Discover more in Life on a Reservation

The Foragers

SEED HARVESTING
When the seeds ripened, Californian women beat them straight from the bushes into their carrying baskets.

From early spring to late autumn, many Native American tribes moved around frequently, searching for things to eat. They foraged for seeds, berries, nuts and roots. In the Great Basin, March cattail shoots were the first fresh vegetables, while September's pine nuts provided needed stores for winter. Occasional grasshopper plagues were tasty feasts. The men caught almost anything, from rats, mice, caterpillars and lizards, to larger game including jack rabbits and deer. They made lifelike decoys from reeds and feathers to lure wild ducks into shallow water. Then they grabbed the birds or shot them down with arrows. In California, foragers hunted in the mountains and on the sea coast. They harvested acorns from oak trees and ground them into flour.

THE RIPE TIME

The summer sun ripened fruits, nuts and berries, and tribes knew where to find them in their region. The foragers dried some of these and put them aside for winter, when food was often in very short supply.

Persimmons

Cranberries

Black walnut

Buffalo berries

Pine nuts

WICKER LARDER
The foragers used woven baskets to collect, store and carry food. These were much lighter than containers made from clay and were not easily broken.

A DIFFICULT CATCH
The sure-footed bighorn sheep that browsed in the desert mountains provided a satisfying meal. Men, women and children banded together to trap the animals.

Discover more in Buffalo Hunters

523

Club

**HALIBUT
FISHING TACKLE**
The Tlingits carved hooks
to catch halibut. They
stunned the heavy fish
with clubs before hauling
them into the boat.

Hook

HARPOONING WALRUSES
Hunting walruses was
dangerous but worthwhile.
They provided meat, skin
for boats and clothing,
and ivory tusks for
ornaments and utensils.

The Fishers

In late spring, salmon begin to race up the inland waterways
to lay their eggs. The Northwestern tribes thought they came
especially to provide their people with food, and caught them
in nets and traps. They offered prayers to the first salmon landed,
roasted it for everyone to taste, and returned the whole skeleton
to the river where, they believed, it would come alive again. The
ocean was also a rich source of food for these people. From it,
they took whales, seals, huge halibut, cod and sturgeon, all kinds
of shellfish and oily candlefish, which gave them fuel for lamps.
Other tribes speared fish by torchlight from their canoes on the
Great Lakes and used nets and traps during the day. In parts of the
Southeast, the people depended on fish, as well as turtles,
alligators and deer, for protein in their diet. The Inuit (Eskimos)
needed meat from sea mammals to survive in winter when they
had no plant food. They caught seals and stalked walruses
among the ice floes.

DID YOU KNOW?
The Inuit (Eskimos) divided their world
into land and sea things. They would not
eat the flesh of land and sea animals in
the same meal or cook caribou meat
over a driftwood fire.

NET GAUGE
Twine to make mesh fishing nets was wound around a rectangular gauge. This one is made from elk antlers.

WHALING IN A UMIAK

Whales were dangerous prey. Arctic tribes went after them in single-sailed rowing boats called umiaks. They were made from whalebone covered with walrus hides and waterproofed with seal oil. Umiaks were much stronger than kayaks. Inuit (Eskimo) fishers sometimes hunted bowhead whales that were twice as long as the boats. They caught smaller beluga whales for their skin and tusked narwhals for their oil.

Whaling float

TRAPPING FISH
The Ingalik people of the Subarctic set wheel fish traps beneath the ice. Large ones, like this, were mainly used for snaring ling, a cod-like fish.

ULU KNIFE
Women butchered meat, scraped skins and cut leather with an ulu. It had a slate blade and a bone, ivory or wood handle.

Discover more in Canoes and Kayaks

Making a Meal

Imagine making a meal with no tap water, no gas or electricity and no refrigeration and supermarkets. Native Americans had none of these, and the women spent hours collecting food and water and preparing things to eat. Their daughters worked alongside them from a very early age. Pueblo women baked corn bread in outdoor ovens. The Northwestern Tlingits filled baskets of closely woven spruce roots with water, meat and vegetables, dropped in hot stones till the water boiled, and made a stew. Plains Indians used buffalo-hide containers in the same way. Metal cooking pots, which lasted much longer than skin or woven ones, became popular after tribes began trading with the Europeans. Native Americans had no set meal times. People ate when they were hungry, after a good hunt or when travelers arrived. Sharing was very important. Plenty of food meant lots for everyone. If provisions were in short supply, the small amounts were divided evenly. Most tribes stored food for the winter, when plants and game were harder to find.

PLENTY ON THE PLAINS
During most of the year, there was plenty to eat on the Plains. These Cheyenne women are pounding wild cherries and cooking with hot stones and metal pots. Turnips and squash will be added to the stew.

DRYING CHILIES
Chili peppers were a popular crop in the Southwest and gave the food a hot, spicy flavor. The pods were strung on plant fibers or cotton cords and hung up to dry.

SUN-DRIED FOODS
When food was plentiful, it could be preserved for winter. Fish, meat, corn, fruit and vegetables were hung on racks or spread out on platforms to dry in the sun.

IN STORAGE
Large storage jars for food were woven from willow wood and sedge roots. They were patterned with darker plants, such as devil's claw, or decorated with brightly colored feathers.

GRINDING CORN

Most days young Hopi women gathered to grind corn. They placed the corn kernels on a rough stone slab called a "metate" and rubbed them with a smaller stone called a "mano." Then they moved the pieces to progressively smoother slabs and crushed them into finer and finer particles. The Hopi cooked cornmeal flour in several ways. They made it into more than 30 different dishes including bread, gruel, pancakes and dumplings.

Discover more in Village Life

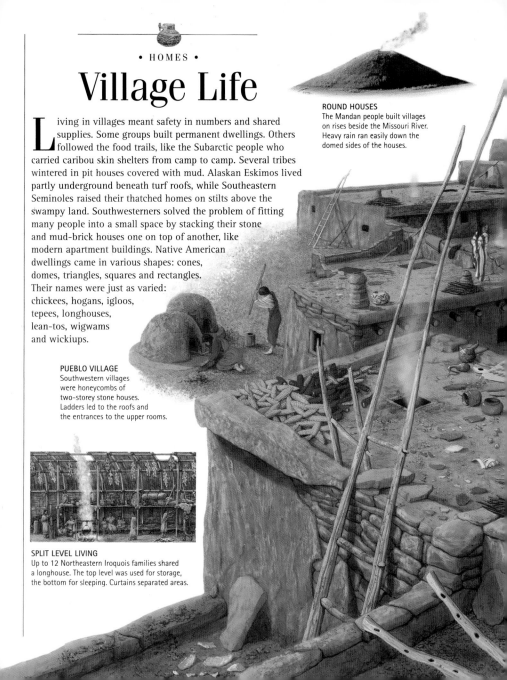

Village Life

Living in villages meant safety in numbers and shared supplies. Some groups built permanent dwellings. Others followed the food trails, like the Subarctic people who carried caribou skin shelters from camp to camp. Several tribes wintered in pit houses covered with mud. Alaskan Eskimos lived partly underground beneath turf roofs, while Southeastern Seminoles raised their thatched homes on stilts above the swampy land. Southwesterners solved the problem of fitting many people into a small space by stacking their stone and mud-brick houses one on top of another, like modern apartment buildings. Native American dwellings came in various shapes: cones, domes, triangles, squares and rectangles. Their names were just as varied: chickees, hogans, igloos, tepees, longhouses, lean-tos, wigwams and wickiups.

ROUND HOUSES
The Mandan people built villages on rises beside the Missouri River. Heavy rain ran easily down the domed sides of the houses.

PUEBLO VILLAGE
Southwestern villages were honeycombs of two-storey stone houses. Ladders led to the roofs and the entrances to the upper rooms.

SPLIT LEVEL LIVING
Up to 12 Northeastern Iroquois families shared a longhouse. The top level was used for storage, the bottom for sleeping. Curtains separated areas.

VILLAGE LAYOUT

In the well-planned Creek villages of the Southeast, airy summer sleepouts were built beside warmer lodges. The largest round council buildings could seat 500 people. The villagers used them for ceremonies, dancing, winter meetings of the tribal elders and to house the homeless and the aged.

FRONT DOOR POLES

Northwest Coast tribes, such as the Haida, lived in wooden buildings. Carved cedar totem poles indicated who lived in each house.

BARK SHELTER

Some tribes built huts from chunks of redwood or cedar bark. The sweet-smelling wood repelled insects.

Discover more in Tepees

Tepees

TEPEE DECORATION
Ornaments were tied to the top of the tepee poles. These are leather thongs wrapped in grass and tipped with yarn tufts.

Family possessions
Families kept everything they owned inside their tepee. It was kitchen, bedroom, playroom, living room and shed all in the one space.

The Plains Indians lived in cone-shaped structures called tepees, made out of buffalo hides sewn together. When tribes needed to move on to find food or to escape from enemies, they could fold the tepees and transport them easily. At first, they carried their belongings on dog travois, so tepees were limited in size to the height of a man. Later, when horses were used for transportation, the tepees became much bigger. The space inside these portable homes was very limited, and the furniture was simple and functional. Buffalo skins made comfortable bedding. Backrests of willow rods laced together with string and supported by poles formed chairs, which could be rolled up neatly. Rawhide saddlebags called "parfleches" doubled as cushions or pillows. Tepees were erected with the steeper, rear side against the westerly winds, and the doorway facing east, towards the rising sun. In the windy areas of the southern plains, the Sioux and Cheyenne supported their tepees with three foundation poles, while further north, the Hidatsa, Crow and Blackfeet used four poles. Ceremonial and larger tepees had more supports.

Symbols of the Blackfeet
Painted symbols, such as rainbow stripes or a buffalo head, protected the tepee owners against sickness and bad luck.

RESERVATION CAMP
When the Plains Indians were forced to move to reservations by the United States government, they took their tepees with them and pitched them in tribal groups. They tried to preserve as much of the way of life from their homelands as possible.

Smoke flaps
These could be opened from the inside with a pair of long poles.

TEPEE CAMPS

Tepee camps were pitched in a C shape with the opening facing east. Behavior inside the tepee followed strict rules. An open door was a sign of welcome. Men entered to the right, women to the left. Younger men remained silent until they were invited to speak. No one walked between the fire and another person. Visitors brought their own bowls, and when the host cleaned his pipe, it was time for visitors to leave.

N

Dew cloth
A decorated lining called a dew cloth was tied inside the tepee. It kept out moisture and helped to insulate the tepee.

Lacing
The tepee was laced from the bottom to the smoke hole with pins carved from flexible willow wood.

Entrance
The door was made from a flap of skin and was oval or V-shaped.

Fireplace
The fire was placed under the smoke hole, and the woodpile was near the door.

Hemline
The hem was pegged to the ground with stakes but raised in hot weather to let in air.

BUFFALO SKIN
Hides were tanned and smoked so that they would be waterproof but still remain soft. Some were decorated for ceremonial tepees or painted with a family clan's symbol.

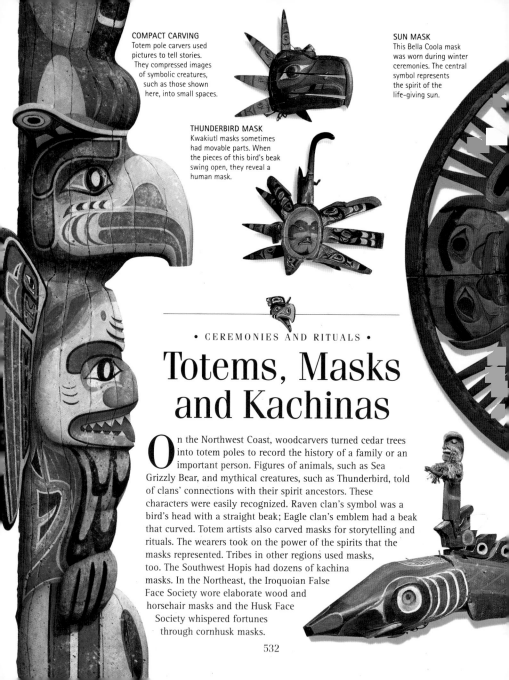

COMPACT CARVING
Totem pole carvers used pictures to tell stories. They compressed images of symbolic creatures, such as those shown here, into small spaces.

SUN MASK
This Bella Coola mask was worn during winter ceremonies. The central symbol represents the spirit of the life-giving sun.

THUNDERBIRD MASK
Kwakiutl masks sometimes had movable parts. When the pieces of this bird's beak swing open, they reveal a human mask.

• CEREMONIES AND RITUALS •

Totems, Masks and Kachinas

On the Northwest Coast, woodcarvers turned cedar trees into totem poles to record the history of a family or an important person. Figures of animals, such as Sea Grizzly Bear, and mythical creatures, such as Thunderbird, told of clans' connections with their spirit ancestors. These characters were easily recognized. Raven clan's symbol was a bird's head with a straight beak; Eagle clan's emblem had a beak that curved. Totem artists also carved masks for storytelling and rituals. The wearers took on the power of the spirits that the masks represented. Tribes in other regions used masks, too. The Southwest Hopis had dozens of kachina masks. In the Northeast, the Iroquoian False Face Society wore elaborate wood and horsehair masks and the Husk Face Society whispered fortunes through cornhusk masks.

KACHINA DOLLS

Southwestern tribes, such as the Zuni and the Hopi, carved kachina dolls out of wood. They clothed them in masks and costumes to look exactly like the men who dressed up as kachina spirits. These dolls were not playthings. They were given to the children to teach them to identify the many different kachinas and the parts they played in tribal ceremonies.

Zuni kachina doll

TODAY'S TOTEM ART
Thunderbird, on top of this modern totem, is the carver's personal crest and shows he belongs to a Kwakiutl warrior group.

DOGFISH MASK
A dancer wearing this dogfish mask pulled a string to make the figure riding the dorsal fin twirl.

SEA OTTER MASK
The curved arms around the sea otter's head whirl around and support birds with flapping cloth wings.

Special Occasions

Native Americans enjoyed many special occasions. They celebrated important times in people's lives, such as reaching the age of puberty, getting married or being successful in battle. Ceremonial clothing, decorated with fur, feathers, quilling and beadwork, was worn for these events, and people made necklaces, earrings and bracelets from animal teeth, bones and claws, shells and stones. Native Americans believed that the sky, the soil, plants, birds, animals, rivers and everything else had spirits that must be respected. These spirits could be reached through dance, song, prayer and other religious rituals. Some tribes also worshipped monsters, such as the dreadful Cannibal Spirit of the Northwest, and danced to show their evil power.

SACRED ROOM
The underground "kiva" was the holy place of Southwest kachina priests. The chamber's floor became an altar when it was sprinkled with cornmeal, sand, ground bark and flowers.

DRESSED TO DANCE
Kwakiutl shamans wore cedar bark costumes and painted wooden masks to represent the birdlike friends of the fearsome Cannibal Spirit.

CANNIBAL SPIRIT DANCERS
The Cannibal Spirit dancer is standing on the left wearing a cedar bark ring around his neck. His fearsome followers squat in front.

FAMILY JEWELRY
Navajo women wore ear pendants before marriage. Afterwards, they attached them to bead necklaces until their own daughters were old enough to wear them.

FEATHER HEADDRESS
Ceremonial costumes made from eagle feathers, like this chieftain's war bonnet, had special importance. Eagles were linked with heavenly spirits and admired for their speed, fierceness and sharp eyesight.

POLLEN RITUAL

A blessing of pollen, the tribe's most powerful "medicine," is a highlight of the four-day puberty ritual for an Apache girl. Baskets of yellow pollen from tule rushes are gathered for the ceremony. The grains are then sprinkled over the girl's hair and face.

Pollen basket

Discover more in Pipes and Powwows

SNOWSHOE DANCER
An Objibwa hunter rejoices after the first winter snowfall. Now his prey will flounder in drifts and be much easier to catch.

FLUTED MELODY
Flute music, called the "wind that breathes life into the heart," accompanied many dances. The six round holes represented the Earth, sky and the four directions: north, south, east and west.

MARKING TIME
This rattle has a dried gourd body, a handle covered in glass beads and streamers made from feathers and strips of sinew.

• CEREMONIES AND RITUALS •

Ceremonial Dancing

Ceremonial dancing was the Native American way of celebrating joyous occasions and praying for health, successful hunting and good harvests. Plains Sioux imitated the sounds and movements of bears before the hunt or whooped in a scalp dance after a battle victory. The Californian Patwin tribe danced in huge headdresses and cloaks made of feathers or grass to encourage the growth of wild crops. Hopi men in the dry Southwest collected snakes for an elaborate ritual. The snake priests, wearing feathered headdresses and kilts patterned with the serpent motif, circled the village square with the reptiles in their mouths. Their companions stroked the creatures with eagle feathers to stop them from biting. The snakes were then returned to the desert where their lightning-like, zigzag movements were supposed to bring pre-harvest rain.

KACHINA SPIRITS

Hopi men impersonated kachinas, important spirits in their religion. They performed dances during the seasons of seed sowing, plant growth and harvesting. Kachina dancers taught young children tribal ways and gave them dolls.

Kachina doll

RATTLING RHYTHM

Inuit (Eskimo) men wore sealskin gauntlet gloves for ceremonial dancing. This pair is decorated with horned puffin beaks and quills from feathers, which rattled to the beat of the drum.

TAPPING RHYTHM

An Inuit (Eskimo) woodcarver made this baton. The woodpecker is attached to the shaft with springy whale cartilage. During the dance, it pecks like a real bird.

537

Pipes and Powwows

Native Americans used solemn pipe-smoking rituals to ask for the spirits' help to make war, peace or rain, to hunt successfully, or to seal a good trade bargain. Pipes were very special and very beautiful. Each one took several weeks to make. The stem was hollow wood, and the bowl was fashioned from soft soapstone, clay or wood. Supernatural powers did not flow through the pipe until these two parts were ceremonially joined. In the past, pipes were smoked at powwows where people gathered to pray for the sick or for the tribe's success in battle or hunting. Today, powwows are joyful events, held at least once a year, to remind people of old customs and to celebrate new ones.

TOBACCO BAG
The pipe bowl, the pipe st
and the smoking mixture
were kept in a quilled and
beaded buckskin pouch.

PIPE CEREMONY
Ceremonial pipe
smoking had special
importance. As one
Sioux tribesman explained:
"This pipe is us. The stem
is our backbone, the bowl
our head. The stone is our
blood, red as our skin."

PIPE TOMAHAWK
Pipe tomahawks with sharp steel blades and inlaid handles were prized ceremonial objects, rather than weapons of war.

IVORY PIPE
Arctic pipes were often made of ivory. This one was carved from the tusk of an Arctic walrus.

PRAYER RITUAL
A Sioux warrior placed a buffalo skull at his feet and pointed his pipe skyward while he prayed for supernatural powers.

TOBACCO PLANT
Tobacco was believed to have magical powers to heal or to hurt, to change people's fortunes, or to call up good spirits and drive away evil.

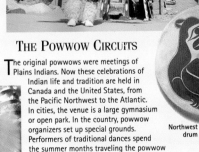

THE POWWOW CIRCUITS

The original powwows were meetings of Plains Indians. Now these celebrations of Indian life and tradition are held in Canada and the United States, from the Pacific Northwest to the Atlantic. In cities, the venue is a large gymnasium or open park. In the country, powwow organizers set up special grounds. Performers of traditional dances spend the summer months traveling the powwow circuit. These cheerful, colorful festivities also involve singers, craftspeople, families, friends and the local community.

Northwest drum

Discover more in The Future

Healers and Healing

C alling a doctor was no simple matter for Native Americans. Often, visits and ceremonial treatments lasted for several days. They were organized by healers, known as shamans or medicine men and women, some of whom had as much power as any chief. Shamans cured the sick with herbs, performed healing rituals, told the future or found missing property. They knew the dances, chants, prayers and ceremonies that would bring good fortune to their tribes and please the spirits. Young people had to take many difficult tests of physical strength before they could become shamans, and few succeeded. They did not receive their special powers until they had seen a vision in the form of a sacred animal or object. Native Americans were generally very healthy until fatal diseases were introduced from Europe.

MEDICINE KIT
A Sioux woman once owned this rawhide bag of medicines. The wrapped bundles contained crushed leaves and powdered bark and roots.

SWEAT LODGES
Skins over a wooden frame made an airtight hut. Inside, water poured on hot stones turned to steam, like a modern sauna. Warriors cleansed their bodies and the sick eased fevers and aching bones.

PRAYER SNAKES
Some tribes thought that snakes caused stomach complaints. To cure such gastric upsets, Navajo medicine men made snake-shaped prayer sticks from wood and feathers.

DRY SAND PAINTING
Navajo healers used powdered rocks to create large pictures on the floor of the patient's dwelling. They believed the sand could absorb evil sickness. After the ritual, they destroyed the painting.

TOOL OF TRADE
Rattles were an essential part of the Tlingit shaman's ritual equipment. This one's wild hair, beard and mustache are made from human hair.

Q: What did Navajos use to cure stomach complaints?

HONORED HEALER
Slow Bull, photographed on the Plains with a sacred buffalo skull, was a well-respected Oglala Sioux medicine man.

HERBS AND FLOWERS
Native Americans used many herbal medicines. From willow bark they extracted a pain-relieving ingredient, used in today's aspirin. Southeastern Cherokees believed every plant would cure a specific sickness. Iris roots ground with suet, lard and beeswax made an ointment for cuts and grazes. Juice of lady's slipper roots eased pain, soothed hysterics and relieved colds and flu.

Wild purple iris

Yellow lady's slipper

FOE KILLED

The Sioux clipped and dyed feathers to count special coups. A red spot indicated the wearer had killed his foe.

SINGLE WOUND

A feather dyed red meant the wearer had been wounded in battle.

MANY COUPS

A jagged edge proclaimed that the wearer had felled several enemies.

MANY WOUNDS
Split feathers were a sign that the wearer had been wounded many times.

FOE SCALPED
A notched feather showed the wearer had cut his enemy's throat and then taken his scalp.

THROAT CUT

The top of the feather clipped diagonally signaled that the wearer had cut his foe's throat without scalping him.

GREAT WARRIOR
Braves who fought well became respected war leaders and were entitled to wear elaborate ceremonial costumes.

• A CHANGING WORLD •

Warriors and Warfare

S ome Native American tribes hated war, but many fought constantly over land and horses, to avenge their people and to rack up battle honors. They carried out hit-and-run raids more often than full-scale warfare. Sometimes they adopted prisoners; other times they tortured or scalped them. Southeastern men wore loincloths and moccasins on the warpath and carried weapons and moccasin repair kits. The men walked in single file, and stepped in the footprints of the warrior in front. Chickasaw scouts tied bear paws to their feet to lay confusing trails. Northeasterners stalked the woodlands, signaling to each other with animal calls, or clashed in canoes on the Great Lakes. Plains warriors considered hand-to-hand fighting more courageous and skillful than firing arrows from a distance. A brave proclaimed his "coup" (French for "blow") score by attaching golden eagle tail feathers to his war bonnet and ceremonial robe.

WAR SHIELDS
Warriors prized their buffalo hide shields. Some were covered with deerskin and decorated with symbolic animals, bells and feathers.

TROPHIES OF WAR

In some tribes, scalps brought honor to the warrior who took them. In others, counting coups or capturing horses from the enemy were much more important. Native Americans believed the scalp contained a person's soul and that spiritual power flowed from the slain warrior to the victorious brave. After a successful battle, many tribes danced through the night around the scalps of their foes. They preserved these war trophies by stretching them over a wooden hoop attached to a stick.

WAR WHISTLE
When a Mandan warrior saw an enemy, he blew a whistle like this, which is made from bone wrapped in porcupine quills.

BOWS AND ARROWS
Warriors carried their tightly strung bows and sharp arrows in tanned leather cases.

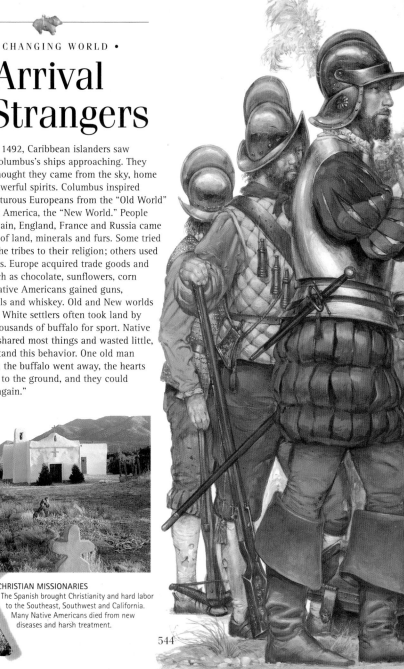

Arrival of Strangers

In 1492, Caribbean islanders saw Columbus's ships approaching. They thought they came from the sky, home of powerful spirits. Columbus inspired adventurous Europeans from the "Old World" to visit America, the "New World." People from Spain, England, France and Russia came in search of land, minerals and furs. Some tried to convert the tribes to their religion; others used them as slaves. Europe acquired trade goods and new foods, such as chocolate, sunflowers, corn and peanuts; Native Americans gained guns, horses, metal tools and whiskey. Old and New worlds did not mix well. White settlers often took land by force and shot thousands of buffalo for sport. Native Americans, who shared most things and wasted little, could not understand this behavior. One old man despaired, "When the buffalo went away, the hearts of my people fell to the ground, and they could not lift them up again."

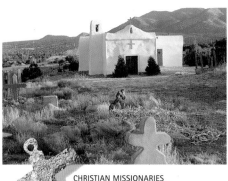

CHRISTIAN MISSIONARIES
The Spanish brought Christianity and hard labor to the Southeast, Southwest and California. Many Native Americans died from new diseases and harsh treatment.

544

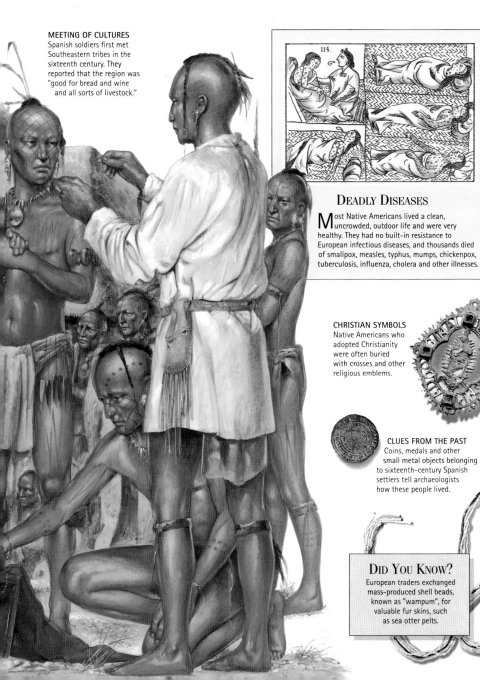

MEETING OF CULTURES
Spanish soldiers first met Southeastern tribes in the sixteenth century. They reported that the region was "good for bread and wine and all sorts of livestock."

DEADLY DISEASES
Most Native Americans lived a clean, uncrowded, outdoor life and were very healthy. They had no built-in resistance to European infectious diseases, and thousands died of smallpox, measles, typhus, mumps, chickenpox, tuberculosis, influenza, cholera and other illnesses.

CHRISTIAN SYMBOLS
Native Americans who adopted Christianity were often buried with crosses and other religious emblems.

CLUES FROM THE PAST
Coins, medals and other small metal objects belonging to sixteenth-century Spanish settlers tell archaeologists how these people lived.

DID YOU KNOW?
European traders exchanged mass-produced shell beads, known as "wampum", for valuable fur skins, such as sea otter pelts.

• A CHANGING WORLD •

Life on a Reservation

When Europeans began to colonize America, they fought bitterly with the Native Americans over land. In 1830, President Jackson passed a law saying that the government could set up areas in the west called reservations. These were exchanged for tribal homelands, which the new settlers wanted to farm. Native Americans were not allowed to argue their case in the law courts. When they resisted being moved, many people, including United States soldiers, died in the struggles to drive Native American families from their homes. The Southeastern tribes were forced to walk the "Trail of Tears" to the west. Some arrived without food, mules, plows or building materials. Thousands of Native Americans perished from hunger and misery. Although reservation schools taught the ways of white people, Native Americans never forgot where they had come from, and children learned their own customs from their parents and grandparents.

GOVERNMENT PROMISES
This letter from the government, dated 1838, granted reservation land to the Cherokees but kept the right to build roads and forts within the area.

THE TRAIL OF TEARS
Bluecoat soldiers marched 16,000 Cherokees along "the trail where they cried." Two thousand died along the way, and another 2,000 died soon after the long journey.

546

RESERVATION LANDS

The dark color marks the reservations set up by the United States government in the west. These small areas often had poor soil and a bad climate. Native Americans were expected to grow their own food, and this caused great hardship because many tribes had never farmed before. They were unable to hunt or to move around freely any more, and their new life often made them feel like prisoners.

COLORING CRAYONS
Some Native Americans who ran reservation stores could not write. They used crayons to draw pictures of what they sold.

SEWING LESSONS
In the reservation schools, children wore uniforms and were taught to read and write English. The girls used sewing machines to make European-style clothing.

547

MICKEY MOCCASINS
Beadwork is a very old
Native American craft.
The designs of some
moccasins now show
modern influences.

• A CHANGING WORLD •

Arts and Crafts

Native Americans made practical and beautifully crafted
objects for everyday use. Their ceremonial clothing and
sacred things were richly decorated. Painting, carving and
embroidery told stories and were linked with the spirits through
designs that had special meanings. Skills such as basketry, pottery
and weaving have been passed down from one generation to
another for many centuries. The Navajo learned silversmithing from
the Mexicans in the 1850s, and then taught the Hopi and Zuni.
Today, many artists and craftspeople use modern materials and tools.
They take ideas from the twentieth century and
blend them with patterns from the past, often
using vibrant colors in their work. Native
American arts and crafts are now
famous throughout the world.

HANDCRAFTED
JEWELRY
Turquoise, mined by
Southwestern tribes, is
the stone of happiness,
health and good fortune.

FETISH BOWL
Southwestern tribes
collected good luck
objects called fetishes.
They kept them in bowls
painted with crushed
turquoise. Some fetishes
were strapped to the outside.

548

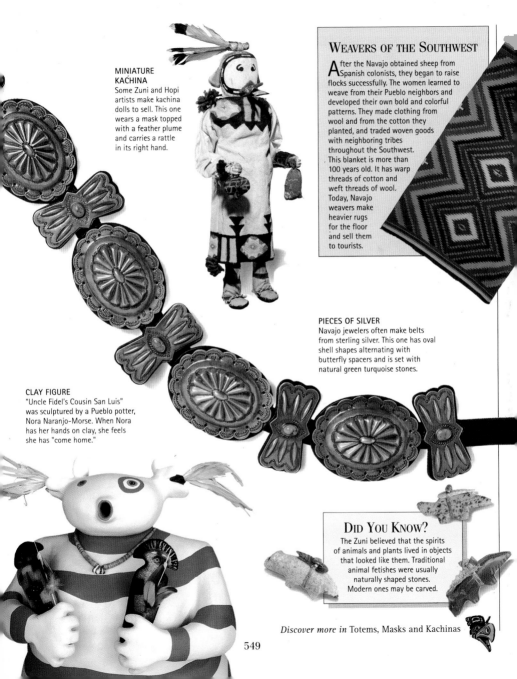

MINIATURE KACHINA
Some Zuni and Hopi artists make kachina dolls to sell. This one wears a mask topped with a feather plume and carries a rattle in its right hand.

WEAVERS OF THE SOUTHWEST

After the Navajo obtained sheep from Spanish colonists, they began to raise flocks successfully. The women learned to weave from their Pueblo neighbors and developed their own bold and colorful patterns. They made clothing from wool and from the cotton they planted, and traded woven goods with neighboring tribes throughout the Southwest. This blanket is more than 100 years old. It has warp threads of cotton and weft threads of wool. Today, Navajo weavers make heavier rugs for the floor and sell them to tourists.

PIECES OF SILVER
Navajo jewelers often make belts from sterling silver. This one has oval shell shapes alternating with butterfly spacers and is set with natural green turquoise stones.

CLAY FIGURE
"Uncle Fidel's Cousin San Luis" was sculptured by a Pueblo potter, Nora Naranjo-Morse. When Nora has her hands on clay, she feels she has "come home."

DID YOU KNOW?

The Zuni believed that the spirits of animals and plants lived in objects that looked like them. Traditional animal fetishes were usually naturally shaped stones. Modern ones may be carved.

Discover more in Totems, Masks and Kachinas

549

The Future

Native Americans suffered badly from the changes caused by European settlement. Many people died and some tribes disappeared altogether. Now, the number of Native Americans is growing as more healthy babies are born each year. Governments are beginning to recognize Native Americans' rights as citizens of the United States and Canada. More than half now live outside reservations. Today, Native Americans try to blend the old with the new and to keep their religious ceremonies, customs and languages alive. Cherokee schoolchildren learn to speak Cherokee and have lessons in Cherokee lifestyle. At powwows throughout the country, young children take part in social and competition dances, performing to the beat of the drums. They dress in tribal costumes and put on face paint for these happy celebrations.

STICK GAMES
Native Americans still play the old games of chance, and gamble with bunches of painted sticks.

SACRED CEREMONIAL CHAMBER
Pueblo villagers meet in the kivas to discuss local government, to train the young and to build altars and pray.

550

MOSAIC NECKLACE
This modern necklace was made from silver, inlaid with semi-precious stones and shells. It depicts the Zuni rainbow god.

ZUNI SACRED MOUNTAIN
Today's Native Americans visit their ancestral sacred sites for spiritual guidance. These places are holy ground where plants, paths, shrines and rocks all have religious meaning and must not be disturbed.

HOPI BASKETS
The Hopi method of making baskets has not changed for hundreds of years.

LOSS OF TRADITIONAL LANDS

1850

1865

1880

1995

The light areas on these maps show the enormous amount of land that was transferred from Native American to white control during the 1800s and 1900s. It was taken away through treaties, purchases, sealed bids, lotteries and theft. The dark areas on the maps show the decreasing areas of land held by Native Americans. Today, many Native Americans are trying to reclaim land that was taken from them in the past.

──People and Events──

Native Americans passed the stories of their people from one generation to the next by word of mouth. After European writing was introduced, a more permanent record was made of many of these traditions. Accounts were written of the chieftains who struggled with the white settlers for possession of their homelands. Other Native Americans made it into the history books as well. The following are a few of them.

Sequoyah
Sequoyah, who lived from about 1773 to 1843, had a Cherokee mother and an English father. He believed that white people had power because they could read and write, so he created a written version of the Cherokee language. There was a separate symbol for each of the 86 syllables.

Geronimo
The Apache chief Geronimo was born in 1829. He was first called "Goyanthlay," which means "One Who Yawns." When Mexican troops killed his family, he became a fierce warrior, feared by both Mexican and American soldiers. Geronimo

eventually surrendered and in 1905 took part in President Theodore Roosevelt's presidential parade in Washington. He died in 1909.

Sitting Bull
Sitting Bull was a respected shaman and a Sioux chief. He led a group at the Battle of Little Bighorn but was too old to take an active part in the fighting. He fled to Canada with his people but later surrendered to United States troops. He was killed by reservation police in 1890.

Red Cloud
Red Cloud, war leader of the Oglala Sioux, counted 80 coups in his lifetime of fighting. He bargained with the government to keep some of the Great Plains for his people. In 1870 he said to officials: "Our nation is melting away like the snow on the sides of the hills where the sun is warm, while your people are like blades of grass in spring when the summer is coming."

"Custer's Last Stand"
In 1868, General George Armstrong Custer and his United States cavalry forces brutally attacked unarmed women and children in a Sioux village. On June 25, 1876, bands of Sioux and Cheyenne warriors trapped and killed Custer and his men at the Battle of Little Bighorn. It was the last major victory of Plains Indians against white soldiers.

Battle at Wounded Knee
In 1890, the United States army attacked unarmed Sioux at Wounded Knee. About 200 Sioux were shot and another 100 who escaped, froze to death in the hills. This was a terrible day in Native American history.

Navajo Code Talkers
The languages of several Native American tribes were used for coding

messages in the Second World War. The Navajo code talkers developed terms to describe military movements and equipment: a submarine was an "iron fish;" a machine gun was a "fast shooter."

N. Scott Momaday
Native American author N. Scott Momaday won the Pulitzer Prize in 1969 for *House Made of Dawn*. The story is about Native American people in modern times.

Siege at Wounded Knee
In 1973, an armed band of representatives from the American Indian Movement occupied the village of Wounded Knee. They said the government had not listened to their requests for fairer treatment. After 71 days, officials from the government negotiated a withdrawal.

Native Americans were organized in different tribal groups, spread out across what is now Greenland, Canada and North America. The Sioux, who pitched their tepees on the Great Plains, were as different from the Tlingits on the Northwest Coast as people from Sweden are from those who live in Greece today.

Tribes spoke their own languages, had particular religious customs and laws, and were different in many other ways. The roles of men and women varied greatly from group to group. There has not been space to put every Native American tribe in this book. This map shows the location of the tribes that are featured in this book and gives an example of one type of dwelling for each region.

Inuit (Eskimo),
Netsilik,
Nunivagmiut,
St Lawrence
Island

Cree, Ingalik,
Ojibwa

ARCTIC

SUBARCTIC

Kalispel,
Wishram,
Nez Perce

Bella Coola,
Haida, Kwakiutl,
Nootka, Tlingit

NORTHWEST
COAST

PLATEAU

GREAT
BASIN

Chumash,
Hupa, Patwin

CALIFORNIA

GREAT
PLAINS

NORTHEAST

Iroquois,
Menominee,
Sauk

SOUTHWEST

SOUTHEAST

Ute,
Paiute

Cherokee,
Chickasaw,
Seminole,
Creek

Apache,
Hopi, Navajo,
Papago, Zuni

Arapaho, Blackfoot,
Cheyenne, Comanche,
Crow, Gros Ventre,
Hidatsa, Mandan, Sioux

Explorers
and Traders

How did the Polynesians navigate the vast Pacific Ocean?

Who was the first to sail around the Cape of Good Hope?

What did traders carry along the Silk Road?

Contents

Early Trade and Exploration

T he world looked very different 20,000 years ago. Great glaciers, caused by an ice age, covered much of the land. The people who lived at this time were hunters and gatherers, and they were always on the move. They followed herds of animals; gathered wild nuts, berries, plants and shellfish; and fished the rivers. They traveled long distances for things they valued, such as flint for making tools and weapons. Gradually, they drifted across much of Europe and Asia, and crossed into North America. At the end of the last ice age, around 10,000 BC, the glaciers thawed and lush forests grew. As the climate changed, so did the way humans lived. Many continued to hunt and gather food, but people in the Middle East planted crops and bred animals. They made pots, wove cloth, and used metals such as gold. Soon they started to trade with other villages for goods they could not produce themselves.

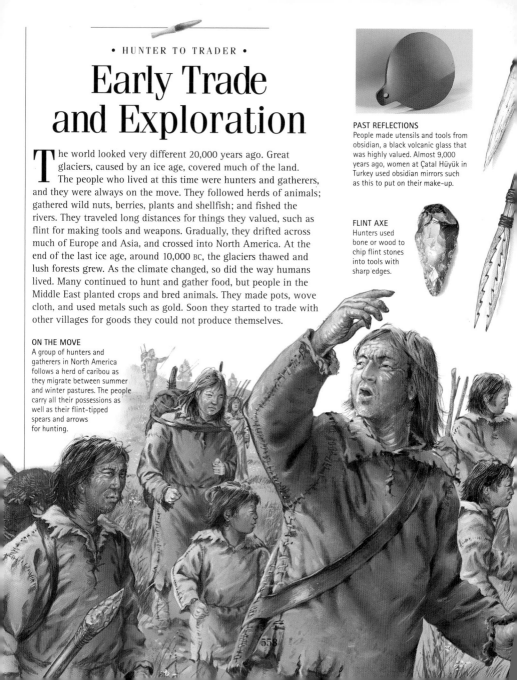

PAST REFLECTIONS
People made utensils and tools from obsidian, a black volcanic glass that was highly valued. Almost 9,000 years ago, women at Çatal Hüyük in Turkey used obsidian mirrors such as this to put on their make-up.

FLINT AXE
Hunters used bone or wood to chip flint stones into tools with sharp edges.

ON THE MOVE
A group of hunters and gatherers in North America follows a herd of caribou as they migrate between summer and winter pastures. The people carry all their possessions as well as their flint-tipped spears and arrows for hunting.

SURVIVAL TOOLS
Hunters and gatherers caught their prey with weapons such as wooden daggers with deer-horn points (far left), harpoons made from wood (middle) and spearheads made from deer bone with flint set in carved grooves (left).

GOLDEN BULL
This bull made of gold came from Bulgaria. People often traded for precious metals, such as gold.

WORKING THE LAND
Early farmers in the Middle East made the first plows and harnessed oxen to them. Thousands of years later, this farmer in central India uses similar tools to plow his land.

VILLAGE LIFE

The town of Çatal Hüyük, in southern Turkey, is one of the oldest towns in the world. People built these mud-brick houses, which were joined together and entered through the roofs, in 7000 BC. Some of the houses were special shrines, decorated with wall paintings, for worshipping the gods. The people herded cattle; grew wheat, barley and peas; and were skilled clothmakers. They had plenty of obsidian and exchanged it for goods from other areas. Çatal Hüyük soon became a busy trading center.

Discover more in Sailing West

559

JEWELS FROM AFAR
Trade in precious stones gave
Sumerian jewelers new materials
to use. This necklace is made from
lapis lazuli from Afghanistan and
carnelian from the Indus Valley.

The First Civilizations

ROYAL SKIN
Egyptian royalty and
priests sometimes wore
the skins of exotic animals.
Princess Nefertiabt,
shown here, wears
the skin of a leopard.

T he first civilizations, Sumer and Egypt, flourished beside great
rivers between 5000 and 500 BC. Sumerian villages prospered on
the flat plains watered by the Tigris and Euphrates rivers, while
the flooding waters of the River Nile left fertile silt for Egyptian farmers
to sow their crops. These large societies organized their resources. They
developed irrigation systems to direct and control the floodwaters,
and store them for later use. They invented the plow and the wheel,
which they used for chariots and to make pottery. People made laws
to govern society and developed their knowledge of subjects such as
mathematics. New groups in society, such as priests and skilled
craftspeople, began to emerge. The Egyptians and Sumerians
exchanged local produce at regional centers, but they also traded
outside their own countries for goods they needed, such as timber.
They began to keep records of their trade, and early systems of writing
developed. The first civilizations were large and successful. They were
the basis for the way society is organized today.

TRADE LINKS
Sumerian and Egyptian
traders often made long
and dangerous voyages.
They obtained goods from
India, Afghanistan, Crete
and Greece. The Egyptians
also traded with Punt.

FOR THE QUEEN'S COURT
The Egyptian Queen Hatshepsut sent a trade
expedition down the Red Sea to the ancient land
of Punt. The Egyptians here are loading their
ships with frankincense trees, elephants'
tusks, ebony, gold, spices and exotic
animals such as panthers.

WIND POWER
By about 3200 BC, the Egyptians had invented sails to power their boats,
rather than relying on oars. This enabled them to explore farther for trade.

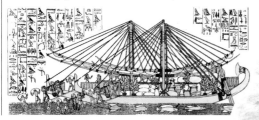

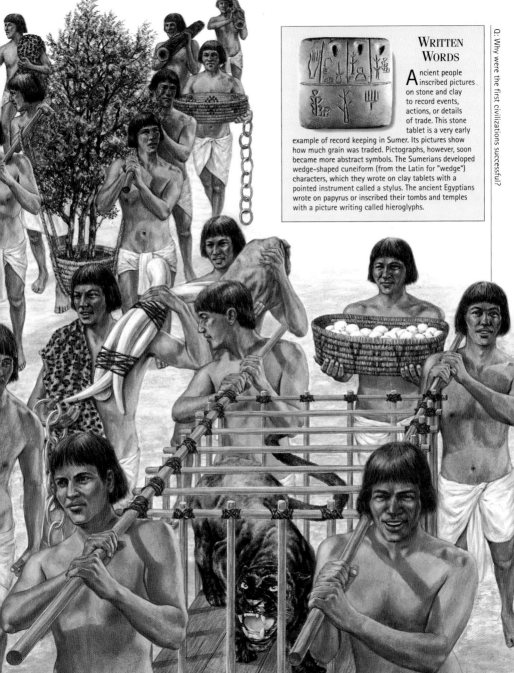

WRITTEN WORDS

Ancient people inscribed pictures on stone and clay to record events, actions, or details of trade. This stone tablet is a very early example of record keeping in Sumer. Its pictures show how much grain was traded. Pictographs, however, soon became more abstract symbols. The Sumerians developed wedge-shaped cuneiform (from the Latin for "wedge") characters, which they wrote on clay tablets with a pointed instrument called a stylus. The ancient Egyptians wrote on papyrus or inscribed their tombs and temples with a picture writing called hieroglyphs.

Mediterranean Trade

STORMY WATER
A Phoenician merchant ship struggle
through wild seas. The crew checks that th
cargo of timber and pottery jars of wine, c
and grain are firmly secured. Such tradir
ships were often lost at sea. Shipwrecks giv
archaeologists valuable clues abou
people and customs of the pas

T he Phoenicians were great seafarers and traders. Around 1000 BC, some set out from their cities on the Lebanon coast to find new lands to trade with and farm. They filled their ships with goods such as carved ivory, glass and cloth that they dyed with purple ink from the murex shellfish (left). Sailing west, they established trading bases in areas that were rich in metals, such as copper and tin. They supplied King Solomon with cedars from Lebanon for his temple in Jerusalem, and the Egyptians with timber to build their ships. These excellent navigators also made the first known voyage around the coast of Africa. The Greeks began to trade in the Mediterranean around 800 BC. With olive oil, pottery and wine, they traded for grain, timber and metal along the northern shores of the Mediterranean and the Black Sea. They also ventured down the Red Sea and sailed on the monsoon winds to India. By AD 110, the Romans had an enormous and wealthy empire that extended into North Africa and the Middle East. They traded only for exotic goods, such as silk from China and wild animals for circuses.

SILVER EXCHANGE
This silver Roman cup was found in Denmark. It may have been traded, given as a gift, or taken as a prize of war.

TWO-FACED VASE
Trade with other countries gave artists ideas for their work. This Greek vase, from 540 BC, shows two of the different races of people in the Mediterranean.

562

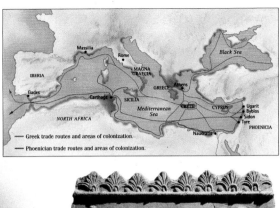

— Greek trade routes and areas of colonization.

— Phoenician trade routes and areas of colonization.

ROMAN SPORTS
The Romans obtained most of the goods they wanted from their own empire. But they did trade for wild animals such as lions, which their gladiators fought in the great amphitheaters.

SMALL CHANGE

Small silver coins appeared in some parts of Turkey as early as the seventh century BC. King Croesus, the ruler of the ancient kingdom of Lydia, introduced both silver and gold coins, such as these, into his realm. A standard system of money made it easier for different countries to trade with each other. When Greece began to use coins, each city-state made its own and stamped them with special symbols. Coins from Athens were stamped with an owl, the sacred bird of Athena, the Greek goddess of wisdom.

Discover more in The Silk Road

Arab Traders

By the ninth century, Arabs were trading extensively by land and sea. They sailed down the east coast of Africa, and caught the monsoon winds to India. Arab traders even reached China. Merchants roped together camels, donkeys, mules or horses and traveled over land in Asia in groups called caravans. They also crossed the Sahara to the gold-mining areas of West Africa. The Prophet Mohammed, the founder of Islam and once a merchant himself, approved of honest trade. Merchants were very respected, and tales of their adventures were the talk of the markets. Some of the stories in the famous *Arabian Nights* tell of merchants and magic carpets, awakened genies and mysterious places. The busy market of Baghdad in Persia was an important meeting place for traders from Europe and Asia. European traders brought furs, cloth and manufactured goods; Indian traders brought gold, jewels and spices; and there were slaves from many nations. After much bartering, traders returned to their own lands laden with goods.

LOCAL TRADE
The Arabs made pottery plates with beautiful glazes and colors, which they traded in the local markets.

AN EYE FOR A BARGAIN
Merchants, such as these from Turkey, checked the merchandise carefully. They bartered and haggled for hours to get the best price.

HEAT AND DUST
The market of Baghdad was a maze of narrow, dusty streets with small, dimly lit shops on either side. Huge canopies deflected the fierce heat of the sun. Different streets were set aside for different goods. Here, the carpet sellers' street meets the spice traders' street.

THE POWER OF ISLAM

In the seventh century, the Prophet Mohammed founded the religion of Islam, which means "submit to the will of Allah (God)." Arab traders carried Islam with them into Africa and across Asia to China. Followers of Islam are called Muslims, and they read the *Koran*, the sacred book of Islam, for advice on how to live their lives. Muslims pray five times a day, facing the Holy City of Mecca where the Prophet Mohammed was born. Every Muslim tries to visit Mecca once in his or her life.

SPREADING ACROSS THE SEAS

The Arabs conquered a vast empire, which stretched from Spain to Persia. They set up trading networks along the eastern coast of Africa and across the seas to China.

SAILING THE SEAS

Arab merchants sailed in vessels called dhows. With lateen (triangular) sails to catch the wind, these ships were light and fast. Many dhows are still built as they were hundreds of years ago.

• KINGS OF THE SEA •

The Vikings

A BOOTY OF SILVER
Silver buckles, coins, necklaces and bracelets were part of a hoard of tenth-century Viking treasure discovered in England.

T he Vikings were proud, skillful sailors. They fished and traded in Scandinavian waters and farmed the land around them. But as the population grew, farm land became scarce. Toward the end of the eighth century, Viking explorers left Scandinavia in strong, swift ships to see the world beyond their jagged coasts. The Vikings from Sweden sailed to eastern Europe laden with walrus tusks, ivory and furs. They set up trade routes in the Baltic and bartered for silver, pearls and Chinese silk. Early in the tenth century, they ventured along the great rivers of Russia to the rich markets of the Near East, such as Baghdad. The Vikings from Denmark and Norway sailed to western Europe. They stormed towns and looted monasteries, and Europeans feared the sight of the fierce Viking warships. The Vikings conquered some of the lands they raided and many settled down as farmers and traders. Others crossed the Atlantic, where they discovered Iceland, Greenland and the New World.

FOR THE AFTERLIFE
This intricately carved wooden post was part of a ship that was buried with a Viking chief in Oseberg, Norway.

GOODS FROM RUSSIA
The Vikings traded with Russia for slaves and luxury goods such as delicate silver arm rings.

INTO THE EAST
Viking traders discovered the wealth of the Middle East in the ninth century. They traded for goods such as this gold statue of Buddha.

EASTWARD TO TRADE
Viking traders row their knorr to the shore of the Baltic coast to find a camp for the night. During the tenth century, Swedish Vikings traveled along the Dnieper and Volga rivers and traded for silver, bronze vessels, pearls and Chinese silk.

VIKINGS ABROAD
Vikings from Sweden traveled east to trade. Vikings from Denmark and Norway went west to invade and settle new lands.

566

A VIKING EXPLORER
Erik the Red, the famous Viking explorer, discovered Greenland. His son Leif Ericson, shown here pointing at land, explored the North American coast in AD 1000. He and his crew spent the winter in what he called Vinland, which may have been Newfoundland.

ALL AT SEA

The Vikings built superb wooden boats that were strong but light. They used small fishing boats for local trade. The wider and slower cargo boats (knorrs) had shallow bottoms so they could come in close to the shore to be loaded with goods and livestock. Raiding parties traveled great distances in sleek, fast longships with shallow hulls. The Vikings named them after spirits of the wind and the sea and carved their prows with fierce dragon heads.

Fishing boat

Knorr

Longship

Across the Pacific

The Pacific Ocean covers one-third of the Earth's surface. Small islands are sprinkled across the enormous blue expanse and often isolated from each other by great areas of sea. The islands that make up Polynesia (meaning "many islands") were settled by seafarers from Indonesia and Malaysia, who spread gradually across the Pacific looking for new lands to settle. Between 2000 BC and AD 1000, they navigated incredible distances in sturdy dugout canoes, reading changes in the swell of the sea, the patterns of the stars and the easterly winds. They brought with them a patterned pottery called Lapita (above left), which has become an archaeological clue to their movements. The settlers adapted to the different environments they found, from the dry atolls to the lush and fertile volcanic islands. They reached Tonga and Samoa by at least 1000 BC and developed their own customs and a society that was ruled by chiefs. By about AD 1000, Polynesians had reached the easterly islands of Hawaii, Easter Island and New Zealand.

DID YOU KNOW?

Gigantic stone statues line the coast of Easter Island. The people carved these guardians of the island from soft volcanic stone, then dragged them to platforms on the cliff edges.

SEEKING NEW LANDS

Polynesian settlers sailed across the Pacific to New Zealand, which the Maoris call "the land of the long white cloud." Their double-hulled canoes were loaded with fruit and vegetables, such as breadfruit and sweet potato; pigs, chickens and dogs; and crops such as taro, yams, bananas and coconuts.

Sails
These were made of matting, woven from a palmlike tree called pandanus.

READING THE SEA AND THE SKY

The Polynesians were expert navigators. They found islands to settle in the vast Pacific Ocean by "reading" the sea and watching for land-based birds such as frigates (above). They traveled to and from these islands using maps they made from palm sticks and cowrie shells (left). The sticks represented the swells and currents of the sea, while cowrie shells marked islands.

Paddles
These added to the power of the sails. The Polynesians steered the canoe with the back paddles.

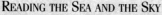

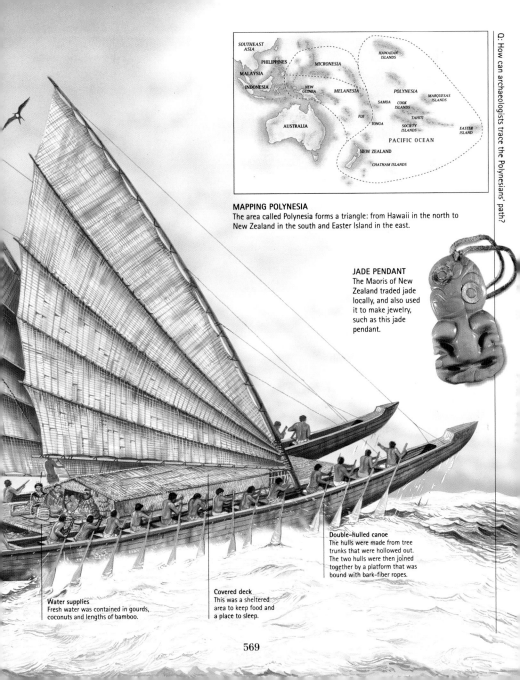

MAPPING POLYNESIA
The area called Polynesia forms a triangle: from Hawaii in the north to New Zealand in the south and Easter Island in the east.

JADE PENDANT
The Maoris of New Zealand traded jade locally, and also used it to make jewelry, such as this jade pendant.

Double-hulled canoe
The hulls were made from tree trunks that were hollowed out. The two hulls were then joined together by a platform that was bound with bark-fiber ropes.

Covered deck
This was a sheltered area to keep food and a place to sleep.

Water supplies
Fresh water was contained in gourds, coconuts and lengths of bamboo.

The Silk Road

VALUABLE HORSES
The Chinese needed strong horses to help them defend their country. Pottery horses were often buried with royalty.

A WINDING ROAD
Chinese traders make camp on their difficult and dangerous journey along the Silk Road. Often, they sold their goods to other merchants on the Silk Road, who then traveled to markets in the Middle East.

T rade routes began to link the civilizations of the ancient world. China had been isolated for centuries by mountains, forests and harsh deserts. The world beyond seemed hostile. Around 138 BC, the Chinese went outside these boundaries to look for allies against invading tribes from the north, and to find strong horses for their heavily armed warriors. They brought back news of other countries and exotic goods. Traders and travelers began to journey to and from China along a network of trade routes called the Silk Road, which crossed the whole continent of Asia. It became a highway for goods, ideas, knowledge, skills and religions. Camel caravans from China carrying silk and porcelain struggled through sandstorms in the Gobi Desert and across high mountains on the way through India and central Asia to the Middle East and Europe. The Silk Road also joined up with sea routes, and local traders filtered goods through to the Indian Ocean and the Mediterranean.

MONEY, MONEY
Ancient Chinese coins had holes in the middle so they could be carried on strings. This made it easy for Chinese traders to keep control of large sums of money.

WORSHIPPING BUDDHA
Chinese travelers brought Buddhism back from India. It is still an important religion in China.

THE SECRET OF SILK

The Chinese were famous for their silk and they guarded carefully the secret of silk making. This delicate and precious fabric was woven from the fibers of the silkworm's cocoon (left). More than 2,500 cocoons must be unwound by hand to make just 7 oz (500 g) of yarn. The Chinese embroidered the silk cloth, painted it or made it into clothes. Many Chinese families bred silkworms so they could spin and weave their own silk. In this tenth-century Tang painting, the women of the royal court are pounding newly woven silk with wooden poles to soften it.

ACROSS A CONTINENT
The Silk Road connected traders and travelers from different countries.

THE HOLY CITY
Christians, Muslims and Jews all believe that Jerusalem is a
holy city. In the First Crusade, which was in 1099, Christian
soldiers assembled at Constantinople (today's Istanbul) and
marched to Jerusalem. They captured and ransacked the
city, but it was eventually retaken by the Muslims.

PILGRIM'S FLASK
Pilgrims had traveled to the Holy Land for
centuries. They made the long and difficult
journey in the hope that God would forgive them
their sins. They brought back reminders of their
pilgrimages, such as holy water in flasks.

• TRAVEL AND TRADE •

The Crusades

In 1095, Pope Urban II called for a holy war, a crusade, to win the
Holy Land (Palestine) from the Muslims. He promised a place in
Heaven to all who took arms. During the next 200 years, kings and
knights from Europe led three main crusades against the Muslims, who
now blocked pilgrims from going to the land where Jesus Christ had
lived and taught. There were also unofficial crusades, such as the
People's Crusade in 1095, which was made up mainly of peasants.
In 1212, more than 30,000 children set out for Palestine. Most
never returned home. While many people fought and died in the
battles for the Holy Land, others saw the crusades as a chance
to trade and to make money. Ship owners in Venice, Pisa and
Genoa, the Italian cities bordering the Mediterranean, took
crusaders to the Holy Land. These cities became important
trading centers where European goods were exchanged for
those from the Middle East, Africa and
Asia. The West learned much from
the East during the crusades,
such as ancient philosophy,
techniques for weaving and
making silk fabrics, and
new architectural styles.

TRADING WITH THE EAST

Ship owners in Genoa, Pisa and Venice took the Christian armies across the Mediterranean. As a reward, they were given trading concessions in Muslim lands and cities, such as Constantinople, which had been conquered during the crusades. They now had direct access to the trade routes of Asia. The Italian cities grew wealthy from their trading links with the Muslims, but a bitter rivalry grew up between them and Constantinople. In 1204, the Venetians were chosen to lead the Fourth Crusade against Alexandria, but they attacked Constantinople instead.

LEAVING HOME
Men left their wives and children to join the crusades. They traveled great distances and fought in terrible battles. Many also died from diseases and lack of food.

LIVING IN THE EAST
The crusaders were impressed with the luxury, lifestyle and learning they found in the East. They enjoyed delicacies such as raisins, sugar and figs. In times of truce, some played games such as chess with the Saracens (Muslim fighters). They returned to Europe with a great desire for the riches of the East.

573

Genghis Khan united the
nomadic Mongolian tribes
and defeated the Turkish
Muslims. He allowed
European Christians to
travel overland to the East.

Venice

Departure
In 1271, 17-year-old Marco
Polo sailed from Venice with
his father and uncle to the
port of Acre in Palestine.

Constantinople
(Istanbul)

Acre

Samarkand

The journey out
From Acre, they rode
camels to the Persian port
of Hormuz, where they saw
beautiful patterned silk.

Baghdad

Kerman

Hormuz

• JOURNEYS OF DISCOVERY •

Traveling East

Close to home
The Polos left the princess
here and traveled overland
and then by sea to Venice.

M arco Polo (left) is one of the most famous travelers in
history. In 1271, he left Venice for the mysterious East
with his father and uncle, who had made the journey
once already. They sailed down the Persian Gulf, then traveled
inland over the high mountains of Afghanistan. They crossed the
perilous Gobi Desert, where night winds wailed like eerie voices.
After three years, they reached the court of Kublai Khan, the
Mongolian emperor of China and grandson of the great Genghis
Khan. Kublai Khan ruled an enormous empire. He welcomed the
strangers and asked Marco Polo to travel around his vast empire. Marco
Polo saw magnificent cities with canal systems, precious jewels, skilled
craftspeople and palaces decorated with gold and silver. When the Polos
returned to Venice, Marco was captured by the Genoese who were at war
with the Venetians. Marco told another prisoner, Rustichello, about his
amazing adventures, and Rustichello published them in a book called
The Description of the World. Many did not believe Marco Polo's
tales, but others dreamed of seeing the lands he described.

——— Traveling to the East
——— Returning to Venice
——— Travels in China
——— Mongolian Empire

BY LAND AND BY SEA
The Polos traveled overland to the East, but
returned by sea 24 years later. They saw lands
and riches unknown to Europeans. Marco Polo
recorded his observations of the people and
the places. Tales of China and India inspired
poets and explorers for years.

574

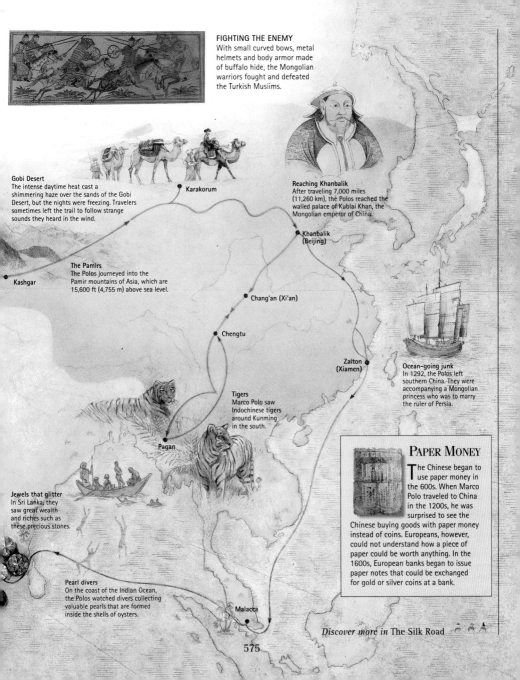

FIGHTING THE ENEMY

With small curved bows, metal helmets and body armor made of buffalo hide, the Mongolian warriors fought and defeated the Turkish Muslims.

Gobi Desert
The intense daytime heat cast a shimmering haze over the sands of the Gobi Desert, but the nights were freezing. Travelers sometimes left the trail to follow strange sounds they heard in the wind.

Karakorum

Reaching Khanbalik
After traveling 7,000 miles (11,260 km), the Polos reached the walled palace of Kublai Khan, the Mongolian emperor of China.

Khanbalik (Beijing)

The Pamirs
The Polos journeyed into the Pamir mountains of Asia, which are 15,600 ft (4,755 m) above sea level.

Kashgar

Chang'an (Xi'an)

Chengtu

Zaiton (Xiamen)

Ocean-going junk
In 1292, the Polos left southern China. They were accompanying a Mongolian princess who was to marry the ruler of Persia.

Tigers
Marco Polo saw Indochinese tigers around Kunming in the south.

Pagan

Jewels that glitter
In Sri Lanka, they saw great wealth and riches such as these precious stones.

Pearl divers
On the coast of the Indian Ocean, the Polos watched divers collecting valuable pearls that are formed inside the shells of oysters.

Malacca

PAPER MONEY

The Chinese began to use paper money in the 600s. When Marco Polo traveled to China in the 1200s, he was surprised to see the Chinese buying goods with paper money instead of coins. Europeans, however, could not understand how a piece of paper could be worth anything. In the 1600s, European banks began to issue paper notes that could be exchanged for gold or silver coins at a bank.

Discover more in The Silk Road

China Ventures Out

T he Chinese rarely explored beyond their sea borders. While other nations set out on voyages to trade, to spread their religion, or to find new lands to settle, the Chinese felt they had all they needed in their own country. They did, however, want to show the rest of the world the power and strength of their empire. In the early 1400s, Ch'eng Tsu, the Emperor of the Ming dynasty, placed Admiral Zheng He in charge of a massive fleet of ocean-going junks. With 255 vessels, 62 enormous ships laden with treasure, and 28,000 men, Zheng He sailed grandly out of Nanking harbor, bound for the southern seas. Between 1405 and 1433, he led seven voyages, and his ships sailed to Southeast Asia, India, the Persian Gulf and Africa. Zheng He gave gifts of tea, silk and porcelain to the people he encountered on his travels. He returned to the Chinese court with foreign diplomats and exotic animals, such as giraffes.

FINDING THEIR WAY
Some of the greatest inventions come from China. Early sea compasses, such as this magnetized needle floating in a bowl of water, allowed sailors to navigate accurately across wide expanses of ocean.

SHIPS OF THE SEAS
Zheng He's junks were five times bigger than Portuguese caravels. Prince Henry of Portugal used some of the design features of the Chinese junk in his ships.

MAPPING ZHENG HE'S VOYAGES
Zheng He's seven voyages introduced many countries to the wealth and splendor of China.

DID YOU KNOW?

Zheng He's ships reached Africa in 1415 and rounded the Cape of Good Hope 70 years before the first Europeans. When Zheng He's voyages were over, the Chinese isolated themselves once more from the rest of the world.

GIFTS FOR THE EMPEROR
In 1414, Zheng He brought back ambassadors from east Africa and unusual presents. Here an attendant shows Ch'eng Tsu one of the gifts—a giraffe.

CHINESE PORCELAIN

Delicate Chinese porcelain was finer than any other pottery in the world. Made from a mixture of clays found only in China, it was highly prized and admired throughout the world. Traders brought Chinese porcelain to Europe in the twelfth century. It was so rare and expensive that only wealthy people could afford to buy it. European manufacturers tried to make porcelain themselves, but for a long time their porcelain could not compete with the high quality of Chinese porcelain.

Discover more in China and Japan Open Up

577

Passage to India

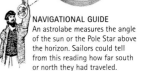

In the 1400s, the Portuguese led the race to find a sea route to the East. Prince Henry of Portugal ("the Navigator") set up a school of navigation, and explorers sailed along the west coast of Africa in small ships called caravels. At first, they did not sail far. They imagined that monsters and boiling seas awaited them in the hot lands to the south. In 1485, Diogo Cão reached Cape Cross and set up a stone marker called a *padrão*. Two years later, Bartolomeu Dias passed Cão's marker. He rounded the Cape of Good Hope and sailed into the Indian Ocean, but his fearful crew begged him to turn back. The explorer Vasco da Gama traded for gold, ivory and slaves in Africa and reached the Indian port of Calicut in 1498. He wanted to trade with the Indian princes, but they were unimpressed with the goods he offered them.

So the Portuguese came with men and guns and used force to set up trading bases in India and Africa. Portugal soon became a powerful empire.

NAVIGATIONAL GUIDE
An astrolabe measures the angle of the sun or the Pole Star above the horizon. Sailors could tell from this reading how far south or north they had traveled.

VALUABLE IVORY
This African salt cellar, showing armed Portuguese on horses, was carved from ivory. The Africans traded ivory with the Portuguese.

LOOKING EAST
Prince Henry never went on a sea voyage himself, but he still inspired many great Portuguese explorers.

SAILING TO THE EAST
The Portuguese ships stayed close to the coast and followed familiar landmarks as they sailed to Africa. The Portuguese were welcomed in some lands, but mistrusted in others. They often overpowered the local people with guns.

PTOLEMY'S WORLD
The Greek astronomer and mathematician Ptolemy drew this map in AD 150. It shows only the top half of Africa because no-one knew how far south it spread. The first Portuguese explorers also thought the world looked like this.

SPICES FOR SALE

Spices were precious in the 1400s. People did not have refrigerators, so spices disguised the taste of old meat. Pepper was so rare that it was sometimes used instead of money. The Portuguese discovered that most of the spices sold in India did not come from there. Cloves, mace and nutmeg, the most valuable spices, came from the Molucca Islands further to the east. They became known as the Spice Islands.

Q: Who was the first European to sail to India?

SAILING TO GLORY
Blessed by King Manuel I and officials from the church, Vasco da Gama left Lisbon in 1498, bound for India. His four ships were stocked with enough supplies for three years. They battled storms and unpredictable currents, and many sailors died from scurvy.

• THE NEW WORLD •

Sailing West

The Italian explorer Christopher Columbus (left) believed he would reach the East if he sailed west. Like many others, Columbus had no idea that the Americas, discovered by the Vikings many hundreds of years before, even existed. He imagined that most of the world was covered by one huge piece of land, made up of Europe, Africa and Asia. Columbus persuaded King Ferdinand and Queen Isabella of Spain to finance a sea voyage. In 1492, he sailed from Spain with his ships the *Niña*, the *Pinta* and the *Santa Maria*. When Columbus saw the Caribbean islands, he thought they were the islands near mainland Asia and he called them the West Indies. He found strange new foods and animals, which he took back to the Spanish court. Columbus made three more voyages across the Atlantic, and explored many of the Caribbean islands and parts of the Central American coast. He always believed he had discovered Asia, but others soon suspected he had found a "New World."

DID YOU KNOW?

The people of the West Indies slept in hanging beds called "harmorcas." Columbus's sailors copied this idea and made these beds above the dirty, wet decks, away from the scurrying rats. They called their beds hammocks.

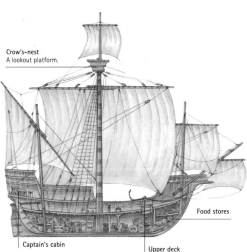

Crow's-nest
A lookout platform.

Food stores

Captain's cabin

Upper deck
The sailors slept here and prepared their meals.

A MODEL FLAGSHIP
The *Santa Maria* was Columbus's flagship. No-one knows what it really looked like, but models have been built that give us some idea. The *Santa Maria* was slower and heavier than the other two ships. It was wrecked off the island of Hispaniola (now Haiti) in 1492.

A MEETING OF WORLDS
Columbus landed on San Salvador in the Bahamas and claimed it for Spain. The people on the island were friendly toward the strangers.

LAND AHOY!
After more than a month at sea, Columbus's crew were weary and disgruntled. They wanted to turn back. But a cry from the crow's-nest soon lifted their spirits. Land was in sight!

BETWEEN TWO WORLDS

Columbus did not find Asia, but he did link the Eastern and Western hemispheres—the Old and the New worlds. Soon, food and animals began to pass between them. The Europeans brought horses, sheep, pigs, wheat, sugar cane, wheeled transport, iron and steel to the New World. They returned to Europe with potatoes, pineapples, avocados, tomatoes, corn, kidney beans, vanilla, peanuts, turkeys, cocoa, tobacco and rubber.

Discover more in The Vikings

Spain and the New World

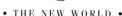

S panish conquistadors (conquerors) set sail for the New World early in the sixteenth century. Lured by thoughts of gold and power, they followed Columbus's route across the Atlantic in sturdy galleons. Catholic priests accompanied them to convert the peoples of the new lands to Christianity. The New World was all they had hoped it would be. The Aztecs of Mexico and the Incas of Peru were highly organized societies that had existed for hundreds of years. The Spanish conquistadors were dazzled by the gold and silver treasures and the magnificent stone cities, pyramids, temples and sculptures, which were built by the people without iron tools or the wheel. In 1521, Hernando Cortés captured Tenochtitlán, the Aztec capital. Francisco Pizarro reached the center of the Inca Empire in the Andes Mountains in 1532. On horseback and armed with guns, the Spaniards defeated the large numbers of Aztecs and Incas. They soon controlled this rich new world.

ISLAND CAPITAL
Tenochtitlán, the spectacular Aztec capital, was built on an island in a lake. It was destroyed by the Spanish conquistador Hernando Cortés in 1521, but was later rebuilt as Mexico City.

BEADS OF PRAYER
The Catholic priests who traveled to the New World wore rosary beads. These help people to concentrate when they are praying and to remember how many prayers they have said.

DID YOU KNOW?
The Europeans brought many infectious diseases, such as measles, smallpox and even colds, to the New World. The people there had no resistance to such diseases, and many died. Their traditions and customs died with them.

TREASURES OF GOLD
The Spanish conquistadors found great treasures in the New World. The Incas used this gold figure, made in the shape of a winged god, in religious ceremonies.

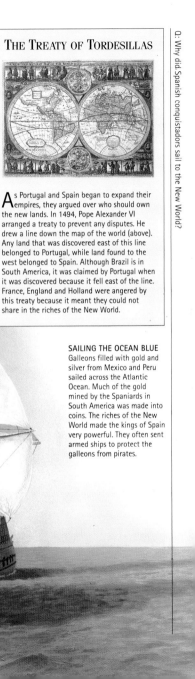

THE TREATY OF TORDESILLAS

As Portugal and Spain began to expand their empires, they argued over who should own the new lands. In 1494, Pope Alexander VI arranged a treaty to prevent any disputes. He drew a line down the map of the world (above). Any land that was discovered east of this line belonged to Portugal, while land found to the west belonged to Spain. Although Brazil is in South America, it was claimed by Portugal when it was discovered because it fell east of the line. France, England and Holland were angered by this treaty because it meant they could not share in the riches of the New World.

SAILING THE OCEAN BLUE

Galleons filled with gold and silver from Mexico and Peru sailed across the Atlantic Ocean. Much of the gold mined by the Spaniards in South America was made into coins. The riches of the New World made the kings of Spain very powerful. They often sent armed ships to protect the galleons from pirates.

STORMY STRAIT
Magellan had discovered a passage to the East, but rowboats had to guide his ships through the rough waters. When Magellan emerged into a calm ocean on the other side, he named it the Pacific—the peaceful sea.

• THE NEW WORLD •

Around the World

As ships began to sail in all directions across the oceans, knowledge of the world grew. Like Columbus, the Portuguese explorer Ferdinand Magellan (left) believed that he could find a westward route to the Spice Islands. He persuaded Charles I of Spain to sponsor his voyage. Magellan set out across the Atlantic in 1519 with five ships. He discovered the treacherous and stormy passage at the bottom of South America, now called the Strait of Magellan. In 1521, he arrived at the Philippine islands. He had reached the East by sailing west. Magellan was killed in the Philippines, but his crew on the ship *Victoria* sailed on to Spain. Fifty-six years later, the brilliant English navigator Francis Drake set out to explore the southern seas and sail around the world. He journeyed up the west coast of South America, raiding Spanish ports and seizing the cargo of Spanish ships. The Spaniards were furious! Relations between Spain and England deteriorated and in 1588 Philip II sent the Spanish Armada to invade England. His warships, however, were unsuccessful.

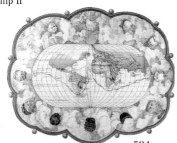

DRAKE'S TRAVELS
In 1577, Drake sailed from Plymouth, England, on his most famous voyage. Three years later he returned a hero. He had circumnavigated the world.

584

THE MANILA GALLEONS

Ferdinand Magellan claimed the Philippines for Spain in 1521. About 50 years later, Spanish galleons began to sail from Acapulco (today's Mexico) across the Pacific to the port of Manila in the Philippines. Manila became a trading center for India, Southeast Asia, Japan and China. Galleons carried European products and Mexican silver to Manila and returned to Spain loaded with riches from the East. Some galleons, such as the *Nuestra Señora de la Concepción*, were shipwrecked on the perilous route. Its cargo of jewels, silks, spices and porcelain sank with the ship to the ocean floor.

Breaking the Ice

In the 1500s, Spain and Portugal controlled the southern routes to the East. Other countries were forced to seek another way to sail from the Atlantic to the Pacific—a northwest passage. In 1497, Italian John Cabot followed the routes of fishermen who caught cod in the cold North Atlantic Ocean and discovered Newfoundland. Frenchman Jacques Cartier found the St. Lawrence River, which became the main route into northern North America, and fur traders soon moved into the area. In 1576, Englishman Martin Frobisher explored Canada's northeast coast and reached Baffin Island. Another Englishman, Henry Hudson, named the Hudson River, the Hudson Strait and Hudson Bay. But these explorers did not have suitable clothes or ships to survive the winter in this vast area of islands and rivers, and they did not discover a northwest passage. Sir John Franklin found this ice-bound route when he explored the Arctic between 1845 and 1848, but no-one sailed through it until 1906.

DANGEROUS JOURNEYS
The cold northern seas challenged explorers. Even when the water seemed calm, long winters and a landscape of ice made their journeys difficult and slow. Ships often became trapped in ice for months. Many men suffered from scurvy and died in the terrible conditions.

IN THE NAME OF THE KING
In the service of Henry VII of England, John Cabot sailed west from Bristol to find a route to the Spice Islands. He opened up Canada to the world.

WALRUS IVORY
The Inuit (Eskimos) lived on the northern coast of Canada as well as in Alaska and Siberia. They traded in furs and ivory.

CAST ADRIFT

Henry Hudson made three journeys to the northern seas. During the bitter winter of 1610, Hudson's crew mutinied when he wanted to continue the voyage north. The mutineers put Hudson, his son and loyal crew members in a boat without oars, water or fresh food. They were never seen again.

EXPLORING NORTH AMERICA

Many places in the north of Canada are named after the explorers who died trying to find a northwest passage.

HUDSON BAY TRADING POST

In 1670, a group of wealthy English merchants and noblemen set up the Hudson's Bay Company. King Charles II of England said the company could trade in all the lands drained by the streams that flowed into Hudson Bay. Trading posts and forts were built all along the bay. The Inuit (Eskimos) traded the skins and furs of wild animals with the Europeans for arms and other goods.

Discover more in Russian Trade

A PIRATE'S PISTOL
A pirate carried several weapons, such as a dagger, an axe, a curved sword called a cutlass and a pistol.

• THE BUSINESS OF TRADE •

Pirates

Pirates had roamed the seas for centuries, attacking ships and stealing their cargoes. In the 1500s, however, the pickings were especially rich. Spanish galleons filled with gold, silver and emeralds from the New World crossed the Atlantic. Ships laden with spices and silks sailed along the trade routes in the Mediterranean and the Indian Ocean. Because England, France and Holland were often at war with Spain, their rulers gave sailors called "privateers" permission to attack Spanish ships and steal their treasures for the government. When peace was made with Spain in the 1600s, many privateers became pirates. They set up bases in the West Indies and Madagascar and attacked ships from every country. The life of a pirate appealed to poorly paid, harshly treated sailors. Henry Morgan was the most successful pirate in the 1650s, while "Blackbeard" Edward Teach terrorized the American East Coast in the 1700s.

DID YOU KNOW?
Pirate flags fluttered menacingly in the wind as pirate ships sailed the seas. At first, pirate captains had their own flags, such as this one. In the 1700s, the Jolly Roger became the common pirate flag.

SHIPMATES, TO ARMS!
Using hooks and ropes to keep the ships together, pirates scramble on board a Spanish trading ship. The crew fight fiercely on deck, but the pirates outnumber them. Pirates often gave defeated crews a choice: join them or die!

SMUGGLERS' COVE

As trading ships began to sail in all directions around the world, trade became more organized. Many governments demanded that traders pay special taxes on goods they brought into the country. They placed agents at sea ports to make sure this money was paid. Smugglers avoided paying the tax by bringing goods into the country secretly. They sailed their ships into deserted coves away from the main ports, and rowed ashore in the black of night to unload their cargoes.

PORTRAIT OF A PIRATE
A pirate robbed ships for himself and was hanged if he was caught. A privateer, however, had a special document from his government that made it legal for him to capture the ships of an enemy.

589

The Dutch Influence

Every country in Europe wanted a share of the East and its riches. The Spaniards and Portuguese dominated trade with the East in the 1500s, but the Dutch, English and French soon began to challenge them. In 1602, the Dutch government formed the Dutch East India Company, which based itself in Batavia, on the island of Java in the East Indies (now Indonesia). This company bypassed the main spice-trade ports in the Indian Ocean and China, which were controlled by the Portuguese. The English also formed their own East India Company and became a great rival of the Dutch. The Dutch, however, were the strongest in the seventeenth century. Between 1618 and 1629, they even drove the Portuguese out of the prized Spice Islands. The Dutch created a vast trading network in the East Indies. They encouraged the islanders to grow new crops such as tea, coffee, sugar and tobacco. Eager to develop other trade links, the Governor of the Dutch East Indies sent Abel Tasman to the South Pacific in 1642. He visited New Zealand, and in 1644 he explored the northern and western coasts of Australia. But the Dutch were unimpressed with the trade opportunities in these distant lands and did not explore them further.

EAST INDIAMEN
The heavily armed Dutch merchant ships were called East Indiamen. They carried cargoes of gold from the Netherlands to India and the Far East, where they loaded up with spices, tea, jade, porcelain and jewelry for the markets in Europe.

NORTH AMERICA
The Dutch also looked west for trading opportunities. In 1612, Dutch explorers settled on an island in the Hudson River, which they called New Amsterdam. The English captured it in 1664 and renamed it New York. This map shows the English fleet in the harbor.

DELFTWARE
Dutch art and crafts flourished in the 1600s. Craftspeople in the town of Delft made a distinctive blue and white earthenware of the same name. It was shipped abroad· and bought by the wealthy.

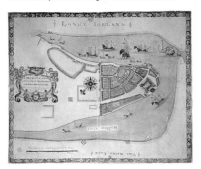

PICKING THE COFFEE BERRIES
Javanese women carry baskets of coffee berries on their heads. Coffee comes from red berries that grow on shrubs. Each berry contains two beans, which are roasted and processed. By 1720, the Dutch were the largest suppliers of coffee to Europe, where coffee houses were very popular.

DUTCH EAST INDIA COMPANY

The Dutch East India Company was very powerful. For nearly 200 years, the Dutch traded between Asia and the Netherlands and ruled the islands of present-day Indonesia. They had well-positioned bases along their trade routes, and offices abroad, such as this one in London. By 1700, the company controlled the cinnamon, clove and nutmeg trade in the East Indies. They also established a trading base and settlement in South Africa.

Slave Trade

A TRIANGLE OF TRADE
Traders loaded their ships in European ports, then sailed to Africa, where they bought slaves. They sold the slaves in the New World. The goods they bought there became the return cargo to Europe.

DID YOU KNOW?

Slaves were crammed side by side on shelves or platforms inside the holds of the ships for up to two months. Amid the filthy conditions on board, diseases spread rapidly. Many slaves died during the voyage to the New World.

In the 1500s, traders from many countries began to buy and sell Africans to work in the New World. The Spaniards there needed workers for their gold and silver mines, because many Native Americans had been killed during the Spanish conquests or had died from European diseases. So the Spaniards looked to Africa. Slave hunters raided inland villages and captured men, women and children. They were taken to the coast where the Europeans bought them from local African chiefs for goods such as guns and copper. The slaves were bound in chains, shipped across the Atlantic and sold at the New World ports. Later, slaves were also sold to European settlers in the West Indies and North America to work on the sugar, tobacco and cotton plantations. The slaves were branded by their new owners and taken to the fields or mines, where they worked from sunrise to sunset. In the 1700s, several million Africans were sent to the New World. Europe and the New World became rich from the slave trade, but Africa lost many of its people and traditions.

SHACKLES
The slaves were shackled in chains such as these during the long voyage across the Atlantic to stop them from jumping overboard.

WORKING IN THE FIELD
Slaves on a Caribbean sugar plantation harvest the crop, pick up the trimmed sugar cane and tie it in bundles.

MAKING SUGAR
The slaves carried the bundles to the crushing machine, which forced a sweet juice from the stalks. This cane juice was used to make sugar, rum and molasses. The owners of the plantations lived in great style and bought more slaves with their profits.

AGAINST SLAVERY

THE LIBERATOR.

The slave trade thrived in the 1700s, but many people were against it. They said that slavery took away the basic right of all humans to be free. William Wilberforce led a movement to stop slavery in the British Empire. When slavery was outlawed there in 1807, he began to campaign against the foreign slave trade. *The Liberator* was an anti-slavery journal published in Boston by William Lloyd Garrison. By the end of the 1800s, most nations had abolished slavery.

Russian Trade

In the 1500s, Russia began to extend its borders. Russian tsars conquered neighboring lands and sent peasant soldiers, called Cossacks, to explore parts of Siberia—a cold, harsh land to the east with many raw materials. The English and Dutch, searching for a northeast passage to the Spice Islands, began to trade with Russia for furs. In 1672, Peter the Great became Tsar of Russia. He was determined to make his country a world power. He traveled to western European countries to study their societies and economies, and their shipbuilding techniques. He expanded Russia's territory to the Baltic Sea and in 1703 founded St. Petersburg to give Russia a port that was close to the west. Because the Russians did not know how far to the east their vast country stretched, Peter the Great hired Danish explorer Vitus Bering. In 1728, Bering sailed between Russia and Alaska and discovered the Northeast Passage. Russian hunters trapped fur seals, sea otters and walruses along the Alaskan coast, and the fur trade flourished.

LIVING ON THE EDGE
The ancestors of this 1890s Cossack were peasant soldiers who lived in the frontier areas of the Russian Empire. They fought for the tsars and were given many privileges. They were the first people to open up Siberia in the east.

FROM THE NORTH POLE
The Russians increased their trade outlets by establishing St. Petersburg and opening up the Northeast Passage from the top of Scandinavia to the Bering Sea.

594

NORTHEAST PASSAGE
Baron Nils A. E. Nordenskjöld was a Swedish polar explorer. In 1879, he became the first person to sail through the Northeast Passage that connected the Atlantic and Pacific oceans. He tells of the journey in his book, *Voyage of the Vega*.

CAPITAL BY THE SEA
St. Petersburg became the capital of Russia in 1712. This beautiful city was an important center for trade with the west.

EXCHANGE OF GOODS
Russian fur trappers from northwest Russia show their polar fox pelts to English traders from the Muscovy Company. The traders would anchor in a White Sea bay and row ashore to meet the trappers. The English exchanged woolen cloth for furs.

BREAKING THROUGH
The Northeast Passage is bitterly cold and frozen with ice. Icebreakers are specially designed, powerful ships that can forge a path through ice. They force their bows up onto the top of the ice until the weight of the ship makes the ice collapse. Since the 1930s, Russian ships have sailed regularly through the Northeast Passage to the Bering Sea, opening up new trade routes.

Discover more in Breaking the Ice

595

India's Wealth

CHURCH AT GOA
The Portuguese made Goa their capital in India. They built churches such as this to try to convert the local people to Christianity.

India seemed a place of great wealth to the Europeans. The ruling Mogul emperors had magnificent palaces and treasures, and the country was rich in gold, jewels, spices and cotton. Vasco da Gama led the Portuguese to India in 1498 and they traded for spices and luxury goods from their coastal base at Goa. London merchants formed the British East India Company in 1600 and made Bombay their headquarters, while the French arrived in India in 1668. When the rule of the Mogul emperors weakened in the middle of the 1700s, the French and British began to fight each other for control of India. The British eventually won, and by the early nineteenth century all Indian states were under British rule. Queen Victoria took the title "Empress of India" in 1877. The British now had unlimited access to the wealth of India. They sent goods and raw materials back to England, where the Industrial Revolution had created radical changes in the way goods were produced and transported. Trade was entering a new stage.

GRACEFUL CLIPPER
Clippers with billowing canvas sails carried goods from India and China on most trade routes until faster steamers took over in the 1860s.

TEA FOR TWO
An official from the British East India Company drinks tea with an Indian prince. Before the British arrived, Indian princes ruled most of India. By the nineteenth century, however, the British East India Company and the British government controlled the country. Many Englishmen made fortunes in India and they lived in great luxury.

INDIAN COTTON

People in India have grown cotton and woven its fibers by hand into cloth for thousands of years. At first, the English bought cloth from India, but then they began to import the raw cotton and weave it themselves by hand. As the demand for cloth became greater, machines were invented to weave large quantities of it. England became one of the largest producers of cloth in the world during the Industrial Revolution.

CURIOUS CREATURES
Indian elephants and rare animals were transported from India all around the world. People flocked to zoological gardens, such as this one in London, to see these unusual creatures.

The Heart of Africa

For hundreds of years, Africa seemed very mysterious to the Europeans. Sailors navigated around the continent but few explorers ventured into the interior. They knew there were jungles, swamps, deserts and vast plains with wild animals. Explorers also died from tropical diseases, such as malaria and yellow fever. Part of West Africa became known as the "White Man's Grave." Because Africa was rich in ivory, gold and slaves, many European countries set up trading ports on the African coast. In 1652, the Dutch East India Company established a supply base for its ships in Cape Colony, South Africa. The Arabs shipped slaves from the east coast of Africa to the Middle East and India, while on the west coast, Europeans took millions of slaves to the sugar and cotton plantations of the New World. Many people were appalled by the slave trade. They formed anti-slavery societies and encouraged Christian missionaries to travel to Africa to try to stop it. Dr. David Livingstone, a Scottish missionary and explorer, journeyed into the heart of Africa in 1849. The records he kept enticed other Europeans to follow his path. When diamonds and gold were discovered in South Africa in the late 1800s, immigrants poured into the country and settled large areas.

DISCOVERING DIAMONDS
Diamond fields were found near Kimberley in South Africa in 1868. As tons of rock must be mined and crushed to find one small diamond, a great deal of money and labor were needed to extract large amounts of diamonds.

SLAVES FOR THE NEW WORLD
The slave trade to the New World was at its peak in the 1700s. African chiefs would capture men and women from enemy tribes and sell them to the Europeans at trading bases on the coast.

EXPLORING AFRICA

Dr. David Livingstone was a famous missionary and explorer. He spent years in Africa, mapping the land and searching for rivers that could be navigated by missionaries and traders (above). When Livingstone disappeared for a few years, Henry Morton Stanley (left), an English reporter from the *New York Herald*, went to Africa to find him. In 1871, Stanley found Livingstone by Lake Tanganyika and greeted him with the famous words, "Dr. Livingstone, I presume?"

THE DUTCH INFLUENCE
The Dutch (Boers) settled in South Africa as farmers. When British settlers came to their colony in the 1830s, the Boers trekked into the interior. They founded the Transvaal and the Orange Free State.

STRIKING IT RICH
Two gold-miners dig painstakingly by candlelight. Miners first found gold in patchy seams in the rock, as seen here. Later, they discovered it in surface deposits and then in deep underground deposits.

• GLOBAL TRADE •

China and Japan Open Up

OPENING UP JAPAN
The Japanese were astonished at the sight of the strange foreigners who sailed into Edo Bay in their black ships. Cautiously, they approached the steamships in small craft. The British, Russians and French soon followed the Americans into Japan. By the 1860s, many foreign diplomats and traders were living in Japan.

By the early 1800s, Europeans had set up trading bases in most countries except China and Japan. The Chinese hated foreign "barbarians" and allowed only Dutch and Portuguese merchants to trade in certain areas. Europeans first ventured into Japan in the 1500s, bringing Christianity with them. But in the 1600s, the ruling Tokugawa shoguns expelled all Europeans, except the Dutch. For the next 200 years, Japan was closed to the rest of the world. In the 1800s, the western powers tried to open up China and Japan for trade. In 1839, Britain went to war with China. Three years later, the Chinese signed a treaty giving Hong Kong to the British and allowing them to trade in other ports. In 1853, four American warships, led by Commodore Perry, sailed into Edo Bay (now Tokyo Bay) in Japan. Perry carried a letter from his president to the Japanese emperor, requesting trade ports. Japan and the United States signed a treaty a year later. The Japanese began to build railways and factories and soon became a major industrial nation.

DUTCH BOY
From the 1600s to the 1850s, the Japanese allowed the Dutch to trade from an island in Nagasaki harbor. Japanese artists included Dutch figures in their art.

OPIUM WARS
The British East India Company began to bring the drug opium into China from India to trade for Chinese tea. But it was illegal to trade in opium, and wars broke out between the Chinese government and the British.

JUST LIKE THE WEST
After 1854, many Japanese, including the royal family, gave up their traditional costumes for western clothes. They wanted their people to be as modern as those in the West.

BOXER REBELLION

Some Chinese hated anything that was foreign. They formed a secret group called Yihequan (Righteous and Harmonious Fists), nicknamed the "Boxers." In 1900, they attacked foreign factories, railways, churches and schools, and besieged diplomats in Peking for 55 days. Many Chinese and foreigners were killed in this rebellion.

Canals Linking Oceans

World trade flourished in the 1800s. Goods were produced quickly, cheaply and in great quantities during the Industrial Revolution. But merchant ships still had to carry their cargoes enormous distances around the world. In 1859, Frenchman Ferdinand de Lesseps began to build the Suez Canal to link the Mediterranean and the Red seas. It made the route between the United Kingdom and India 6,000 miles (9,700 km) shorter, and the world seemed much smaller. The Panama Canal, which links the Atlantic and Pacific oceans, was opened in 1914. Ferdinand Magellan and those who followed him had to sail around South America to reach the Pacific Ocean, but now ships could save 7,900 miles (12,600 km) by cutting through Central America. The Panama Canal was a great engineering achievement. For ten years, thousands of laborers cut through jungles, hills and swamps (above left), many suffering from tropical diseases. It is now the busiest canal in the world.

OPENING THE WAY
The 118-mile (190-km) long Suez Canal was opened by Empress Eugenie of France in 1869. Spectators lined the banks and steam boats formed a grand procession through the canal that would alter travel and trade across the world forever.

BIGGER AND BETTER

The Suez Canal has been widened several times to handle bigger ships and more traffic. Today, it is 36 ft (11 m) deeper than when it was first built, 230 ft (70 m) wider at the bottom and 512 ft (156 m) wider at the surface.

1869

1939

Today

LOCKED IN

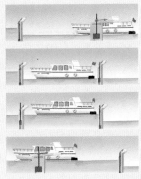

Canals can be built across land that is not level. These canals have water-filled chambers, called locks, which separate two sections that are at different heights. To pass from one level to a higher level (right), a boat enters the lock. The gates close and water from the higher level flows in to raise the boat to that level. When the gates at the higher end are opened, the boat moves forward. To lower a boat, water from the lock flows out. Vessels traveling along the Panama Canal use the Gatún Locks (above left).

Q: How did the Suez and Panama canals affect travel and trade?

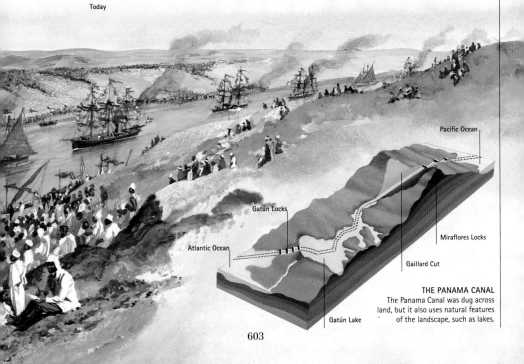

THE PANAMA CANAL

The Panama Canal was dug across land, but it also uses natural features of the landscape, such as lakes.

Pacific Ocean

Gatún Locks

Miraflores Locks

Atlantic Ocean

Gaillard Cut

Gatún Lake

Changes in Transportation

T rade is not possible without transportation. At first, our ancestors could trade only what they could carry. Then with pack animals, sleds and wheeled carts, they began to transport heavier goods. Eventually, sailing ships carried them beyond their shores, extending their world. Ideas, religions, skills, knowledge and goods passed between different peoples. Towns and cities grew, and civilization spread. At the end of the 1400s, ships became bigger and faster and navigation improved. The first engine-powered vehicles were invented in the early 1800s, and transportation has changed rapidly ever since. Steam engines powered trains and ships, and goods were carried in greater quantities. As traveling times became shorter, new items, such as fresh fruit and flowers, were transported across the world. In industrial countries today, goods are loaded onto trains, trucks, airplanes and ships. Jet airliners take travelers to destinations all around the world, while millions of people travel in cars and buses every day. Changes in transportation have revolutionized the way we trade and travel.

ON THE ROAD
Huge semi-trailers carry large amounts of cargo over long distances. Roads have to be very strong and well maintained to cope with such heavy vehicles.

LOADING UP
Cargo is loaded into the nose of a 747 freighter. Because there is limited space for cargo on airplanes, it is the most expensive way to send goods.

DID YOU KNOW?

Ice is one of the oldest ways to keep food fresh. The Chinese used it as long ago as 1000 BC. Technology has now made it possible to carry fresh foods long distances in refrigerated trucks, railway cars and the compartments of ships.

SHIPPING CARGO
The cheapest way to move general cargo is by water. Cargo ships travel mainly across oceans and on bodies of water linked to oceans, such as the Mediterranean Sea and the Baltic Sea.

SEEING THE WORLD

Traveling is an ancient pastime. The Arab traveler Ibn Batuta spent 24 years moving from country to country, while Marco Polo was one of the first Europeans to travel through the East. Today, transportation has made the world accessible to travelers. Tourists travel across the globe to see glimpses of the ancient world, such as the temple of Ramses II in Egypt (above).

CROSSING CONTINENTS
This 2-mile (3.8-km) long train stretches across the landscape in Western Australia. Its cargo of iron ore is being taken to the coast to be shipped to Japan. Freight is often carried by more than one form of transportation on the way to its final destination.

Shipping News

Egyptian trading boat
3000 BC to 500 BC.
Propelled by oars and sail.
Traders navigated by the stars,
currents, clouds, birds, fish
and mammals' movements.

Phoenician trading boat
1000 BC to 300 BC.
Hull covered with copper
for extra strength.

Viking knorr
AD 800 to AD 1070.
Overlapping planks
strengthened the hull.
Steered by single stern oar.

Polynesian canoe
Early AD onward.
Two hulls made the
canoe very stable.

Hanseatic cog
1350–1450.
Enclosed quarters. Traders
navigated with astrolabe.

Explorers and traders have sailed across the world in all kinds of ships. From the ancient trading boats, which relied on muscle and wind power, to the invention of the steamship, ship-building techniques and ways of navigating have changed dramatically. As boats became faster and better, explorers were able to journey farther. They discovered new lands and peoples, and opened up new trading routes. Slowly, the map of the world began to grow as land and sea borders were defined and claimed by different countries. The speed of travel also became important as traders competed for the limited number of markets around the world. The history of exploration and trade is the history of explorers and ships. These pages map the routes of some of the explorers in this book and show a few of the ships that took traders on their voyages.

ALASKA

CANADA

NORTH
AMERICA

Tenochtitlan

PACIFIC OCEAN

Panama
Canal

SOUT
AMER.

Cu

Caravel
1430–1520.
Streamlined and strengthened
by internal frame.
Able to venture into shallow
waters.

Zheng He's junk
Early to mid-15th century.
Big sails stiffened by bamboo
slats for support.
Zheng He navigated with
a magnetic compass.

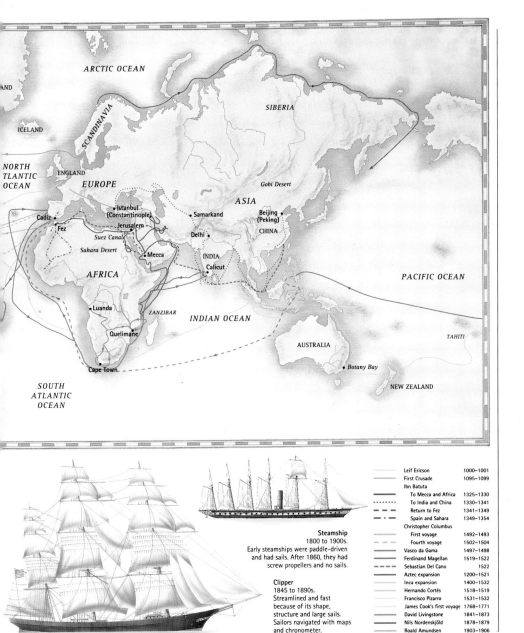

ARCTIC OCEAN

ICELAND

SIBERIA

SCANDINAVIA

NORTH
ATLANTIC
OCEAN

ENGLAND

EUROPE

ASIA

Gobi Desert

Istanbul
(Constantinople)

Samarkand

Beijing
(Peking)

Cadiz

Fez

Jerusalem

Suez Canal

Delhi

CHINA

Sahara Desert

Mecca

INDIA

Calicut

AFRICA

PACIFIC OCEAN

Luanda

ZANZIBAR

INDIAN OCEAN

Quelimane

AUSTRALIA

TAHITI

Cape Town

Botany Bay

SOUTH
ATLANTIC
OCEAN

NEW ZEALAND

Steamship
1800 to 1900s.
Early steamships were paddle-driven
and had sails. After 1860, they had
screw propellers and no sails.

Clipper
1845 to 1890s.
Streamlined and fast
because of its shape,
structure and large sails.
Sailors navigated with maps
and chronometer.

Leif Ericson	1000–1001
First Crusade	1095–1099
Ibn Batuta	
To Mecca and Africa	1325–1330
To India and China	1330–1341
Return to Fez	1341–1349
Spain and Sahara	1349–1354
Christopher Columbus	
First voyage	1492–1493
Fourth voyage	1502–1504
Vasco da Gama	1497–1498
Ferdinand Magellan	1519–1522
Sebastian Del Cano	1522
Aztec expansion	1200–1521
Inca expansion	1400–1532
Hernando Cortés	1518–1519
Francisco Pizarro	1531–1532
James Cook's first voyage	1768–1771
David Livingstone	1841–1873
Nils Nordenskjöld	1878–1879
Roald Amundsen	1903–1906

Glossary

Barcode scanner

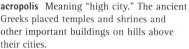

Binoculars

Pan flute

Viking space probe

Vacuum cleaner

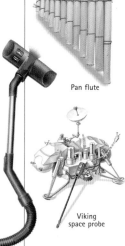

acropolis Meaning "high city." The ancient Greeks placed temples and shrines and other important buildings on hills above their cities.

acupuncture An Chinese treatment for illness involving inserting needles under the skin at certain points on the body.

AD An abbreviation for the Latin *anno Domini*, meaning "in the year of our Lord." AD indicates the number of years since the supposed date of Christ's birth.

Aeneid Virgil's long story poem about the adventures of the Trojan prince Aeneas after the fall of Troy.

aerodynamics The science that deals with air and how aircraft fly.

Aeschines (c.389–c.322 BC) A soldier, actor and clerk who later became one of the most famous orators in Athens.

Aeschylus (c.525–c.456 BC) Regarded as the founder of Greek tragedy. He wrote roughly 60 plays. Seven of them survive.

agora The open space in the middle of a Greek city used for meetings and markets.

ailerons Movable controls fixed to the wings used to make an airplane bank.

air pressure The force of the air inside a container or in the Earth's atmosphere.

airflow The flow of air past a moving aircraft.

airfoil A structure such as a wing, a tailplane or a propeller blade that develops lift when moving quickly through the air.

airplane A powered, heavier-than-air aircraft.

airship A lighter-than-air aircraft that is driven by an engine and able to be steered.

airstream A current of moving air.

altar A block with a flat top or table where offerings are made to a god.

altitude An aviation term for height.

amateur A player who does not receive any payment for playing a sport.

amber yellow or yellow-brown fossilized resin that comes from coniferous trees that are now extinct. Ancient Roman jewelers used it as a gemstone.

ambrosia The food of the gods that made them immortal and prevented them from growing old.

amphitheater An oval arena, surrounded by a seating area, for mass entertainment.

amphora A large pottery jar with a narrow neck. Goods such as wine, olive oil and fish sauce were transported in them.

amplitude The size of a wave. A sound wave with a large amplitude is louder than a sound wave with a small amplitude.

amulet A charm or piece of jewelry worn as protection against evil.

ancestor A member of your family who died a long time ago.

anesthetic A drug that keeps the body from feeling pain and other sensations.

angle of attack The angle at which the wing meets the airstream.

ankh A symbol that meant life. It was carried by gods and pharaohs. Later, the ankh was adopted as a Christian symbol.

anonymous Without a name. Artisans in ancient China were anonymous.

antenna A device that receives or transmits radio waves. An antenna connected to a radio receiver changes radio waves into electrical signals. An antenna connected to a radio transmitter changes electrical signals into radio waves.

aqueduct A channel built for moving water across long distances. It can also be a bridge that carries such a channel across a valley or river.

Arabian Nights A collection of folk tales dating from the tenth century. The stories come from Arabia, Egypt, India, Persia and many other countries.

arch A curved structure built over a doorway or window.

archaeologist Someone who studies the cultures of the past using the things people left behind them as clues, for example, bones, tools, clothing and jewelry.

Archimedes (c.287–c.212 BC) A mathematician, inventor and engineer. He studied in Alexandria, Egypt.

Archimedes' screw A device used to raise liquids. It consists of a screw inside a pipe. As the screw is turned, the water rises up inside the pipe.

architect A person trained to design and oversee the construction of buildings.

architecture The art of planning, designing and constructing buildings.

area rule A special way of designing the shape of an airplane to reduce drag when it flies at supersonic speeds.

Aristophanes (c.448–c.380 BC) A Greek playwright who made fun of the political situations of his day.

Aristotle (c.384–c.322 BC) A Greek philosopher, scientist and writer. Aristotle is one of the most important people in the history of Western thought.

assembly The main governing body of a democratic city-state. All citizens had the right to attend the assembly.

assembly line Part of the mass-production method of manufacturing. Workers fit one part to a product as it moves past them on a conveyor system.

astrolabe A navigational instrument used by early sailors to measure the position of stars and planets to find out how far north or south they had traveled.

atoll A circular coral reef or string of coral islands that surround a lagoon.

atom A particle of matter. It is the smallest existing part of an element.

atrium A small courtyard completely surrounded by the rooms of a house, or a walled courtyard in front of a church.

autogiro An airplane that gets lift from an unpowered rotor.

automaton A mechanical figure or toy that moves by itself. It was originally powered and controlled by clockwork.

autopilot A system that keeps an airplane flying automatically at an altitude (height) and heading (direction) chosen by the pilot.

auxiliaries Soldiers recruited from men who were not Roman citizens.

axial combine harvester A small grain harvester that separates grain in a spinning cage.

Aztecs The people who established a great empire in Mexico during the 1400s and early 1500s. Their civilization was very advanced, and they built huge cities. The Aztec Empire was destroyed by Spanish invaders.

Ba An Egyptian word for a person's spirit, similar to the word soul.

Stethoscope

Fax machine

SLR camera

Cement truck

Bicycle

Mythical Garuda bird

Aerial Steam Carriage

Butterfly

Wilbur and Orville Wright

Space shuttle

backhand A stroke used in racket sports, such as tennis, squash and badminton, and played with the back of the hand turned towards the ball.

bactericide A chemical that kills bacteria.

bailey An open area inside the walls of a castle. A large castle may have more than one bailey.

bails Two small wooden crosspieces set on top of the stumps in a game of cricket. In many cases, the bails have to be knocked from the stumps for the batsman to be dismissed.

ballast A weight used to stabilize a ship, submersible or submarine.

ballistic missile A missile that is launched into the air by an explosive force then continues to fly by itself.

balloon An unsteerable aircraft that is lighter than air.

balloon basket This holds the pilot, passengers and flight instruments.

bamboo A treelike grass with a hollow, woody stem. The ancient Chinese wrote on long strips of bamboo.

banking When the pilot lowers one wing and raises the other during a turn.

barbarians A term used by the ancient Chinese, Greeks and Romans to describe foreigners and people who did not sure their customs.

barcode A pattern of lines containing information about a product. The information is read by pointing a laser at the barcode to detect the varying reflections.

bartering A system of trade by which goods of the same value are exchanged, and money is not used.

bases The four fixed contact points a baseball batter must run to and touch on the way to scoring a run.

basilica A building that contained government offices and law courts.

BC Before Christ. Used for the measurement of time, BC indicates the number of years before the supposed date of Christ's birth.

beam A long, squared piece of wood, stone or metal that has a support under each end. The beams between two walls hold up the ceiling of a room or the floor of the room above.

bi disk A round shape, made of jade or some other material.

biotechnology The process of changing living things to make new products.

biplane A fixed-wing airplane with two sets of wings.

Black Land The fertile soil in the River Nile valley and delta. The ancient Egyptians called this land kemet.

black–figure ware A style of decorating pottery in ancient Greece that featured black figures on a red background.

block and tackle Several pulleys linked together to make it easier to lift a heavy weight. The top set of pulleys is the block and the lower set is the tackle.

board game A game that uses markers on a flat surface or board to indicate progress.

Book of the Dead Sheets of papyrus paper covered with magic texts and pictures. These were placed with the dead in ancient Egypt to help them pass through the dangers of the underworld.

bow The front of a boat—pronounced to rhyme with "cow."

bow drill A tool used for drilling holes in bone and shell—bow rhymes with "no."

bracket A piece of wood, stone or metal that sticks out from a wall or column to support a heavy object above it.

brick A block of clay baked in an oven so that it hardens and becomes waterproof.

bronze An alloy (mixture) of copper and tin and sometimes lead.

Bronze Age The period of time when a civilization first discovered how to make bronze and then used it for tools, weapons and other things.

bubonic plague A contagious disease that is carried by flea-infested rats. It spread through Europe in the Middle Ages.

Buddhism One of the major religions in the world. It began in India more than 2,000 years ago and is based on the teachings of Siddhartha Gautama, who became known as the Buddha.

buffeting The shaking and bumping of an airplane as it nears the speed of sound (Mach 1.0) and is affected by shock waves.

bullboat Circular craft used by native Americans made from a framework of willow branches covered with buffalo hide.

buttress A structure that pushes some part of a building inwards to prevent it from moving outwards.

Byzantine Empire The Eastern Roman Empire from the time of Constantine I until the fall of Constantinople in 1453.

c. An abbreviation of circa, which means "about." Used with dates to mean "at the approximate time of."

cable A strong, thick rope made from hemp or wire.

calculator A machine that can add, subtract, multiply and divide numbers.

cameo A picture carved into the colored layers of semi-precious stone or carved into colored layers of glass.

camphor A sharp-smelling substance taken from camphor trees. The ancient Chinese used it to flavor sweet dishes. Today it is often used as an ingredient in mothballs, ointments for injured joints and to treat colds.

Grob 115

Macchi M.52

DC.3 Dakota

Flying lizard

canal An artificial waterway. Canals in ancient China were dug by hand.

Cannibal Spirit An unfriendly spirit in the stories of the Kwakiutl tribe of the Northwest Coast of North America.

canopic jars Small jars used in ancient Egypt for storing the organs of a dead person when the body was mummified. '

caravel A sailing ship with two or three masts. The Spaniards and the Portuguese sailed in caravels in the fifteenth and sixteenth centuries.

carbon paper Paper that has been coated with carbon dust. It may be placed between two pieces of plain paper so that anything written on the top piece of paper is copied onto the bottom piece.

Carnarvon, Lord (1866–1923) England's wealthy fifth Earl of Carnarvon who contributed the funds for Howard Carter's expeditions to excavate Egyptian tombs.

carnelian A red or reddish-orange kind of quartz used as a gemstone in jewelry.

carrier wave A radio wave used to carry information, such as a radio program.

Carter, Howard (1874–1939) An English Egyptologist who found the tombs of Hatshepsut, Tuthmosis IV and Tutankhamun in the Valley of the Kings.

Float plane

cartonnage A material made from scraps of linen or papyrus glued together with plaster or resin. Cartonnage was light and strong and often used in ancient Egypt for mummy cases instead of wood.

cartouche An oval shape containing the pair of hieroglyphs depicting the name of a pharaoh.

caryatid A column carved in the shape of a young woman.

castle A fortified building designed to hold off enemy attacks.

catapult A powerful, steam-powered device that launches airplanes from aircraft carriers.

catcher The person who, in baseball and softball, crouches behind home base and catches any balls missed by the batter.

cattail A tall marsh plant with reedlike leaves. Many Native Americans ate the pollen and roots.

cavalry A unit of soldiers on horseback.

CD-ROM An abbreviation for "compact diskread-only memory". This refers to a compact disk used with a computer system.

centurion The officer in charge of a unit of foot soldiers in ancient Rome.

century A period of time lasting 100 years. It also refers to a unit of the Roman army, first containing 100 men, but later reduced to 80 men.

ceramics Articles produced from clay and other substances that have been fired in a kiln.

chalice A gold or silver cup used in the Christian church to hold the wine at Mass or Communion.

Champollion, Jean-François (1790–1832) An expert in oriental languages and French founder of Egyptology who deciphered the hieroglyphic inscription on the Rosetta Stone in 1822.

chapel A room in a large building where religious services are held.

charcoal The carbon remains of burned materials. It was used as an ingredient in gunpowder.

chickee An open-sided dwelling, thatched with palm branches and raised above the ground on stilts. Used in swampy areas in the Southeast of North America.

Chinese parsley (coriander) An herb used by the ancient Chinese for medicinal purposes. It was eaten to relieve indigestion and made into a lotion to bathe patients with measles.

citizen A freeborn Greek or Roman man entitled to take part in the government of his city-state.

city-state An independent city that has its own laws and government. Athens and Rome were the most famous city-states in ancient times.

civilization A human society that has developed social customs, government, technology and culture.

clan A group of people related to each other by ancestry or marriage.

clippers Fast ships with narrow hulls and large sails on tall masts. They were built in the United States in the mid-1800s and were used on many trade routes.

clockwork The cogs, wheels, gears, shafts and springs which make mechanical clocks work.

cockpit The compartment where the pilot or crew sit to control the aircraft.

code A system of symbols where each symbol represents a different piece of information.

Quill

Navigational astrolabe

Chopsticks

Box "brownie" camera

Robot

cogs The type of trading ships used in Europe by the Hanseatic League in the 1300s and 1400s.

collective pitch The control that makes a helicopter climb and descend.

colonnade A set of evenly spaced columns that often supports a roof.

column A tall, thin cylinder with a capital at the top and often a wide, round base at the bottom. Columns are used to support a roof or the upper story of a building.

compact disk (CD) A disk on which music or other sounds are recorded as a pattern of microscopic pits.

computer A machine that automatically performs calculations according to a set of instructions that are stored in its memory.

concave lens A lens that is thinner in the middle than at the edges.

concrete A synthetic building material made from a mixture of cement, lime, sand, small stones and water.

conquistadors From a Spanish word for adventurers or conquerors. It is used especially to describe the Spaniards who conquered the New World in the sixteenth century.

contraceptive A chemical or device designed to prevent women from becoming pregnant.

convex lens A lens that is thicker in the middle than at the edges.

corbel A piece of stone or wood that sticks out from a wall. A corbel is held in place at one end by pressure from above and below.

corbeled vault A stone ceiling made up of rows of corbels that rise from two walls and meet in the middle.

Corinthian column A grooved pillar, built of sections of stone, used in the Corinthian

order. The top was decorated with carved acanthus leaves.

court An enclosed area, or one specially marked out for a game. Games played on a court include tennis, croquet, basketball, netball and squash.

courtyard An area surrounded on several sides by walls or buildings and open to the sky.

cox or coxswain A small, lightweight member of a rowing crew whose job is to keep the rowers in time and to steer the craft.

Coyote The mischievous wolf spirit of North America that loved playing tricks.

creases In cricket, these are the lines marked on the ground near each wicket that mark the playing positions for the batsman and the bowler.

creation stories Native Americans told different stories to explain where they came from.

cremate To burn a dead body.

crook A stick with a curved top carried by a god or pharaoh to symbolize kingship.

cue ball In billiards, snooker and pool, the cue ball is white. It is struck by the cue to make contact with the other balls.

cuneiform A system of writing invented by the Sumerians and then used by later peoples of Mesopotamia. The word means wedge-shaped, and describes the shape of the characters used in cuneiform.

Cush A country to the south of Egypt, probably where North Sudan is now.

cyclic pitch The control that makes a helicopter move in a horizontal direction.

dandy horse A two-wheeled "bike" pushed along with the feet. It was made

Padlock

Dandy horse

Electric iron

String telephone

Pitchfork

613

Corinthian capital

Church of St Charles, Vienna

SkyDome, Toronto

Vatican City stamp

Eiffel Tower, Paris

popular in the early 1800s by fashionable men called "dandies."

data Information. Computer data may include numbers, text, sound or pictures.

deben A metal ring used by traders for measuring weight. One deben weighed about 4 oz (90 gm).

decathlon An athletics contest of ten different events taking place over two days.

democracy In ancient Greece, a system of government in which all citizens who were entitled to vote could have a say.

demotic script A form of writing that developed from hieratic script from 700 BC onwards. It was used for administration and business.

dhows Arab ships with triangular sails and one or two masts. Arab traders sailed along the coast in dhows hundreds of years ago. They are still used in the Middle East.

diamond This refers to the marking of a baseball field, which looks like the outline of a diamond.

dihedral The angle at which wings or tailplanes are attached to the fuselage. It helps to keep the airplane stable.

disk drive The part of a computer that is used to read or write data.

dolerite A hard rock used by the ancient Egyptians to grind and crush other rocks.

dome A curved stone roof that covers an area the shape of a circle. Most domes are made from arches. Domes built with corbels are called corbeled domes.

Doric column A grooved pillar with a plain top, built of sections of stone, used in the Doric order.

drag A force that tries to slow down an object, such as an

airplane or boat, moving through air or water.

dragon A mythical creature. The ancient Chinese worshipped the spirits of dragons. The five-toed dragon became the First Emperor's special symbol.

dribbling The process of moving the ball forwards with slight touches of hands, feet or sticks. Basketball, soccer and field-hockey players all dribble the ball.

dugout canoe A small boat made by hollowing out a log.

dyke A barrier to stop water from flooding. The ancient Chinese tried to control a flooding river by building a dyke.

dynasty A ruling family. Members of a dynasty were related by birth, marriage or adoption.

Eastern Hemisphere The part of the globe that contains Europe, Asia and Africa.

Egyptologist A special kind of archaeologist who finds out about how people lived in ancient Egypt by studying the things they left behind.

ejection seat A rocket-powered seat that fires (or ejects) the pilot out of an airplane. The pilot then parachutes to safety.

electrocardiograph A machine that records the electrical activity of the heart.

electromagnetic wave A wave of energy made of vibrating electric and magnetic fields. Light, radio and X-rays are examples of electromagnetic waves.

electron A particle of matter with a negative electric charge found in an atom. Electrons moving in the same direction form an electric current.

electronic circuit The pathways and connections followed by electrons to control computers, robots and modern domestic appliances.

electronic television The modern television system that uses an electron gun to scan images and reproduce them on a cathode-ray tube.

electrum A mixture of gold and silver. This occurred naturally in Asia Minor and was made into coins throughout ancient Greece.

elevator A movable control attached to the tailplane that makes an airplane climb or descend.

elimination match A competition match in which the losing players leave the competition at the end of each round, until only the winner remains.

elixir A magical mixture that is supposed to enable people to live forever.

embalmer A person who treats a dead body with spices and oils to prevent it decaying.

emblem A decorative mark or symbol that means something special.

emery A hard, grayish-black rock used for smoothing and polishing.

empire The people and territories under the rule of a single person or state. Great Britain, the Netherlands, France and other countries had trading empires throughout the world.

emulsion The light-sensitive coating on photographic film.

environmentally friendly Machines, appliances and materials that do not damage the natural resources and features of the Earth.

escapement The part of a clock or watch that regulates its speed.

Eskimo (see Inuit) Dwellers in the Arctic region. Eskimo means "eaters of raw meat," but they cooked at least part of their daily ration.

Euripides (c.480–c.406 BC) A writer of tragedies. He wrote between 80 and 90 plays, but only 18 complete ones survive.

excavate To uncover an object, a skeleton or even a whole town by digging.

eyepiece The lens or group of lenses nearest the eye in a microscope, telescope or pair of binoculars.

Far East The countries of East Asia, such as China, Japan and Indonesia.

fax machine A facsimile machine. A machine used to send and receive documents by telephone.

Ferghana horses Horses imported from central Asia. The ancient Chinese could not breed horses easily themselves.

fetishes Natural objects, such as an unusually shaped stone, a perfect ear of corn or a shell. The Southwestern tribes of North America believed they brought supernatural power.

fiber-optic cable A cable made from hair-thin strands of pure glass.

filament A thin wire inside a bulb. The filament glows when an electric current flows through it.

fin The fixed, vertical part of the tail unit that helps keep an airplane flying straight ahead.

flagship The main ship in a fleet of ships. The commander has his quarters there.

flail A tool carried by a god or pharaoh in ancient Egypt to symbolize kingship and the fertility of the land.

flax A plant from which thread can be made and woven into linen.

flint A grayish-black form of quartz that was used by ancient peoples to make tools and weapons. It was often traded for other goods.

Church of the Nativity, Novgorod

Kandariya Mahadeo floorplan

Colosseum, Rome

Lloyd's of London

615

Bowls

Fireman's carry

Trimaran

Fencing equipment

Badminton racquets and shuttlecock

float plane A seaplane that is supported on the water by floats.

fluorescent tube A tube that glows when an electric current passes through the gas inside.

flying boat A seaplane that is supported on the water by its fuselage.

flying buttress An arch in a Gothic building that connects a buttress outside the building to an arched vault inside.

focus The point where light rays bent by a lens come together.

fodder The feed grown for livestock.

forehand A stroke played in racket sports with the palm of the hand turned towards the ball.

fossil fuels The remains of animals and plants left in the earth that form coal, oil and natural gas over millions of years.

freedman/freedwoman A man or woman who has been freed from slavery.

frequency The rate of vibration of any wavelike motion including light, radio or water waves.

fresco A painting made with watercolors on wet plaster.

frieze A strip of painting or carving on a temple or a tomb wall that told a story.

fuselage The body of an aircraft.

galleons Large sailing ships with three or more masts. They were used as warships and trading vessels by countries such as Spain from the fifteenth to the eighteenth centuries.

gallery A long, narrow room that is open on at least one side.

gargoyle A waterspout that is carved as a grotesque face or creature. It drains water from the gutters around the roof of a building.

garlic An herb used by the ancient Chinese to treat colds, whooping cough and other illnesses.

gas burners Burners that are fed by propane gas to provide the heat that lifts a hot-air balloon.

gauntlet glove A glove with a long cuff.

gild To cover in gold leaf.

ginseng An herb used by the ancient Chinese for medicinal purposes. It was prescribed for poor appetites, some forms of upset stomachs and coughs, excessive perspiring and forgetfulness.

gladiator A man, or sometimes a woman, trained to fight in public to provide entertainment.

glaze A liquid glass that is baked onto the surface of materials made from clay, such as pottery, bricks and tiles, to give them a hard, shiny and waterproof surface.

glider An unpowered, heavier-than-air aircraft.

goalkeeper or goal-tender In soccer, field hockey and ice hockey, this person has the task of defending the goals to stop the ball or puck from passing between the posts.

goals In some ball sports, the goals are special areas at which you aim the ball to score points. In various codes of football and hockey, the goal is a pair of posts through which a ball or puck is kicked or struck. In basketball and netball, it is a ring through which the ball is thrown.

gold leaf A very thin sheet of beaten gold.

gondola The passenger and crew cabin of an airship.

gourd The fruit of a species of plant that includes squashes. The dried shells were used as rattles and containers.

616

graphite A soft, grey-black form of carbon used in pencils.

gravity The force that pulls us down to the ground and also keeps the Earth and the other planets circling the sun.

green In golf, this is the smooth putting area in which the hole is cut. In lawn bowling, it is the section of lawn used for a game.

groma An engineer's surveying tool. It was used to make sure that roads and buildings were straight.

guided missile An airborne bomb that can chase a moving target. It may be guided by wires, radio controls, laser light or infrared sensors.

gunpowder A mixture of potassium nitrate, sulfur and charcoal used as an explosive and in fireworks.

helicopter An aircraft that gets its lift from a powered rotor.

hemp A plant with tough fibers that can be woven into fabric and rope.

heptathlon A track and field competition of seven separate events over two days.

Herodotus (c.485–c.425 BC) A Greek writer who is regarded as the father of history. He based his reports of the Persian Wars on eyewitness accounts and facts.

Hesiod (8th century BC) One of ancient Greece's earliest poets.

hieratic script A faster form of writing than hieroglyphs used in ancient Egypt. It was always written from right to left, unlike European languages, which are written from left to right.

hieroglyphs The ancient Egyptian form of writing in which pictures or symbols represented sounds, objects and ideas.

Hippocrates (c.460–c.377 BC) A Greek doctor, teacher and writer on medicine

who is regarded as the father of medicine.

hippodrome A course used by ancient Greeks and Romans for the staging of chariot and horse races.

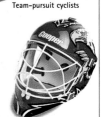
Team-pursuit cyclists

hogan A Navajo dwelling built of timber and poles covered with bark and dirt.

Homer (8th century BC) A Greek poet who composed long story poems called epics.

hoplite race A foot race held at the ancient Olympic Games. The competitors had to run naked except for helmets and leg armor, and carry their shields.

husky A breed of dog. They were the Inuits' (Eskimos') only domestic animal.

hydroponics The growing of plants by placing their roots in water enriched with nutrients rather than in soil.

Ice-hockey face mask

Ice Age A time when large parts of the Earth were covered with glaciers.

igloo A temporary winter dwelling built by some Arctic tribes from packed snow.

Iliad Homer's long story poem about the war between the Greeks and the Trojans that was said to have lasted ten years.

Incas The people who lived in Peru in South America from about 1100 to the 1530s. They had a great empire, which was destroyed by the Spaniards.

Golf caddie

incense An aromatic substance extracted from resin. Incense was used in ritual ceremonies.

Darts

Industrial Revolution A process that began in Great Britain in the mid-1700s and spread to other parts of Europe and North America in the 1800s. It involved new methods of transport and new machines, driven by coal, electricity or other power sources.

Canoeing safety equipment

Faience make-up holder

Tomb painting

Cartouches

Ceremonial
ax

Wooden
porter

infantry A unit of foot soldiers.

infrared missiles Rocket-powered missiles that are guided to their target by an infrared homing system.

inlay A form of decoration in which one material is inserted into prepared slots or cavities in the surface of another material to form a pattern or a picture. The ancient Chinese often inlaid bronze with silver.

innings A team's turn at batting in cricket, softball, baseball and rounders.

insulation Material used to stop heat, electricity or sound from passing through a surface.

internal combustion engine An engine in which the fuel is burned inside the engine, such as in a car or jet.

Inuit (see Eskimo) Many tribes in the Arctic prefer to be known as Inuit. This is their word for "people."

invention An original or new product or process.

Ionic column A grooved pillar, built of sections of stone, used in the Ionic order. It was taller and more slender than the Doric column. The top was usually decorated with two swirls called volutes.

iron A strong metal used to make tools and parts of some buildings. It is found in certain types of rocks and is removed by very high temperatures.

irrigation The system of supplying the land with an artificial water supply using pumps, pipes and channels.

jade An extremely hard mineral. The jade in ancient China was nephrite, a white stone streaked with reds or browns. The more familiar green jadeite was not known in China until much later.

javelin A light spear thrown by a competitor in a track and field event.

Josephus (c.AD 37–c.100) A Jewish soldier and historian who was born in Jerusalem.

joyflights When people pay to fly in small airplanes to experience the excitement of flying.

junks Chinese ships with square sails.

Ka An Egyptian word for a person's life force created at birth and released by death.

kachinas Supernatural spirits who guided tribes of the Southwest of North America.

keel The lowest part of a sailing boat. It is often lined with lead or other heavy material to weight the boat and help keep it upright.

keep A tall tower that is a castle, or the tall tower within a castle where the defenders retreat. The entry is usually high in a wall so it can only be reached by a ladder.

kite A tethered glider that is lifted by the wind. A kite was the first heavier-than-air aircraft.

kiva A place of worship and a council chamber for native Americans, usually built below ground.

knorrs Shallow-bottomed ships used by the Vikings to carry cargo.

lacquer A natural plastic varnish that resists heat, moisture, acids, alkalis and bacteria, and is made from the milky sap of the lacquer tree. The Chinese heated and purified this substance, and painted many very thin coats onto objects made from bamboo, silk, wood and other materials.

laminate To press or cement together several layers of materials, such as wood, glass or plastic, so that they become a single sheet.

landing gear The name for the wheels that support an airplane on the ground.

lapis lazuli A brilliant blue mineral that is used as a gemstone.

laser A very intense light of one wavelength and frequency that can travel long distances. It is used to cut materials, carry television transmissions, print onto paper and guide machines.

lean-to A framework of poles covered with branches to form a simple shelter.

legion A unit of the Roman army. Its size varied during the history of the Empire.

lever A bar that rests at one point on a raised support or fulcrum. When the long end of the lever is pushed down, the short end of the lever on the opposite side will lift a heavy object a short distance.

li A Chinese unit of length. One li equals about 3/10 mile (0.5 km).

lift The force that enables an aircraft to fly. Lift is produced when an aircraft's wings or a helicopter's rotor blades cut through the air.

light Electromagnetic waves that the human eye can detect. Different wavelengths are seen as different colors, red is the longest and violet is the shortest.

linear Made up of lines. The linear designs on Stone Age (Neolithic) pottery were made with a soft brush.

Linear A An early form of writing used by the Minoans.

Linear B A form of writing that the Mycenaeans adapted from Linear A.

liquid crystal display (LCD) The type of display, or screen, used by electronic calculators and digital watches.

Livy (59 BC–AD 17) Writer of a history of Rome in 142 books, which became textbooks in Roman schools.

locomotive A self-powered vehicle that runs on a railway track.

loincloth Cloth or skin draped between the legs and looped through a belt. Worn by many Native Americans living in a warm climate.

longships The swift, sleek warships that carried invading parties of Vikings to western Europe.

loom A wooden frame used to hold the threads during the process of weaving.

Mach number The speed of an airplane compared with the speed of sound.

magistrate A Roman official who judged the local criminals and looked after the affairs of a district.

maglev A magnetic levitation train that floats above a special track. Magnetic fields support the train's weight and move it along the track.

magnetron The part of a microwave oven and radar equipment that produces microwaves.

malachite A copper ore. When crushed, it was used as green eye paint and symbolized fertility.

marble A popular building stone found in many colors. It can be polished until it is as smooth and shiny as glass.

Martial (c.AD 40–c.104) A Roman poet, known for his wit. He wrote 1,561 poems.

mass production A method of manufacturing large quantities of goods, often using a number of machines. Each worker or machine in a factory works on just one part of a product.

medicine man/woman A shaman with powers to heal and contact the spirits.

memory The part of a computer that stores data and instructions.

microphone A device for changing sound into a varying electrical current.

Tutankhamun's gold funeral mask

Hair comb and toys

Pyramids at Giza

Rosetta Stone

Crocodile mummy

Shang bronze statue

microsurgery Surgery conducted using specially designed microscopes and tiny instruments to repair the smallest parts of the body.

Middle Ages The period of European history from about 1000 to the 1400s.

migrating When people or animals move from one country or place to another.

milestone A column of stone beside Roman roads that was inscribed with the distances to the nearest towns.

minaret A tall tower built outside a mosque that has a staircase inside and a platform at the top.

minister A high-ranking Roman official who advised the emperor and helped to see that laws were obeyed.

moat A ditch filled with water outside the walls of a castle. It is intended to keep attackers away from the walls.

Mogul A member of the Muslim dynasty of Indian emperors that began ruling India in 1526.

molecule A group of atoms linked together. Chemical substances are made from molecules.

monoplane A fixed-wing airplane with one set of wings.

monsoon winds Winds that change direction with the season.

mosaic A picture or design made by mounting small pieces of colored stone or glass on a wall, ceiling or floor.

motor A device for changing electricity into movement, such as the spinning motion of a shaft.

Mt. Vesuvius A volcano in southern Italy that erupted in AD 79 and buried the towns of Pompeii and Herculaneum.

mummification A process of drying and

embalming that preserves the dead body of a person or an animal.

mutineers People who rebel against those who have authority over them. Sailors sometimes mutinied against their captains.

mythology Stories passed down by word of mouth from one generation to another.

natron A natural salt from the desert that absorbs moisture. The ancient Egyptians used it for mummification.

Near East Another term for the Middle East. It describes the countries round the eastern end of the Mediterranean Sea.

nectar The drink of the gods.

Neolithic A period of time when humans made tools and weapons of flint and stone. This is the final stage of the Stone Age.

New World A name used by the Europeans for the land area that included Canada, North America and South America.

nomads People who wander from place to place, usually in search of game to hunt or grazing land for their flocks.

nomarch An official in charge of a province or region called a nome. There were 42 nomes in ancient Egypt.

nymph A spirit of nature in Greek myths.

oasis (singular), oases (plural) A fertile area in the desert with its own water supply.

obelisk A tall, thin monument that is pointed at the top and usually square at the bottom.

objective lens The lens, or group of lenses, which forms the image in a microscope, telescope or a pair of binoculars.

obsidian A dark glass made by a volcanic explosion. It is formed when lava hardens quickly. Obsidian is sharp and cuts well. It was valuable to ancient peoples who made utensils from it.

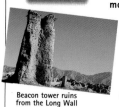

Beacon tower ruins from the Long Wall

Terra-cotta granary

Zhou jade disks

Odyssey Homer's long story poem about the adventures of the Greek hero Odysseus.

Old World The part of the world that was known by Europeans before the discovery of Canada and the Americas. It included Europe, Asia and Africa.

omen A sign or warning about the future that indicated a happy or disastrous event.

optical fiber A communications cable made of solid glass or plastic fiber. It transmits light from one end to the other, even when the cable is curved or bent.

optical illusion When your eyes are tricked into seeing something that is not actually there. For example, you may be able to see a deep room in a flat painting.

oracle A message spoken by a priest or priestess on behalf of a god.

orbit To circle the Earth, another planet or a star in space.

order A style of architecture. The three ancient Greek orders were Doric, Ionic and Corinthian.

ostracon (singular), ostraca (plural) A fragment of pottery or large flake of stone used by students to practice their writing.

Ovid (43 BC–c.AD 17) A popular Roman poet. He wrote love lyrics and verses about mythology.

oxygen The gas that is essential for life and also for combustion. Oxygen makes up 20 percent of the air around us.

palette A flat piece of wood or stone on which artists mix paints.

papyrus A kind of paper made from the papyrus plant. It was used by the ancient Egyptians, Greeks and Romans.

parallel bars Two bars of equal length supported by uprights used in gymnastic performances.

Paralympics An international sporting festival for athletes with a disability. Like the Olympic Games, the festival is held every four years.

parfleche A saddlebag made from rawhide in the pattern of an envelope. This word comes from the French *parer une flèche*, meaning "to turn an arrow."

patent A law that guarantees inventors the exclusive rights to perfect, build, sell and operate their inventions for a number of years.

Bronze and turquoise bird

patricians Aristocratic, well-born members of Roman society who held important positions in government.

pediment A wall shaped like a triangle that closes the end of a building between two sloping roofs. It is also the triangular decoration above a window or door.

Han bronze seal

pendentives Supports that make it possible to build a dome over a square room.

percussion instrument A musical instrument that produces sound when it is struck by the hand or a hammer.

Pericles (c.490–c.429 BC) An Athenian statesman and general. He was also a powerful public speaker who influenced people's opinions.

persistence of vision The illusion of movement produced when viewing a film or television. Our eyes see a series of still pictures moving as one.

pharaoh The ruler of ancient Egypt. The name comes from the ancient Egyptian word *per-ao*, meaning "the great house." It referred to the palace where the pharaoh lived.

Tang pottery dancer

Pheidias (5th century BC) Considered to be the greatest sculptor in ancient Greece. He designed sculptures for the Parthenon and the statue of Athena.

Tang pottery camel

Rosemary

Mycenaean fresco

Miniature wine jug

Voting tokens

Terra-cotta statue

philosopher A person who searches for knowledge and wisdom. Philosophers study the natural world and human behavior.

photodiode A light-sensitive device that varies the size of an electric current flowing through it according to the amount of light that falls on it.

pictographs Pictures or symbols that represent words. Pictographs were one of the earliest forms of writing.

pigment A powder that is mixed with a liquid to make ink or paint.

pilgrim A person who travels to a holy place. A pilgrim's trip is called a pilgrimage.

pinnacle A small peak that stands on top of a wall or a buttress.

piston A movable, solid cylinder that is forced to go up and down inside a tube by the exertion of pressure.

pitch A marked playing area, such as a soccer or hockey field, and the area used for bowling and batting in a game of cricket. Pitch also means to throw or fling a ball, such as in baseball. Pitch is alos how high or low a musical note is when compared to other musical notes.

pitcher The player in a baseball or softball game whose task is to throw the ball at the batters.

pitching The aerodynamic term to describe an airplane's nose moving up and down.

plaque A wall ornament in the form of a plate or tile, made from porcelain, wood or other material.

plaster A mixture including lime, sand and water, which is spread on walls and ceilings and left to dry.

Plato (c.427–c.347 BC) An Athenian philosopher. Plato's ideas for running an ideal state are still studied today.

plebeians Common, low-born members of Roman society. The plebeians were the poorest and most numerous class of Roman citizens.

Pliny the Elder (AD 23–79) A Roman writer, cavalry officer and government official. He wrote about war, weapons and natural and human history.

Pliny the Younger (c.AD 62–c.113) Nephew of Pliny the Elder and a writer, lawyer and public speaker. His letters to Trajan and the historian Tacitus fill ten volumes.

pok-ta-pok A game, similar to basketball, which was played in the tenth century BC by the Mayan and Toltec peoples of Central America.

porcelain A thin china made from fine kaolin clay that was found in the Chinese mountains.

porch The covered entrance to a temple or other large building with its roof supported by columns.

porcupine An animal covered with long protective quills. Native American women softened and dyed the quills and used them for embroidery.

Portland Vase A blue and white vase that is the best known example of Roman glass. It is named after the Dukes of Portland, who once owned it.

poultice A medical pack or dressing placed on part of the body to cure a sickness or soothe an inflammation.

primary feathers A bird's outermost wing feathers that provide the thrust for flight.

prism A wedge-shaped block of glass used to refract or reflect light.

production line The process of working in a team to produce something. The tasks are divided up so that workers become very fast and good at what they do.

professional An athlete or sporting person who competes for prize money or receives payment for playing a chosen sport.

program Instructions given to and stored in the memory of a computer so that it can carry out a particular task.

propeller A set of blades driven by an engine that pull or push an airplane through the air. It is sometimes called an airscrew.

province Any area outside the city of Rome (later outside Italy) that was controlled by the Romans.

prow The front part, or bow, of a boat.

P-type silicon A form of silicon that has been treated so that it contains fewer electrons.

puck A rubber disk used in ice hockey.

puffin A diving bird with a brightly colored beak. Arctic tribes ate them and used their beaks and feathers for decoration.

pulley A wheel with a groove around the rim. A rope can be threaded around the rims of several pulleys to make it easier to lift a heavy weight.

pumice A lightweight stone that comes from a volcano.

Punt An important trading country in the ancient world. It was particularly valued for its incense. This country was probably situated in what is now Somalia.

punt A type of football kick in which the kicker drops the ball from the hands. The kicker strikes the ball with the foot before the ball touches the ground.

pylon racing Air racing close to the ground around a course marked by painted pylons.

Pythagoras (c.580–c.500 BC) A philosopher, mathematician and teacher. Pythagoras developed many theories of geometry.

qilin A mythical animal also called a Chinese unicorn. It symbolized long life, greatness, happiness and wise government.

racket A wooden, metal or synthetic bat strung with nylon or other material. It is used to hit the ball in such sports as tennis, squash and badminton.

radar Stands for "radio detecting and ranging." This is a way of locating an object by measuring the time and direction of a returning radio wave.

radiation Energy given out by an object in the form of particles or electromagnetic waves.

radio telescope An instrument designed to detect radio waves from distant parts of the universe.

radio waves Invisible electromagnetic waves that carry information such as Morse code "beeps" and the human voice.

Raven A spirit worshipped by the Native Americans.

reconnaissance plane An aircraft that is designed to take pictures and sometimes to spy over enemy territory.

Red Land The desert that lay beyond the river valley and delta. The ancient Egyptians called this land deshret.

red-figure ware A style of decorating pottery that featured red figures on a black background.

relic A possession of a person who was considered very holy, or a part of that person's body that is kept after death.

relief A picture carved into a flat slab of stone.

Renaissance The period in Europe between 1300 and 1500 when science, invention, art and education were strongly encouraged.

republic A form of government in which the people elect representatives.

Roman god Vulcan

Theatrical mask

Minoan double ax

Discus

Archimedes' screw

The emperor Caracalla

Garden fresco

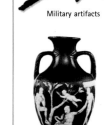

Military artifacts

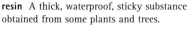

The Portland Vase

Etruscan
sarcophagus

resin A thick, waterproof, sticky substance obtained from some plants and trees.

rink A marked section of ice used for sports such as ice hockey, curling and figure skating. A rink is also a strip of lawn used for lawn bowling.

ritual A set pattern of behavior for a religious or other kind of ceremony.

rivet A metal pin or bolt used for holding two or more pieces of a material together.

rocket A vehicle propelled by burning a mixture of fuel and a substance containing oxygen.

rolling The aerodynamic term used to describe an airplane banking- when one wing lifts and the other drops down.

Roman mile An ancient Roman unit of length, 1,000 paces long (5,000 Roman feet), which was equivalent to 1,620 yards (1,481 meters).

Rosetta Stone An inscribed slab of granite that gave the major clue to deciphering hieroglyphs. It was found by a French soldier near the village of Rosetta in 1799. The stone is 3 ft 7 in (114 cm) high, 2 ft 3 in (72 cm) wide, 11 in (28 cm) thick and weighs 1,684 pounds (762 kg). The Rosetta Stone is now in London's British Museum.

rotor The part of a machine that rotates.

rotors Two or more long narrow wings (called blades) that provide lift for a helicopter or an autogiro.

rounders A team game of bat and ball in which players run the round of the bases. It is an older and more basic form of baseball.

rudder A flat piece of wood attached to the stern of a boat below the waterline, which is used for steering. The rudder was a Chinese invention.

A rudder is also a movable control fixed to the fin that helps control the direction of an airplane.

runners A long set of blades, usually made of steel, on which a sled, toboggan or luge rides over ice.

saguaro A giant desert cactus with a sweet red fruit.

sailplane A high-performance glider designed specially to soar on thermals.

salon A French word for a room in a great house where guests are received and entertained.

saltpeter (potassium nitrate) A chemical ingredient of gunpowder.

sarcophagus (singular), sarcophagi (plural) A large stone box that enclosed a mummy's coffin. The surfaces were usually carved in relief.

satellite An object that orbits a star or planet. Satellites may be natural (moons) or artificial (spacecraft).

satyr A mythical being half man, half goat.

scaffolding Metal or wooden platforms set up along walls or under roofs where workers and artists stand while constructing, decorating or repairing buildings.

Schliemann, Heinrich (1822–1890) A German archaeologist. At 46, he began devoting his time to excavating the sites of Mycenae and Troy.

scrimmage A play in American football that begins with two teams lined up opposing each other and the ball placed on the ground between them.

scrum or scrummage A grouping of rugby players. Players from one team bind together on one side and push against a similar grouping from the opposing team to win possession of the ball.

scurvy A disease, caused by lack of Vitamin C, which was common to early sailors. They did not have fresh fruit and vegetables on their long sea voyages.

Sea Grizzly Bear A spirit creature carved on masks and totem poles.

Seneca (c.4 BC–c.AD 65) A writer and philosopher. Seneca tutored the emperor Nero when he was a boy.

senet A game played by the Egyptians with a board and counters. It had lucky and unlucky squares and was a little like the modern game of checkers.

Seven Wonders of the World The seven most impressive structures of the ancient world: the Pyramids of Egypt, the Hanging Gardens of Babylon, the temple of Artemis at Ephesus, Phidias' statue of Zeus at Olympia, the mausoleum of Halicarnassus, the Colossus of Rhodes and the Pharos (lighthouse) of Alexandria.

shabti A model figure that acted as the servant of the deceased and carried out all the work required in the afterlife.

shaduf An irrigation device made from a bucket and a counterweight. It transferred water from the Nile into canals.

shaman A Native American medicine man or woman with special powers to heal and contact the spirits.

shogun Originally a military title. It was used by the men who controlled the government of Japan from 1192 to 1867.

shorthand writing A fast type of handwriting where simple strokes represent parts of words.

shrine A structure people build over a place or object they consider sacred.

shutter The part of a camera that opens for a fraction of a second when you take a picture to let light fall on the film behind it.

shuttlecock Weighted at one end, a shuttlecock is made of cork or plastic with a ring of feathers attached and is struck back and forth in the sport of badminton.

silicon A common substance found in sand and clay. It is used in computer chips and solar cells.

sinew An animal tendon used by Native Americans as sewing thread.

Skywoman A character in Iroquois creation stories.

slalom In skiing or canoeing, a slalom event is conducted over a course that has a number of flags or artificial obstacles.

slip A thick mixture of water and clay that was used to coat pottery.

soapstone A variety of soft stone that could be easily worked with a knife or hand drill.

Socrates (c.470–c.399 BC) An Athenian philosopher. He taught his students by questions and answers.

software Another word for computer programs, the instructions that make computers work.

solar cell A device that converts sunlight directly into electricity.

sonar Stands for "sound navigation and ranging." A device that sends sound through water and then detects echoes as they bounce off the sea floor and other objects.

sonic boom A sound like a thunderclap that is sometimes heard on the ground. It is caused by the shock waves from a supersonic aircraft.

soothsayer A man or woman who predicts future events.

Sophocles (c.496–c.406 BC) An Athenian playwright who wrote more than 100

Janus

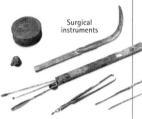
Surgical instruments

Cameo

Dying Gaul

Fresco of Bacchus

Thunderbird mask

Inuit grass socks

Rawhide
playing cards

Inuit snow goggles

Saddle bag

works for the theater. Seven of his tragedies survive.

Sosigenes (birth and death dates unknown) A Greek astronomer and mathematician who helped Julius Caesar to reform the calendar.

sound barrier An invisible, aerodynamic barrier that was once thought to prevent airplanes from traveling faster than the speed of sound.

space shuttle A reusable aircraft that is used to travel into space.

Spanish Armada The fleet of heavily armed warships sent by Philip II of Spain to attack England in 1588. The Spaniards were defeated in the English Channel.

speed of sound At high altitudes, this is 662 miles (1,065 km) per hour; at sea level the speed of sound is 760 miles (1,223 km) per hour.

sphairistike The first name given to the new sport of lawn tennis. It came from the Greek word *sphaira*, which means "ball."

sphinx A mythical creature usually depicted with the head of a woman and the body of a lion.

Spice Islands A group of islands in the East that were the source of precious spices in the 1400s. Later, they became known as the Moluccas. Through the years, traders from many countries tried to gain control of these prized islands.

spire A tall, thin tower in the shape of a cone or a pyramid, which stands on top of a building.

stability A plane needs to be stable when it flies. Its wings, fuselage and tailplanes make it easy, safe and smooth to fly.

stadium An enclosed sports ground, usually with banks of seats for spectators.

stalling This happens when a plane flies too slowly. Its wings are unable to produce enough lift and it loses height.

star anise An herb used as a medicine by the ancient Chinese, for example, to relieve stomach pain and vomiting.

stator The part of a rotary machine that does not move.

steam engine A machine that changes steam into the energy used to power equipment and tools.

steel A strong metal made from iron and carbon melted together at a very high temperature.

sterilize To kill germs on instruments and in the air.

stern The back of a boat.

Stone Age The time when humans made tools and weapons from stone.

streamlining This gives an aircraft a smooth shape to reduce its air resistance, or drag.

strike In basketball, a strike is an accurate pitch that the batter misses or chooses not to hit. In tenpin bowling, a strike is the knocking down of all the pins with one bowl.

striker A soccer player whose main task during the course of a game is to score goals.

stucco A smooth plaster applied to walls, which can be dyed, molded and polished until it shines.

stumps Three upright wooden pegs that form a set that is called a wicket in a game of cricket.

stylus A writing tool. The pointed end was used to mark the covering of wax on a wooden tablet. The blunt end of the stylus was used to smooth out mistakes from the wax.

626

submarine A large craft that can travel underwater for long distances unaided by any other craft.

submersible A small craft that can dive to great depths underwater. A submersible is much smaller than a submarine, and it is carried to and from its dive location on the deck of a ship.

subsonic Flying at less than the speed of sound.

sulfur A chemical ingredient of gunpowder.

Super Bowl The grand final game to decide the annual National Football League championship of the United States. The word "bowl" is used to describe the large bowl-like stadium in which the game is played.

supernatural Relating to things belonging to the world of the spirits.

supersonic Flying faster than the speed of sound.

sweat lodge An airtight hut filled with steam—the Native American equivalent of a sauna.

swivel A part that kept the harness on a dog sled from becoming tangled.

symbol A decorative mark in a painting or carving that means something special.

synthesizer A machine that electronically creates and amplifies musical sounds.

synthetic Something that is made by humans and does not exist naturally.

Tacitus (c.AD 55–c.120) The greatest of the Roman historians. He wrote many books.

tailplane The fixed, horizontal part of the tail unit that helps to keep an airplane stable (also called the horizontal stabilizer).

tanning The process for turning animal skins into leather.

tee A cleared space of land from which the ball is struck at the beginning of each hole in a game of golf. A tee is also the name of the small plastic or wooden holder used to support the ball.

tension The result of pulling or stretching an object. Building materials such as stone and concrete break easily under tension while wood, iron and steel remain strong.

terra-cotta A hard, unglazed clay used to make tiles and for modeling small statues.

terrace A level space that is raised above its surroundings. A building may stand on top of one or more terraces. The top of a building is said to be terraced if it looks like a series of steps.

thatch A covering for roofs made of bundles of straw, reeds or leaves.

thermal A column of rising air used by gliders and birds to gain height.

throttle Like the accelerator of a car, this controls the speed of an airplane's engine.

thrust The force developed by a propeller or jet engine that drives an airplane through the air.

Thucydides (c.460–c.395 BC) An Athenian historian who wrote an account of the Peloponnesian War.

Thunderbird A spirit worshipped by the tribes of the Northwest Coast.

thwart A crosspiece inside a canoe that pushes out the sides and forms a seat for the people who row the boat.

tile A thin slab of baked clay. Tiles are used to cover roofs and floors. They are often glazed. Glazed tiles were first used to cover buildings that water would damage.

War shield

Net gauge

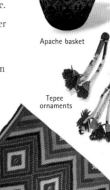

Apache basket

Tepee ornaments

Navajo blanket

Incan god

Turkish merchants

Jade pendant

Map of Drake's travels

totalizator A machine for calculating, recording and indicating a system of betting on horse races.

touchdown In American football, a team scores a touchdown by carrying or receiving the ball over the opposing team's goal line.

tracery Frames of curved stone that hold the stained glass in the windows of Gothic buildings.

trade colony A settlement of merchants in a foreign land who trade with their home country.

trance A hypnotic state resembling sleep. At Greek shrines, priests and priestesses often fell into a trance before speaking oracles.

transgenic Containing genetic features that have been artificially transferred from a different species.

transonic Flying through the sound barrier.

travois A platform for baggage formed by poles, joined together by a web of ropes. It was tied to the back of a horse or a dog.

tributary A stream of water that flows into a larger stream of water. The Egyptian Nile's main tributaries were the Blue Nile and the White Nile.

trireme A Greek warship with three rows of oars.

try In rugby union and rugby league, players score a try by touching the ball down over their opponents' goal line.

Tuareg tribesmen The largest group of nomads living in the Sahara Desert. They herd camels, goats, sheep and cattle and move about finding pasture for their animals. They are sometimes called the Blue Men

of the Desert because they wear robes that are dyed indigo and blue marks are often left on their skin.

tule A species of cattail found in the lakes and marshes of western America.

tule pollen Fine, powdery yellow grains produced by cattail flowers as they turn to seed.

tundra A treeless area with permanently frozen subsoil that lies between the ice of the Arctic and the timber line of North America and Eurasia.

turbine A wheel with many blades that is made to turn by a gas such as steam, or a liquid such as water. It is used to power machines or to generate electricity.

turbogenerator An electricity generator driven by a turbine.

turquoise A blue-green stone imported from the Sinai Peninsula.

type A piece of rectangular metal with a raised letter or symbol on one side.

ultrasound A high-pitched sound that humans cannot hear.

umiak A rowing boat made from whalebone covered with walrus hide and waterproofed with seal oil. It had a single sail and was used by the Inuit to hunt whales.

V/STOL Vertical and/or short take off and landing.

valve A device used to regulate the flow of a gas or a liquid, or to turn it on and off.

vane A blade of feathers or plastic used to keep arrows and darts on a straight course.

vault A curved roof or ceiling made of stone, which uses either arches or corbels.

velocipede An early form of bicycle that looked like the dandy horse.

velodrome An oval-shaped, steeply banked track used for cycling events. A banked track has curved sides that slope steeply upwards.

villa A wealthy family's country house, which was often part of a country estate. Villas were also grand houses located near the seaside.

Virgil (70–19 BC) The greatest poet of the Augustan period. He spent the last ten years of his life writing the *Aeneid*.

virtual reality An artificial environment produced by a computer that seems very real to the person experiencing it.

virus A microorganism that can only reproduce inside a living cell.

vision A religious experience that brought a person into contact with the spirits.

vision quest An expedition made alone in order to see a vision. Native American boys went on their first quest at puberty to test their courage. The boys sometimes stayed alone for several days at a time in an isolated place without food.

vizier Chief adviser to the pharaoh and second only to him in importance. At times there were two viziers.

wampum White and purple disk-shaped beads made from shells. Native Americans exchanged them as ceremonial gifts. Wampum began to have monetary value after the Europeans arrived.

warp The threads arranged vertically on a weaving loom. The weft threads are woven through the warp threads.

waveguide A hollow tube used to channel microwaves from one place to another.

weft The threads woven across the fabric (horizontally) on a weaving loom.

weight The heaviness of an object, caused by gravity pulling on it.

Western Hemisphere The part of the world that contains the Americas.

wicketkeeper In the game of cricket, the wicketkeeper stands behind the batsman and the stumps and catches the balls that go past the batsman.

wickiup A dwelling used by Native Americans in the South-west made of poles covered with branches.

wigwam A shelter used by tribes in the Northeast and Great Lakes areas of America. They bent four saplings towards a center and coveredd them with long strips of bark sewn together.

wing flap A hinged section of the wing that is lowered when landing and taking off to increase lift at low speed.

wing slat A small airfoil that forms a gap at the front of the wing to increase lift at low speed.

woolly mammoth A large, hairy animal that became extinct at the end of the Ice Age. It had trunk like an elephant and curved tusks.

Xiongnu Nomadic people who lived to the north of China.

X-rays Electromagnetic waves that can pass through soft parts of the body. X-rays are used to create images, on photographic film or a computer screen, of the inside of the body.

yawing The aerodynamic term that describes an airplane's nose swinging from side to side.

yucca An indigenous plant used by Native Americans. The chewed stem made an excellent paintbrush.

zither A stringed instrument. The seven-stringed zither was one of the oldest instruments in ancient China.

Weaving cotton in India

Roman gladiators

Greek vase

Ancient weapons

Ivory walruses

Index

frescoes 199
furniture
  Egyptian 314–15, *314–15*

G–I

gadgets 126–7
games 136–7, *136–7*, 315
  Native American 514–15, *514–15*
  sporting 220–1. *See also sport*
genes 171
glass 469, *469*
gliders 80–1, *80–1*
global positioning system (GPS) 59
golf 250–1, *250–1*
  caddies 251, *251*
  Jack Nicklaus 280, *280*
gravity 65
Greece 396–7
  afterlife 412–13, *412–13*
  agora 422–3, *422–3*
  Alexander the Great 332–3, 436–7, *436*
  archaeology 442–3, *442–3*
  architecture 184–5, *184–5*, 432–3, *432–3*
  art 444–5, *444–5*
  Athens 404, 434
  city-states 404–5
  clothing 418–19, *418–19*
  colonization 402–3
  crafts and craftspeople 420, *420*
  Cycladic period 398
  Delphi 305
  democracy 406–7
  education 416–17, 440
  Egypt, invasion of 332–3, 436–9
  family life 414–15, *414–15*
  festivals 232–3, 426–7
  food 424–5, *424–5*
  games 136, 258, 414, *414*
  gods 408–9, *408–9*
  Hellenistic world 438–9
  jewelry 419, *419*
  law 406–7
  Macedonians 436–7, *436–7*

medicine 428–9, *428–9*
metalwork 430–1
Minoans 398–9
music 424
Mycenaeans 398–9, 400–1
myths and legends 400–1, 410–13
Persians 434, 435
pottery 430–1, *430–1*
Roman rule 440–1
school 134
sculpture 438, 444–5, *445–5*
slaves 420–1, *420–1*
Sparta 405, 434
trade 402–3, 422–3, 562
travel 396–9
tyrants 406
war 434–5, *434–5*
writing 417, *417*, 442
greenhouses 23, *23*
Guggenheim Museum, New York 218–19, *218–19*
guitar 152, *152*
gunpowder 163, 384
gymnastics 264–5, *264–5*, 326
Hadrian's Wall 486, *486–7*
hair and hairstyles
  Egypt 316–17, *317*
  Native Americans 512, *512*
  Rome 463, *463*
Herakles 410, *410–11*, 412
hieroglyphs 146, 318–19, *318–19*
high jump 262, *262*
Hinduism 188–9, 190
hockey 240–1, *240*
horses
  China 350, *350*, 374, *374*
  Native Americans and 520–1, *520–1*
  riding 278–9, *278–9*
houses
  ancient 178, 414
  energy efficient 22–3, *22–3*
  grass 178–9, *178–9*
  Native American 528, *529*, 530–1, *530–1*
  stone 179

hunters and gatherers 558–9, *558–9*
  Native Americans 522–5, *522–5*
hydroelectric power 18–19, 161
ice hockey 240–1, *240–1*
ice-skating 276–7, *276–7*
  Sonja Henje 281
Incas 180–1
  Spanish invasion 582
India
  architecture 188–9
  English rule 596–7
  passage to 578–9
  religion 200–3
Industrial Revolution 130
infrared technology 165
insects
  flight 76, *76–7*
Instrument Landing System (ILS) 100
Inuit 506
  clothing 508, *508*
  explorers, contact with 586–7
  food 524–5, *524–5*
  houses 528
  kayaks 516, *516*
  travel 518–19, *518–19*
inventions 124–7, 172–3, 384–5
iron
  bridges 214, *214*
  casting 380
irrigation 18, 132, 363
Islam 200–1, 565
  spread 202–3
Ivan the Terrible 198

J–M

Jainism 188
Japan
  architecture 194–5, *194–5*
  trade 600, *601*
javelin 262, *262*
jewelry
  China 366–7, *366–7*
  Egypt 316, 317, *317*, 322, *322*
  Greece 419, *419*, 430, *430*

N–P

U–Z

Credits

ILLUSTRATORS

Paul Bachem; Graham Back; Kenn Backhaus; Andrew Beckett/Garden Studio; David Boehm; Gregory Bridges; Colin Brown/Garden Studio; Leslye Cole/Alex Lavroff & Associates; Lynette R. Cook; Christer Eriksson; Alan Ewart; Nick Farmer/Brihton Illustration; Rod Ferring; Chris Forsey; John Crawford Fraser; Tony Gibbons/Bernard Thornton Artists, UK; Kerri Gibbs; Greg Gillespie; Mike Golding; Mike Gorman; Ray Grinaway; Terry Hadler; Helen Halliday; Langdon G. Halls; Lorraine Hannay; Adam Hook/Bernard Thornton Artists, UK; Christa Hook/Bernard Thornton Artists, UK; Richard Hook/Bernard Thornton Artists, UK; Keith Howland; Gillian Jenkins; Janet Jones/Alex Lavroff & Associates; David Kirshner; Mike Lamble; Robyn Latimer; Alex Lavroff; Connell Lee; Kent Leech; Ulrich Lehmann; Steinar Lund/Garden Studio; Chris Lyon/Brihton Illustration; Martin Macrae/Folio; Avril Makula; Shane Marsh/Linden Artists Ltd; David Mathews/Brihton Illustration; Iain McKellar; Peter Mennim; David Nelson; Paul Newman; Steve Noon/Garden Studio; Matthew Ottley; Darren Pattenden/Garden Studio; Evert Ploeg; Tony Pyrzakowski; Oliver Rennert; John Richards; Ken Rinkel; Trevor Ruth; Michael Saunders; Stephen Seymour/Bernard Thornton Artists, UK; Nick Shewring/Garden Studio; Ray Sim; Mark Sofilas/Alex Lavroff & Associates; Kevin Stead; Roger Stewart/Brihton Illustration; Sharif Tarabay/Garden Studio; C. Winston Taylor; Rodger Towers/Brihton Illustration; Steve Trevaskis; Ross Watton/Garden Studio; Rod Westblade; Ann Winterbotham; David Wood; David Wun.

PHOTOGRAPHS

(t = top, c = center, b = bottom, l = left, r = right, i = icon) AA&A=Ancient Art and Architecture Collection; ADL=Ad-Libitum; AMNH=American Museum of Natural History; APL=Australian Picture Library; AUST=Austral International; BA=The Bridgeman Art Library; BPK= Bildarchiv Pressischer Kulturbesitz; BPL=Boltin Picture Library; BM=The British Museum; CMD=C. M. Dixon; GC=The Granger Collection; IB=The Image Bank; JJ= Jerry Jacka; JL= Jürgen Liepe; LPA=Luna Park Amusements Pty Ltd, Australia; MEPL=Mary Evans Picture Library; NA= National Anthropological Archives; NMNH=National Museum of Natural History; PR=Photo Researchers, Inc; RHPL=Robert Harding Picture Library; SI=Smithsonian Institution; SM=The Science Museum; TPL=photolibrary.com); WF=Werner Forman Archive 14 tl ADL (LPA). 16 bl Bruce Coleman Limited (M. Ide/Orion Press). 17 tr ADL (LPA). 18 bl PR (J. Steinberg). 19 tr ADL (LPA); br Panos Pictures (J. Dugast). 21 tr ADL (LPA). 23 tc Tom Stack & Associate (G. Vaughn). 24 cl ADL (LPA); tl RHPL. 25 br ADL (LPA); c ADL (LPA); tr TPL (M. King). 29 cr ADL (LPA). 30 cl ADL (LPA). 31 tr ADL (LPA). 32 bl ADL (LPA). 33 br ADL (LPA); tr ADL (LPA). 34 bl ADL (LPA); tl TPL (C. Bjornberg/PR). 37 tr TPL (NRAO/SPL). 36/37 AFL (M. Smith/Retna Pictures). 38 bl AUST (FPG International). 39 tr TPL (SPL). 41 cl IB (A. Pasieka). 42 tl ADL (LPA). 43 cr ADL (LPA). 44 cl, tcl, tl Heather Angel. 45 tr ADL (LPA). 46 bl ADL (LPA). 47 cr International Photographic Agency (SuperStock). 49 tr APL (M. Smith/Retna Pictures); bl TPL (J. Burgess/SPL).

50 b ADL (Australian National Maritime Museum); cl ADL (Australian National Maritime Museum). 53 tr ADL (Australian National Maritime Museum). 54 bl, tl ADL (Australian National Maritime Museum); bc TPL (C. Secula). 61 br International Photographic Agency (SuperStock). 62 tl TPL (NASA/SPL). 64 bct, br ADL (S. Bowey). 70 cl AKG. 71 tl Terry Gwynn-Jones (Royal Flying Doctor Service). 72 tr Auscape International (H. & J. Beste); l NHPA (G. I. Bernard); cl TPL (J. Burgess/SPL). 74 bl Auscape International (J. P. Ferrero); tl Oxford Scientific Films (S. Osolinski). 75 tr Auscape International (J. P. Ferrero). 76 br CSIRO, Division of Plant Industry; tl NHPA (S. Dalton). 77 NHPA (S. Dalton); br Imax Corporation. 78 tl WF (Denpasar Museum); bl GC. 79 bl Enrico Ferorelli; tr, br GC; cr North Wind Picture Archives. 80 br Auckland Museum. 81 tr APL (B. Desestres/Agence Vandystadt); bl SM (Science & Society Picture Library). 80/81 c Chris Elfes. 82 bl AKG. 83 bl Australian Geographic (P. Smith); c Royal Aeronautical Society. 84 tl AUST (L'Illustration/Sygma). 85 t AUST (L'Illustration/Sygma); cr Image Select (Ann Ronan Picture Library); br SM (Science & Society Picture Library). 86 tr APL (Bettmann). 87 tl Terry Gwynn-Jones (Royal Flying Doctor Service); r SI (#A42363C); br SI (#A4943). 88 tr AUST (Rex Features Ltd). 89 cr GC; br Terry Gwynn-Jones (Royal Flying Doctor Service); cl SI (#75–12199); tr SI (#89–923). 90 tl GC; tr Terry Gwynn-Jones (Royal Flying Doctor Service). 91 cr TPL (Hulton Deutsch); tr Topham Picture Source; tl Visions (M. Greenberg). 90/91 c SI (#79–763). 92 tl Budd Davisson; bl Quadrant/Flight. 93 br APL; tr Quadrant/Flight; cr SI (#93–5512). 94 bl APL (UPI/Bettmann); c SI (#77–11329). 95 br Check Six (M. Fizer); tr SM (Science & Society Picture Library). 97 bl APL (G. Hall); br Check Six (M. Fizer). 98 tr Check Six (G. Hall). 99 tl The Boeing Company; tr TPL (J. King-Holmes/SPL). 100 c IB (T. Bieber); tr IB (S. Proehl). 102 tr TRH Pictures (US Navy). 104 tr Terry Gwynn-Jones (Royal Flying Doctor Service); bl AUST (P. Schofield/Shooting Star.); cr Check Six (J. Towers). 107 tr Terry Gwynn-Jones (Royal Flying Doctor Service). 108 c Check Six (G. Hall); br Lockheed. 110 l International Photo Library; bl Picture Media (R. Richards/ Liaison/Gamma). 111 cr Sextant Avionique. 112 tr Check Six (G. Hall); bl tr APL (ACME/ Bettmann); bl NASA; bcl NASA; br NASA; tcr TPL (J. Fitzgerald); tr TPL (Hulton Deutsch); r TPL (SPL/NASA); c TPL (SPL/NASA); cl TPL (SPL/NASA); cr RHPL; tl TRH Pictures (NASA). 115 tl TPL (GEO-SPACE/SPL). 114/115 c TPL (SPL/NASA). 116 tr The Boeing Company; bl The Boeing Company. 117 cr NASA. 122 tl, bl, it, ic, ib, tr ADL (S. Bowey); i Nobel Foundation. 123 itc, ic, ibc, ib, tl ADL (S. Bowey); tr Air Portraits (D. Davies); i IB (L. Reupert); br TPL (Science Photo Library/B. Blokhuis). 124 i, l, r Nobel Foundation. 125 br APL (E. T. Archives); cr Culver Pictures. 124/125 c ADL (S. Bowey). 126 i ADL (S. Bowey). 127 tr ADL (S. Bowey); tl AKG (Wurttembergisches Landesmuseum, Stuttgart); cr TPL (W. Stacey). 124/125/126/127 ADL (S. Bowey); Photo manipulation Richard Wilson studios. 128 i ADL (S. Bowey); bl Victa. 129 cr APL (Bettmann Archive); br Caroma Industries; tr Hoover Co, Nth Canton, Ohio. 130 i, c ADL (S. Bowey); bl APL (Archive Photos); tr Crown Equipment; tl PR (A. Green). 131 i ADL (S. Bowey); tl RHPL (R. Evans). 132 i, tl, tr ADL (S. Bowey). 133 tc, tl ADL (S. Bowey); cr Bruce Coleman Limited (D. & M. Plage); br Spectrum Colour Library. 132/133 c Stock Photos Pty Ltd (P. Steel). 134 i, tl ADL (S. Bowey). 135 tr APL (Bettmann Archive); i IB (L. Reupert); br, cr Xerox Corporation. 134/135 c ADL (S. Bowey). 136 i, c, cl ADL (S. Bowey); bl Mattel Toys. 137 br Brilliant Images; tr Lego Australia; tl National Museum of Roller Skating, Lincoln, Nebraska. 138 i ADL (S. Bowey); bl Coo-ee Picture Library (R. Ryan); i IB (L. Reupert). 139 br ADL (S. Bowey); br Brownel & Wight Car Company 1890; cr PR. 140 i ADL (S. Bowey); tr Polperro Pictures (B. Geach). 141 i ADL (S. Bowey); tr MEPL; cr Peraves Ltd. 142 i ADL (S. Bowey); tr Jenny Mills; tl SM; bl Stock Photos Pty Ltd (D. Madison). 143 cr ADL (S. Bowey/ Australian National Maritime Museum); tr APL; br Spectrum Colour Library. 144 i ADL (S. Bowey); br, cr NASA; tl SI; cl TPL (Science Photo Library). 146 l ADL (S. Bowey); i IB (L. Reupert); tr Powerhouse Museum, Sydney; tc WF (BM). 147 c ADL (S. Bowey); tl Mansell Collection Ltd. 148 bl ADL (S. Bowey); cl E.T. Archives (SM); i IB (L. Reupert); cl Image Select (Ann Ronan Picture Library); tl Powerhouse Museum, Sydney. 149 bl APL (M. Adams). 150 tc, l ADL (S. Bowey); tr British Film Institute (BFI Stills); bl Coo-ee Picture Library (R. Ryan); i IB (L. Reupert). 151 br, cr, tl ADL (S. Bowey); bl AKG (E. Bohr); tr RHPL; bc TPL. 150/151 c British Film Institute (BFI Stills). 152 i ADL (S. Bowey); bl BA (BM); i IB (L. Reupert); br PR (S. McCartney). 153 cr AA&A; c Black Star; br BA (British Library); tr Powerhouse Museum, Sydney; tl Yamaha Japan. 152/153 c ADL (S. Bowey). 154 i ADL (S. Bowey); l RHPL; tr Loren McIntyre. 155 tr, br ADL (S. Bowey); cr Black Star (M. Balderas); tl, bl SM. 156 i, l ADL (S. Bowey); cr ADL (N. Vinnicombe); bl TPL (W. & C. McIntyre); tr TPL (U.S. Dept of Energy). 157 br SM (D. Exton). 158 i ADL (S. Bowey); cl ADL (S. Bowey/ Kellett's Museum of Laundry Irons, Melbourne); tl APL; bl Powerhouse Museum, Sydney. 159 br Edison National Historic Site USA; br SM. 160 i ADL (S. Bowey); tc Bruce Coleman Limited (J. Cancalosi); tl SI; tr TPL (Science Photo Library/J. Mead). 161 t RHPL; cr IB (L. J. Pierce); r TPL (W. Kent); TPL (Science Photo Library/B. Blokhuis). 160/161 c APL (D. & J. Heaton); c TPL (N. Hong); c TPL (R. Smith); c Wheels Magazine (W. Kent);

photo manipulation Richard Wilson Studios. 162 i ADL (S. Bowey); l Australian Museum (C. Bento); tr (Hulton-Deutsch). 163 c ADL (S. Bowey); tr MEPL; br MEPL. 164 i ADL (S. Bowey); tr International Photo Library; bl Military Picture Lib-rary. 165 cr Aviation Picture Library (J. Flack); cr Military Picture Library; tr Check Six (J. Benson); br Department of Def-ence; cl Department of Defence. 166 bl ADL (S. Bowey); i ADL (S. Bowey); tl SM (Science & Society Picture Library). 167 cl TPL (Science Photo Library/O. Burriel); br Deutsches Museum Munich. 168 i ADL (S. Bowey); tl AUST (Rex Features London/The Sun); cl TPL. 169 tl Royal North Shore Hospital, Sydney; tr TPL (R. Chase); br TPL (PR /S. Camazine); br TPL (Science Photo Library/A. Tsiaras). 168/169 c Lennart Nilsson; c TPL; Photo Manipulation Richard Wilson Studios. 170 i ADL (S. Bowey); tl Bruce Coleman Limited (J. Foott); bl Bruce Coleman Limited (H. Reinhard); tr TPL (P. Hayson). 171 c ADL (S. Bowey); tl Bruce Coleman Limited (H. Reinhard); r TPL (Science Photo Library); c TPL (Tony Stone Worldwide). 170/171 Photo Manipulation Richard Wilson Studios. 172 br MEPL; cr MEPL; cl Illustrated London News Picture Library; tr Illustrated London News Picture Library. 173 cr MEPL; tl MEPL; br Illustrated London News Picture Library; c Illustrated London News Picture Library. 179 tr Christine Osborne Pictures; cr RHPL (G. Campbell); tl RHPL (A. Woolfitt). 180 cl IB (M. Martin); tr RHPL (R. Frerck). 181 br IB (G. Covian); tc TPL, (R. Frerck/ TSW); cr TPL, (D.N. Green). 182 tl AKG (H. Bock); cl APL (D. & J. Heaton). 183 tc CMD (Berlin Museum). 184 tr TPL (N. van der Waarden). 185 tl WF (BM). 186 t Robert Estall; bl WF (BM). 188 cl APL (D. Ball); tc IB (D. Heringa); bl Scala; br Scala. 189 tr RHPL (J. H. C Wilson); c Sonia Halliday Photographs (J. Taylor). 190 bl Magnum (P.J. Griffiths). 191 tl Magnum (M. Franck); br Jenny Mills; cr RHPL (G. Campbell). 190/191 c George Gerster. 192 bl APL (K. Haginoya); tl TPL (TSW). 193 tr BM. 194 tl APL (S. Vidler). 195 cr TPL (TSW); tr Michael S. Yamashita. 196 cl AKG (E. Lessing); bc Sonia Halliday Photographs (L. Lushington). 197 tr RHPL (G. Campbell); br RHPL (A. Woolfitt). 198 cl IB (H. Sund); tr RHPL (G. Campbell); br SCR Library (D. Toase). 199 cr Arcaid (M. Fiennes); tr Arcaid (M. Fiennes). 200 bc BA (Courtesy of the Board of Trustees of the Victoria & Albert Museum); bl Sonia Halliday Photographs (L. Lushington). 201 tl RHPL (J.H.C Wilson). 202 tr AKG (H. Bock); tl AUST (Camerapress/C. Osborne); tcl APL (B. Holden). 203 cr CMD (Berlin Museum); tl WF (BM); tr WF (BM). 205 tc Angelo Hornak Library (Courtesy of the Dean and Chapter of Wells Cathedral); cr CMD (Berlin Museum); c Scala; tr Scala. 206 tr Angelo Hornak Library (Courtesy Dean and Chapter of Wells Cathedral); cl BA (J. Bethell); bl TPL (TSW). 208 tl APL (D. Ball); tc APL (R. Price/West Light). 210 br BA (Lauros-Giraudon); cl CMD (Berlin Museum); tl CMD (Berlin Museum). 211 br AKG (E. Lessing); c BA (Johnny van Haeften Gallery, London); bl IB (G. Faint). 210/211 t Scala. 212 tl TPL (R. Smith). 213 bc AKG (E. Lessing); tr Arcaid (N. Barlow); br Bildarchiv Foto Marburg. 214 bl RHPL (G. Campbell); tl RHPL (G. Campbell). 215 c Angelo Hornak Library (Courtesy Dean and Chapter of Wells Cathedral); tl Angelo Hornak Library (Courtesy Dean and Chapter of Wells Cathedral); tcr APL (Bettmann). 216 bl A.G.E. Fotostock; tr AKG (E. Lessing); tl Bilderberg (M. Horacek). 217 tl A.G.E. Fotostock; cr Bilderberg (E. Grames); tr Arcaid (R. Bryant). 218 tl Arcaid (S. Francis); bl Timothy Hursley. 219 tc IB (G. Colliva). 220 br Arcaid (R. Bryant); bl APL (D. Ball). 222 bl Angelo Hornak Library (Courtesy Dean and Chapter of Wells Cathedral). 223 tl APL (D. Ball); tr RHPL (G. Campbell). 224 bl Arcaid (I. Lambot); tr IB (A. Becker); br IB (R. Lockyer). 225 c NEC Corporation, Tokyo. 230 bl ADL (S. Bowey); tcr ADL (S. Bowey); tl ADL (S. Bowey); tr ADL (S. Bowey). 231 tr ADL (S. Bowey). 232 bl TPL (Hulton Deutsch). 233 cr GC; tr IB (Duomo). 234 br ADL (S. Bowey); cl Sporting Pix; tc Sporting Pix. 235 cr China Stock (C. Liu). 236 tl ADL (S. Bowey); bl APL (Allsport/J. Gishigi). 237 tc Action-Plus (P. Tarry). 238 tl MEPL; tr MEPL; bl Sport The Library. 239 cr ADL (S. Bowey); tr Sport The Library. 240 tl Action-Plus (G. Kirk); cl ADL (S. Bowey). 241 cr Action-Plus (D. Davies); tc GC; tcr GC. 242 tl Sport The Library (J. Wachter). 243 br ADL (S. Bowey); tr Colorsport; bl D. Donne Bryant Stock (A. Zaloznik). 244 tr APL (Allsport/M. Powell); bl IB (Duomo/D. Madison). 245 tr Action-Plus (S. Bardens). 246 bl ADL (S. Bowey); br MEPL (R. Mayne); tcl Sport The Library (D. Braybrook); tl Sport The Library (T. Nolan). 247 tr BA (Marylebone Cricket Club); tc Marylebone Cricket Club. 248 bl ADL (S. Bowey); c ADL (S. Bowey); tl APL (S. Powell); bc TPL (J. McCawley). 249 br Action-Plus (N. Haynes). 250 tl MEPL; br RHPL (A. Woolfitt & R. Tonk-inson). 251 tl Action-Plus (S. Bardens); bl ADL (S. Bowey); br ADL (S. Bowey). 252 cr Action-Plus (M. Hewitt). 253 br ADL (S. Bowey); tc ADL (S. Bowey); bc Oliver Berlin; tl MEPL. 254 c ADL (S. Bowey); tl ADL (S. Bowey); tc ADL (S. Bowey); tr ADL (S. Bowey); tl ADL (S. Bowey); cl APL (E.T. Archive); 255 cr Oliver Berlin; tl Oliver Berlin; tr BA. 254/255 c TPL (TSW/D. Leah). 256 bc Action-Plus; bl ADL (S. Bowey/Benson Archery, Sydney); tr ADL (S. Bowey/Benson Archery, Sydney). 257 bc Action Images; br ADL (S. Bowey); cr Sporting Pix (Popperfoto). 258 tl ADL (S. Bowey). 259 cr Art Resource (E. Lessing/The Louvre). 258/259 c APL (Agence Vandystadt/C.H. Petit); bl IB (Duomo/D. Madison); bc MEPL. 260 bl ADL (S. Bowey); tl APL (Agence Vandystadt/Allsport/G. Planchenault). 261 tr APL (ZEFA). 262 tr APL (Allsport/M. Powell); tc Sport The Library (S. Perkins); bl Sport The Library (N. Schipper); bc Sporting Pix (B. Thomas). 263 tc Sport The

638

Library; tr Sport The Library. 264 tl APL (Agence Vandystadt/Allsport); bl IB (Duomo/W. Sallaz); tcl Sport The Library (D. Callow). 265 tr Action-Plus (C. Barry); tl Sporting Pix (B. Thomas). 264/265 b APL (Agence Vandystadt/Allsport). 266 b ADL (S. Bowey). 267 cr APL (Bettmann); tr Live Action (V. Acikalin). 268 tl ADL (S. Bowey); bl, tr APL (Allsport). 269 bl APL. 268/269 c APL (Allsport/S. Bruty). 270 tr APL (Agence Vandystadt/Allsport); tl TPL (TSW/B. Torrez); bl Sport The Library (J. Crow). 271 br APL (Allsport/P. Rondeau). 270/271 c Dean Wilmot. 272 tl ADL (S. Bowey); c APL (Allsport/B. Martin); cl Live Action (V. Acikalin); tr TPL (N. Green). 273 tr RHPL (A. Evrard). 274 bl Action-Plus (G. Kirk); bc IB; c Live Action (SIPA-PRESS); tcl TPL (TSW); tl Sport The Library (J. Crow). 276 tcr BA (New York Historical Society); tr MEPL. 277 tr APL (Agence Vandystadt/R. Martin); cr APL (Allsport/A. Want); tl Sport The Library (J. Crow). 278 bl Action-Plus (R. Francis); cl ADL (S. Bowey). 279 tl APL (Agence Vandystadt/P. Vielcanet); tr APL (Allsport/D. Cannon). 280 bl Sport The Library (J. Crow); tl Sport The Library; tc Sporting Pix (Popperfoto); br Sporting Pix. 281 br Sport The Library (B. Frakes); tc Sporting Pix (T. Feder); bl Sporting Pix (Popperfoto). 280/281 Sporting Pix (B. Thomas). 288 cl BM; tl A. J. Spencer. 289 br, tl BM; bcl National Museums of Scotland; cbr WF (BM); bc WF. 290 bl, br, bc JL. 291 br BPK (Vatican Museum, Rome). 292 bc BPK (Staatliche Museen, Berlin); cl Geoff Thompson. 293 br JL. 294 bl JL. 295 tc JL. 296 cl, tr BM; bcl JL; tl WF (BM). 297 br, tc, tr BM; bl RHPL. 298 tr BM; bc Continuum Productions Corporation (R. Wood). 299 cr BM; br RHPL. 300 tl BPK (Egyptian Museum, Cairo); tc Griffith Institute, Ashmolean Museum. 301 tr Griffith Institute, Ashmolean Museum; cr BM; br St. Thomas' Hospital (S. W. Hughes). 302 tl, tr BM; bl Thames & Hudson Limited. 303 br BPK (Egyptian Museum, Cairo); tr BM; tl Enrico Ferorelli. 304 c BM; cr JL; tr WF (Royal Musem of Art and History, Brussels). 305 cl, cr, t BM; c JL. 304/305 b BM. 306 bc WF (Royal Musem of Art and History, Brussels). 307 br, tl JL. 308 tc APL (D. & J. Heaton); tr John Rylands University Library of Manchester; bl RHPL. 309 tl AUST (Colorific!/Terence Spencer). 311 br BPK (Egyptian Museum, Cairo/M. Büsing); cr BM. 312 bl WF (E. Strouhal). 313 br, tr BM. 314 tl JL; tr WF (E. Strouhal). 315 cr BM. 316 bl WF (The Louvre). 317 br, tr BM. 318 tr BM. 319 bc BM; tr WF (BM); c WF (Royal Musem of Art and History, Brussels). 320 ct Ashmolean Museum, Oxford; c BM; tcl Musées Royaux D'Art; bl Ny Carlsberg Glyptotek; tr WF (Egyptian Museum, Cairo). 321 tr WF (BM); br Werner. 322 bc, cl, tl BM; c WF (Petrie Museum, University College, London). 323 br BM; cr, tr JL. 324 cr, tl BM; bc, cl JL. 325 bl, br JL. 326 br WF; tr Hirmer Fotoarchiv, Munich; tl JL; bl WF(BM). 327 tr APL. 328 tl BM; cl Manchester Museum (Neg. no. 2138). 329 cr BM; br JL; bl Giulio Mezzetti. 330 bl, br JL; c RHPL (F. L. Kenett); tl RHPL (F. L. Kenett). 331 tc, tr JL. 332 tc BM; bc C. M. Dixon; tl RHPL (T. Wood). 333 br, tr BM. 334 tl BM; bc National Geographic Society (O. L. Mazzatenta). 335 tr RHPL. 336 br, bc, cr, cl, tr JL. 337 bc JL; tc BPK (Egyptian Museum, Cairo/J. Liepe); r BM. 340 b Cultural Relics Publishing House; tr WF (Private Collection, New York). 341 br BM. 343 cr Asian Art and Archaeology; tcr BM; tr Museum of Far Eastern Antiquities, Stockholm, Sweden (E. Cornelius); br Scala (Giuganno Collection, Rome). 344 bl Cultural Relics Publishing House. 345 r, tr Asian Art and Archaeology; tc Academia Sinica, Taipei. 346 cl Cultural Relics Publishing House; b, c Daniel Schwartz, Zurich. 347 c Daniel Schwartz, Zurich. 348 tr RHPL (G. Corrigan). 349 tc RHPL (G. Corrigan); tl Asian Art and Archaeology. 350 bl The Nelson-Atkins Museum of Art, Kansas City, Missouri (Purchase: Nelson Trust, 49-40); cl WF (Idemitsu Museum of Arts, Tokyo); tl WF (René Rivkin, Sydney). 351 tr Cultural Relics Publishing House. 352 t China Pictorial; bl Cultural Relics Publishing House; 353 cr Asian Art and Archaeology; tr China Stock; tr MEPL. 354 bl, t Asian Art and Archaeology; br WF (Myron Falk, New York). 356 tl China Pictorial; tc Cultural Relics Publishing House; bl WF (Eskenazi Ltd, London). 357 br China Pictorial. 358 bl BM. 359 cr BM. 360 bc WF (Victoria & Albert Museum, London). 362 tc Asian Art Museum of San Francisco (B60 P130+, The Avery Brundage Collection); tr, tl Cecilia Lindqvist, Stockholm. 364 cl China Pictorial; tl Nelson-Atkins Museum of Art, Kansas City, Missouri (Purchase: Nelson Trust, 35-125/1). 365 tr BM; tl Giraudon (Lauros/Musée de la Ville de Paris, Musée Cernuschi). 366 tc BM; bc China Pictorial. 367 bl BM; tl Cultural Relics Publishing House; c, cl WF (Eskenazi Ltd, London). 368 bc Asian Art and Archaeology; bl Asian Art and Archaeology; c, tl China Pictorial. 370 cl Giraudon (Bonora/Musée Guimet, Paris); bl WF (National Palace Museum, Taipei). 371 br Cultural Relics Publishing House; cr Royal Ontario Museum; tr WF (Sotheby's, London). 373 cr China Stock; br National Museums of Scotland. 374 tr The Seattle Art Museum (P. Macapia/Eugene Fuller Memorial Collection). 375 tc BM; tl The Commercial Press (Hong Kong) Ltd (From *Chinese Textile Designs*); tl The Nelson-Atkins Museum of Art, Kansas City, Missouri (Gift of Earle Grant, 59-63). 376 bl Cultural Relics Publishing House. 377 cr The British Library (S. 3326); t China Pictorial; br WF (Eskenazi Ltd, London). 378 tl The British Library (Or 8210/P2); tr Cultural Relics Publishing House; cr WF (Brian McElney Collection, Hong Kong); tcr WF (Private Collection, New York). 380 bl BM; cl Giraudon (Lauros/Musée Guimet, Paris); c WF (Earl Morse, New York). 381 br, tcr, tr BM; cr Giraudon (Lauros). 382 cl BM; bl The Palace Museum, Beijing. 383

c Giraudon (Lauros). 385 cr Asian Art and Archaeology; br SM, London (Science & Society Picture Library). 386 cl BM; bl Giraudon (Lauros/Musée Guimet, Paris). 387 cl BM; br Giraudon (Bonora/Musée Guimet, Paris); bc IB (P. & G. Bowater). 388 bl New World Press, Beijing; tr RHPL (G. Corrigan). 389 tl, tr BM. 388/389 b China Pictorial. 390 tr Asian Art and Archaeology; l, br China Pictorial (Lauros). 391 tr Giraudon (Lauros/Musée de la Ville de Paris, Musée Cernuschi); l Daniel Schwartz, Zurich; br WF (Idemitsu Museum of Arts, Tokyo). 397 bcr BA (Fitzwilliam Museum, University of Cambridge); br BA (Freud Museum, London); cr BA (The Louvre); tcr, tr GC. 398 tl BM; c GC. 399 c, tl AKG (Archaeological Museum, Herakleion/E. Lessing); tr AKG (The Louvre/E. Lessing); cl APL (ET Archive). 398/399 b Scala (National Museum, Athens). 400 bl AKG (National Archaeological Museum, Athens); tl BA (National Archaeological Museum, Athens). 401 br APL (D. Ball); cr BA (National Archaeological Museum, Athens); bcr, tc GC. 402 bc, tl, tr BM. 403 cr APL (ET Archive/ Archaeological Museum, Ferrara); tc BM. 404 bc AKG (BM); bl APL (ET Archive/Staatliche Glyptothek, Munich). 405 bl AA&A; bc BA (City of Bristol Museum & Art Gallery). 406 br AA&A (R. Sheridan); tc GC. 407 bl AKG (American School of Classical Studies, Athens). 408 tr WF (BM); cl WF. 409 tcl AKG (Archaeological Museum, Nauplion); tc AKG (Museo Nazionale Romano delle Terme, Rome/E. Lessing); tl BA (Freud Museum, London). 410 cl AKG (Badisches Landesmuseum, Karlsruhe); bl AKG (The Louvre/E. Lessing); tr BA (BM); tl GC. 411 bl AKG (National Museum of Archaeology, Naples/E. Lessing). 412 bl, cl BM. 413 bc AKG (Staatliche Antikensammlungen und Glyptothek, Munich/E. Lessing); bl BA (The Louvre); br BA (National Archaeological Museum, Athens). 414 bl, c AKG (The Louvre/E. Lessing); br, tc BM. 415 bl WF (BM). 416 bl AKG (National Archaeological Museum, Athens/E. Lessing). 417 tl BA (Giraudon/The Louvre); br BM; tr GC. 418 bl, c BM; tl WF (Acropolis Museum, Athens). 419 tr BM; cr BA (BM). 420 bl, cl, tl BM. 421 br BA (BM). 422 cl AKG (Kunsthistorisches Museum, Vienna/E. Lessing); tr AA&A. 423 br AKG (BM/E. Lessing); tc BA (BM); tl WF (Acropolis Museum, Athens). 424 bcl AKG (Archaeological Museum, Herakleion/E. Lessing); bl AKG (The Louvre/E. Lessing); tc AUST (M. Friedel); tl WF (BM). 425 bc AKG (The Louvre/E. Lessing). 426 bc AKG (Akademie der Bildenden Kuenste, Vienna/ E. Lessing); tc BA (The Louvre); bl BM; cl GC; tl National Archaeological Museum, Athens. 427 br National Archaeological Museum, Athens. 428 tl AKG; cl BM. 429 tc AKG (The Louvre/E. Lessing); cr AA&A. 430 tl APL (ET Archive/Archaeological Museum, Ferrara); cl BA (Freud Museum, London); bl BM; tr C. M. Dixon. 431 br BA (BM); bc BA (Vatican Museums & Galleries). 432 bc AKG (J. Hios); tl APL (D. Ball); bl BA (BM). 433 tr BA (BM). 435 tl AKG (Musée Vivenel, Compiegne); cr C. M. Dixon (Argos Museum, Greece); cl WF (BM). 436 tl AKG (National Museum of Archaeology, Naples/E. Lessing); tc AA&A (R. Sheridan); b APL (ET Archive/ Archaeological Museum, Istanbul); cl BA (Giraudon/The Louvre). 438 bl AKG (Israel Museum, Jerusalem/E. Lessing); c AKG (Pergamon Museum, Berlin); cl BA (Giraudon/The Louvre). 439 cr APL (ET Archive). 440 br AKG (Museo Nazionale Romano delle Terme, Rome/E. Lessing); bl APL (ET Archive/BM). 441 cr APL (ET Archive/Antalya Museum); tr APL (ET Archive/BM); bc BA (BM); bl IB (J. Zalon). 442 bc APL (ET Archive/BM); br APL (D. & J. Heaton); bl GC; tl IB (S. Dee); cl WF (Bardo Museum, Tunisia). 443 tr AKG (Palazzo Salviati, Florence/E. Lessing). 444 bl BM; cr, tr National Archaeological Museum, Athens; br, cl Scala (National Museum, Athens). 445 br BM; bl, cl BA (BM); br BA (Lauros-Giraudon/The Louvre); tl BA (National Archaeological Museum, Athens). 450 br AA&A (R. Sheridan); bl CMD. 450 bl CMD; tr Scala (Cortona, Museo del l'Accademia Etrusca); tl Scala (Museo Archeologico, Firenze); bc Scala (Museo Civico Piacenza); br Scala; cr (Museo di Villa Giulia, Roma); br Scala (Museo Gregoriano Etrusco, Vaticano). 452/453 r Scala. 454 tl BA (Giraudon); bl BM; br Scala. 455 bl BA (Kunsthistorisches Museum, Vienna); br Scala (Musei Capitolini, Roma). 456 cr BA (Villa dei Misteri, Pompei); tr Bulloz; cl Scala (Casa dei Vetti, Pompei); bl Scala (Museo Pio-Clementino, Vaticano); tr Scala. 457 br AA&A (R. Sheridan). 458 bl BA (Metropolitan Museum of Modern Art, New York); cr BM; cl Scala (Museo Nazionale, Napoli); tl (Villa dei Misteri, Pompei). 459 b Peter Arnold Inc. (M. Cooper); cr BM. 460 tl Scala (Vatican Museums); tr CMD; br Reunion des Musées Nationaux (Louvre, Paris). 461 tl BA (Lauros-Giraudon); cr Nasjonalgalleriet, Oslo (J. Lathion/Ethnological Museum, Oslo); c WF (Scavi di Ostia). 462 tl BA (Archaelogical Museum, Naples); cl CMD (BM). 463 cl BA (BM); tc BM; tr, br CMD (Capitoline Museum, Rome); cr Michael Freeman (Intl. Perfume Museum, Grasse); bl Scala (Museo Nazionale, Roma). 464 cl AKG (E. Lessing/Museo Ostiense); tc AA&A (R. Sheridan); bc, bcl, tl CMD. 465 tr AA&A (R. Sheridan); bl BA (Metropolitan Museum of Modern Art, New York); tr Scala (Museo Nazionale, Napoli). 467 b Scala (Museo Archeologico, Venezia). 468 bc Scala (Musei Capitolini, Roma); l Scala (Museo Nazionale, Napoli). 469 br Peter Arnold Inc. (L. J. Amos); bl, c BA (BM); bc CMD (Capitoline Museum, Rome); tr Scala (Museo Archeologico, Adria). 470 bc RHPL; bl, bc Scala (Museo Archeologico, Aquileia). 472 tl CMD; tr GC; cl RHPL (Rado Museum, Tunis); bc Scala (Arcivescovado, Ravenna). 474 tl BM; tr Scala (Villa Romana del Casale, Piazza Armerina). 475 tr, cr CMD; br Scala (Museo Nazionale Atestino, Este). 476 tcl, bcl CMD; tr Scala (Museo Gregoriano

639

Profano, Vaticano); bl, cr Scala (Museo Nazionale, Napoli). **477** br Scala (Museo Nazionale, Napoli). **478** bl AA&tA (R. Sheridan); tr BA (BM); tl Scala (Museo Archeologico, Taranto); tl, bc Scala (Museo Gregoriano Profano, Vaticano); cl Scala (Museo Nazionale, Napoli). **479** tr Scala (Musei Capitolini, Roma). **480** tl CMD (National Architecture Museum, Naples); c Scala (Museo della Civilta' Romana, Roma). **481** tl Sonia Halliday Photographs; tr RHPL (E. Rooney). **482** br CMD (Vatican Museums); tl Scala (Museo Archeologico, Palestrina); bl Scala. **483** tr Sonia Halliday Photographs. **484** tl AA&tA (R. Sheridan); c BA (Louvre, Paris); tr CMD; cl Michael Freeman; cr RHPL (BM). **486** tl, bl BM; cl Michael Freeman; tr JL. **497** tl RHPL. **488** tl BA (Dorset County Museum); br CMD (Vatican Museums); cl Sonia Halliday Photographs (A. Held); bl Scala. **489** br CMD (Vatican Museums); tr RHPL (P. Scholey). **490** tr BA; tl Scala (Museo delle Terme, Roma); tc Rheinisches Landesmuseum Trier. **491** tl Scala; r Scala. **493** br AA&tA (R. Sheridan); tr BA (BM); tl Scala (S. Apollinare Nuovo, Ravenna). **494** tl BA (Pushkin Museum, Moscow); tr CMD (Heritage Museum, Leningrad); bl CMD; c Scala (Museo Benaki, Athenai). **495** br CMD; tr WF (Topkapi Palace Library, Istanbul). **496** br AA&tA; bl CMD; tl Scala. **497** bl AA&tA; br GC; cr Scala (Casa dei Vetti, Pompei); tr WF. **496/497** c AUST (Pictor Uniphoto). **498** br AA&tA (R. Sheridan); tr Scala. **499** tr BA; br BA; br BA; bc CMD (Vatican Museums); tcl Scala (Galleria degli Uffizi, Firenze); tc Scala (Museo Pio-Clementino, Vaticano); bl, bcl Scala; tcr WF (Barber Institute of Fine Arts, Birmingham University). **502** bi ADL (S. Bowey); br, cli BPL; tr NMNH/SI (Arctic Studies Center, W. Fitzhugh); tli Peabody Museum, Harvard University (H. Burger). **503** br AMNH (#3180[2]); cli Thomas Burke Memorial Washington State Museum (#2.5E1605) (Eduardo Calderon); bli JJ (Heard Museum); tli National Museum of the American Indian/SI (#12/8411); tr NMNH/SI (Arctic Studies Center, W. Fitzhugh); i Peabody Museum, Harvard University (H. Burger); tr Rosamond Purcell. **505** i ADL (S. Bowey); r BPL; r NMNH/SI (Arctic Studies Center, W. Fitzhugh); r Rosamond Purcell; tl Turner Publishing, Inc (A. Jacobs). **506** br AMNH (E. Curtis, #32999); b The Brooklyn Museum, Museum Collection Fund (1905 Expedition Report, Culin Archival Collection); bl Fred Hirschmann; t Library of Congress; c NA/ SI (#1564); l New York Public Library, Astor, Lenox and Tilden Foundations; l Peabody Museum, Harvard University (H. Burger); c San Diego Museum of Man (E.S. Curtis). **507** tc Library of Congress; i National Museum of the American Indian/SI (#12/8411); cr NMNH/SI (#517B); b (#53887); tr PR (J. Lepore); c Provincial Museum of Alberta (Ethnology Program, #H67.251.19). **508** c NMNH/SI (#81-4283); i Peabody Museum, Harvard University (H. Burger). **509** t AMNH (A. Anik, #3563[2]); i BPL. **510** r AMNH (J. Beckett, #4633[3]); tr (J. Beckett, #4634[3]); l NA/ SI (#30951-B); i Peabody Museum, Harvard University (H. Burger). **511** r AMNH (D. Finnin/C. Chesek, #4632[2]); r NA/ SI (#45217C). **512** b CMD; i Peabody Museum, Harvard University (H. Burger); r San Diego Museum of Man (E. S. Curtis); t San Diego Museum of Man (E.S. Curtis). **513** l Peabody Museum, Harvard University (H. Burger); b The Brooklyn Museum, Museum Collection Fund (Museum Expedition 1905); c JJ (Heard Museum); b Museum of New Mexico (C. N. Werntz, #37543); t New York Public Library, Astor, Lenox and Tilden Foundations; l Peabody Museum, Harvard University (H. Burger). **514** tl National Museum of the American Indian/SI (#6/4597); bl NMNH/SI (Arctic Studies Center, W. Fitzhugh); i Peabody Museum, Harvard University (H. Burger). **515** r Denver Art Museum; t Library of Congress; r NMNH/SI (#81-11343); c NMNH/SI (Arctic Studies Center, W. Fitzhugh). **516** bl BPL; i BPL; l NA/ SI (#81-13423). **517** cr Australian Museum (H. Pinelli); r NA/ SI (#75-14718); c NMNH/SI (Arctic Studies Center, W. Fitzhugh); t NMNH/SI (Arctic Studies Center, W. Fitzhugh); r Peabody Museum, Harvard University (H. Burger). **518** i BPL; l Phoebe Hearst Museum of Anthropology, University of CA; br San Diego Museum of Man (E. S. Curtis). **519** cr AMNH (D. Eiler, #4962[2]); t Australian Museum (K. Lowe, #E.12665); r NA/ SI (#10455-A-1); i National Museum of the American Indian/SI (#12/8411); br WF (W. Channing). **520** i BPL; tl Buffalo Bill Historical Centre, Cody, WY (Gift of Mr & Mrs I. H. L. Larom); tr Buffalo Bill Historical Centre, Cody, WY (Gift of A. Black); l Detroit Institute of Arts, Founders Society, Flint Ink Corporation. **521** r, bl Buffalo Bill Historical Centre, Cody, WY (Gift of Mr & Mrs I.H. L. Larom); i JJ (Heard Museum); c Peabody Museum, Harvard University (H. Burger); tr John Running. **522** i ADL (S. Bowey); t NA/ SI (#75-14715). **523** r ADL (S. Bowey); r, cr BPL; r Robert & Linda Mitchell; r PR (C. Ott); r PR (A. G. Grant); br PR (T. & P. Leeson). **524** i ADL (S. Bowey); i AMNH (#3180[2]); t National Museum of the American Indian/SI (#35421). **525** i BPL; t Thomas Burke Memorial Washington State Museum (#2-3845); r Thomas Burke Memorial Washington State Museum (#222b); r NA/ SI (#10455-L-1); tr New York Public Library, Astor, Lenox and Tilden Foundations; br Peabody Museum, Harvard University (H. Burger). **526** i ADL (S. Bowey). **527** cr BPL; r JJ (Heard Museum); i National Museum of the American Indian/SI (#12/8411); br New York Public Library, Astor, Lenox and Tilden Foundations. **528** tr National Geographic Society (R. Madden); i National Museum of the American Indian/SI (#12/8411). **529** r AMNH (#330387); i National Museum of the American Indian/SI (#12/8411); br John Oram. **530** l The Bettmann Archive; t Buffalo Bill Historical Centre, Cody, WY

(Gift of A. Black); i National Museum of the American Indian/SI (#12/8411). **531** b Peabody Museum, Harvard University (H. Burger). **532** c AMNH (L. Gardiner, #4597[3]); tc (L. Gardiner, #4596[2]); r (L. Gardiner, #4595[2]); i Thomas Burke Memorial Washington State Museum (#2.5E1605) (Eduardo Calderon). **533** c AMNH (S. S. Myers, #3837[3]); b AMNH (L. Gardiner, #4588[2]); r Australian Museum (H. Pinelli, #E.81968); tr The Brooklyn Museum, Museum Collection Fund (Museum Expedition 1907). **534** i Thomas Burke Memorial Washington State Museum (#2.5E1605) (Eduardo Calderon); Tom Till. **535** i Thomas Burke Memorial Washington State Museum (#2.5E1605) (Eduardo Calderon); b Arizona State Museum (H. Teiwes); c The Brooklyn Museum, Museum Collection Fund (Museum Expedition 1903); t San Diego Museum of Man (E. S. Curtis); br P. K. Weis; WF (BM). **536** l AMNH (C. Chesek, #4394[3]); t The Brooklyn Museum, Museum Collection Fund (Museum Expedition 1911); i Thomas Burke Memorial Washington State Museum (#2.5E1605) (Eduardo Calderon). **537** tr Museum of New Mexico (T. Harmon Parkhurst, #3895); t John Running; r The University Museum, University of Pennsylvania (#T4.367c3); br The University Museum, University of Pennsylvania (#T4.369c2). **538** i Thomas Burke Memorial Washington State Museum (#2.5E1605) (Eduardo Calderon); r Detroit Institute of Arts, Founders Society, Flint Ink Corporation. **539** tl AMNH (#335500); t Detroit Institute of Arts, Founders Society, Flint Ink Corporation; cr Mark E. Gibson; cl Grant Heilman Photography (J. Colwell); i JJ (Heard Museum); br Jack Parsons; r Peabody Essex Museum, Salem, MA; b Stephen Trimble. **540** bl AMNH (E. Mortenson, #4730[2]); br The Brooklyn Museum, Museum Collection Fund (Museum Expedition 1903); i Thomas Burke Memorial Washington State Museum (#2.5E1605) (Eduardo Calderon); bl NA/ SI (#53401-A). **541** r AMNH (#19855); br, b Daniel J. Cox & Associates; tr NMNH/SI (Arctic Studies Center, W. Fitzhugh); **542** i JJ (Heard Museum); t NA/ SI (#3409-A); br National Museum of the American Indian/SI (#22/8539). **543** br AMNH (E. Mortenson, #4724[2]); t, tr Linden Museum, Stuttgart (U. Didoni). **544** i JJ (Heard Museum); bl, tl National Geographic Society (B. Ballenberg); cl Tom Till. **545** r National Geographic Society (B. Ballenberg); tr AMNH (C. Chesek/J. Beckett, #4051[4]); c Florida Division of Historical Resources; br NA/ SI (#93-2051). **546** tl Cherokee National Museum, Tahlequah, OK; i JJ (Heard Museum); t Rosamond Purcell. **547** r Chief Plenty Coups State Park Museum, Pryor, MT, (From *The American Indians: The Mighty Chieftains*, M. Crummet, 1993 Time-Life Books); b NA/SI (#53372-B). **548** bl, br Four Winds Gallery (S. Bowey); i JJ (Heard Museum); tl National Museum of the American Indian/SI (#2268); l John Running. **549** br Four Winds Gallery (S. Bowey); t The Brooklyn Museum, Museum Collection Fund (Museum Expedition 1903); i Thomas Burke Memorial Washington State Museum (#2.5E1605) (Eduardo Calderon); br Lowe Art Museum, University of Miami (Gift of A. I. Barton, #57.158.000); bl Stephen Trimble. **548/549** c Four Winds Gallery (S. Bowey). **550** i JJ (Heard Museum); br Jerry Jacka; l John Running; r John Running. **551** tl JJ (Heard Museum); tc JJ (Heard Museum); c JJ; r Stephen Trimble. **552** tr APL (Bettmann Archive/UPI); br APL (Bettmann Archive/UPI); tcr NA/SI (#76-7905); bl National Archives, Washington D.C.; tcl National Archives, Washington D.C.; bc National Archives, Washington D.C.; tl Philadelphia Museum of Art (Artists: Lehman and Duval, Gift of Miss W. Adgar). **558** cr GC; tr WF (BM). **559** tl AA&tA; cl RHPL (A. Woolfit). **560** tl Ashmolean Museum, Oxford; bl The Mansell Collection; tcl WF (The Louvre). **561** tr GC. **562** bl APL; cl BA (Nationalmuseet, Copenhagen); tl BA. **563** cr AKG (E. Lessing); br AA&tA; cr APL (E.T. Archive/Correr Museum Library, Venice). **565** cr Auscape International (M. Freeman). **566** tl AKG; cr AA&tA; tr BA (BM); cl WF (Statens Historiska Museum, Stockholm). **567** tr GC. **568** bc Auscape International (T. de Roy); cl Auscape International (J. P. Ferrero); tl Australian Museum; bl Wave Productions (O. Strewe). **569** cr WF (BM). **570** tl WF (Courtesy Sotheby's, London). **571** tr Auscape International (K. Atkinson); cr GC; tc RHPL (J. H. C. Wilson). **570/571** t APL. **572** tl BA (Bibliotheque Nationale, Paris); tc BA (The Louvre). **573** tr AKG (E. Lessing); tl BA (British Library). **574** cl GC; tl GC. **575** br AA&tA; tl APL. **577** br WF (Asiatische-Sammlung Collection, Bad Wildungen, Germany). **578** cr BA (Nationalmuseet, Copenhagen); tr GC; tl RHPL (P. Scholey). **579** tcr APL (L. Meier); tl BA (University of Witwatersrand, Johannesburg); tl North West Picture Archives. **580** tl AKG (Metropolitan Museum of Art, New York). **581** br ADL (S. Bowey); tr AKG. **582** tl ADL (S. Bowey); c AKG; bl IB; tr MEPL. **583** tr MEPL. **584** tl AKG; bc GC; cl GC. **585** tr APL (E.T. Archive). **586** bl BA (University of British Columbia); cl GC. **587** tr APL; br MEPL. **588** tc BA (Tower of London Armouries). **589** tr BA (Fitzwilliam Museum, University of Cambridge); br MEPL. **590** tl APL (E.T. Archive); bl BA (British Library); c BA (Fitzwilliam Museum, University of Cambridge); b BA (O'Shea Gallery, London). **591** br MEPL. **592** bc BA (British Library); cl MEPL. **593** br MEPL. **594** tl MEPL. **595** tr AKG; br APL (J. Carnemolla); tl MEPL. **596** bl BA (British Library); bc BA (Christie's, London); tl RHPL (M. Joseph). **598** bl Black Star (F. Ward); br BA (Stapleton Collection); bc CMD; cr MEPL. **599** bl APL (E.T. Archive). **600** c BA (National Maritime Museum, London); bl Orion Press; tl WF (BM). **601** br GC. **602** tl Auscape International; tr APL (D. & J. Heaton). **603** tc Black Star (J. Lopinot). **604** cr IB (M. St. Gil). **605** tc IB (P. Kaehler); tr IB (M. Melford).